# Collins

NEW GCSE SCIENCE

# Science

for Specification Modules B1–B3, C1–C3 and P1–P3

# OCR
Twenty First
Century Science

Series Editor: Ed Walsh

Authors: Peter Ellis,
Jo Foster, Nicky Thomas

**Student Book**

William Collins' dream of knowledge for all began with the publication of his first book in 1819. A self-educated mill worker, he not only enriched millions of lives, but also founded a flourishing publishing house. Today, staying true to this spirit, Collins books are packed with inspiration, innovation and practical expertise. They place you at the centre of a world of possibility and give you exactly what you need to explore it.

Collins. Freedom to teach

Published by Collins
An imprint of HarperCollinsPublishers
77–85 Fulham Palace Road
Hammersmith
London
W6 8JB

Browse the complete Collins catalogue at:
**www.collinseducation.com**

© HarperCollinsPublishers Limited 2011

10 9 8 7 6 5 4 3 2 1

ISBN-13 978 0 00 741528 1

The authors assert their moral rights to be identified as the authors of this work.

All rights reserved. No part of this publication may be reproduced, stored in a retrieval system, or transmitted in any form or by any means, electronic, mechanical, photocopying, recording or otherwise, without the prior written permission of the Publisher or a licence permitting restricted copying in the United Kingdom issued by the Copyright Licensing Agency Ltd., 90 Tottenham Court Road, London W1T 4LP.

British Library Cataloguing in Publication Data
A Catalogue record for this publication is available from the British Library

Commissioned by Letitia Luff
Project managed by Jane Roth
Contributing authors: Ed Walsh; John Beeby; 'Bad Science' pages based on the work of Ben Goldacre
Typesetting, design, layout and illustrations by Ken Vail Graphic Design
Design manager: Emily Hooton
Edited by Anne Trevillion and Tony Clappison
Proofread by Anna Clark
Photos researched by Caroline Green
Production by Kerry Howie
Cover design by Julie Martin

Printed and bound by L.E.G.O. S.p.A. Italy

Acknowledgements – see page 336

# Contents

How to use this book — 6

## B1 You and your genes — 8

What your genes do • You're different • How genes work together • Variation in families • Genetic crosses • Sex determination

Preparing for assessment: Applying your knowledge — 22

Disorders caused by a single gene • Carrying genes and passing them on • More about genetic testing • Cloning • Multipurpose stem cells

Checklist — 34
Exam-style questions — 36

## B2 Keeping healthy — 38

Microbes and your body • Defending against disease • Vaccination • Making vaccination safe

Preparing for assessment: Applying your knowledge — 48

Mutation and resistance • Human guinea pigs • Your amazing heart • Keeping your heart healthy • Get your pulse racing • Blood pressure and lifestyle factors • Homeostasis • Water in, water out • Your kidneys

Checklist — 68
Exam-style questions — 70

## B3 Life on Earth — 72

Species and adaptation • Changes and challenges

Preparing for assessment: Evaluating and analysing evidence — 78

Chains of life • Recycling nutrients • Environmental indicators • Variation, mutation and evolution • The great competition of life • Evolution has the answers • Evidence from fossils and from DNA • We need diversity • A sustainable future • Thinking ahead

Checklist — 100
Exam-style questions — 102

## C1 Air quality — 104

The air around us • Changing air • Humans and the air • Air quality and health • Burning fuels • Rearranging atoms • Reactants and products • Sources of pollutants • Removing pollutants • Improving power stations • Reducing carbon dioxide • Improving transport

Preparing for assessment: Evaluating and analysing evidence — 130
Checklist — 132
Exam-style questions — 134

## C2 Material choices — 136

Using materials • Choosing materials

Preparing for assessment: Applying your knowledge — 142

Natural and synthetic materials • Crude oil • Separating hydrocarbons • Making polymers • Better materials • Polymer properties • Improving polymers • Nanotechnology • Nanoparticles • Making use of nanoparticles • Staying safe with nanoparticles

Checklist — 166
Exam-style questions — 168

## C3 Chemicals in our lives – risks and benefits — 170

Moving continents • Useful rocks • Salt • Salty food – risks and benefits • Alkalis • Reacting alkalis • Making alkalis • Chlorine in water – benefits and risks

Preparing for assessment: Evaluating and analysing evidence — 188

Electrolysis of brine • Industrial chemicals • PVC • Life Cycle Assessment

Checklist — 198
Exam-style questions — 200

## P1 The Earth in the Universe — 202

Our solar system • Observing stars

**Preparing for assessment: Evaluating and analysing evidence** — 208

Distances to stars • Fusion in stars • The expanding Universe • The Big Bang • Rocks on Earth • Continental drift • Tectonic plates • Earthquake waves • What a wave is • The wave equation

Checklist — 230
Exam-style questions — 232

## P2 Radiation and life — 234

Electromagnetic radiation • Radiation intensity • Ionisation • Effects of ionising radiation • Microwaves • Ozone • The greenhouse effect • Carbon cycling • Global warming • Electromagnetic waves for communication

**Preparing for assessment: Evaluating and analysing evidence** — 256

Digital signals • Storing digital information

Checklist — 262
Exam-style questions — 264

## P3 Sustainable energy — 266

Energy sources • Power • Buying electricity • Energy diagrams • Efficiency

**Preparing for assessment: Applying your knowledge** — 278

Generators • How power stations work • Waste from power stations • Renewable energy sources • The National Grid • Choosing the best energy source • Dealing with future energy demand

Checklist — 294
Exam-style questions — 296

Carrying out practical work in GCSE Science — 298
Controlled assessment in 21st Century GCSE Science — 311
How to be successful in your GCSE Science exam — 312
Data sheet — 319
Bad Science for Schools — 320
Glossary — 326
Index — 331

# How to use this book

## Welcome to Collins New GCSE Science for OCR 21st Century

### The main content

Each two-page lesson has three sections:

- ◐ **The first section** outlines a basic scientific idea
- ◐ **The second section** builds on the basics and develops the concept
- ◐ **The third section** extends the concept or challenges you to apply it in a new way.

The third section can also provide extra information that is only relevant to the Higher tier (indicated with 'Higher tier only')

Each section contains a set of questions that allow you to check and apply your knowledge.

Look for:

> 'Did you know?' boxes

> internet search terms (at the bottom of every page)

> 'Watch out!' hints on avoiding common errors

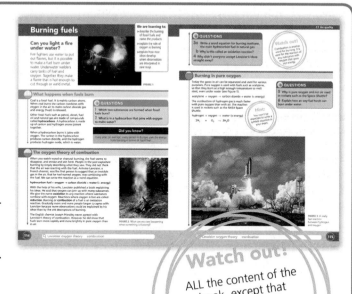

**Watch out!**
ALL the content of the book, except that marked 'Higher tier only', will be assessed at both Foundation and Higher tier.

### Module Introduction

Each Module has a two-page introduction.

Link the science you will learn in the coming Module with your existing scientific knowledge.

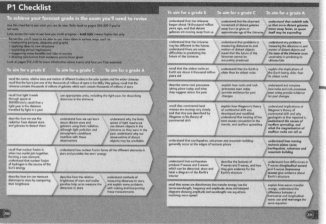

### Module Checklists

At the end of each Module is a graded Checklist.

Summarise the key ideas that you have learnt so far and check your progress. If there are any topics you find tricky, you can always recap them!

# How to use this book

## Exam-style questions

Every Module contains practice exam-style questions for Foundation and Higher tier. There is a range of types of question and each is labelled with the Assessment Objective that it addresses.

Familiarise yourself with all the types of question that you might be asked.

## Worked examples

Detailed worked examples with examiner comments show you how you can raise your grade. Here you will find tips on how to use accurate scientific vocabulary, avoid common exam errors and improve your Quality of Written communication (QWC), and more.

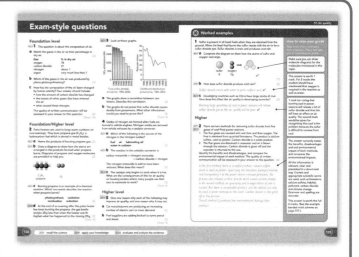

## Preparing for assessment

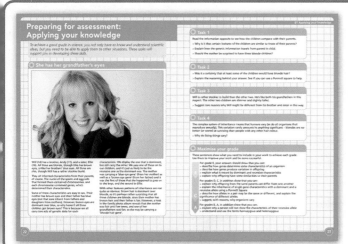

Each Module contains a Preparing for Assessment activity. These will give you practice in tackling the Assessment Objectives and will help build the essential skills that you will need to succeed in your Controlled Assessment tasks and your exam.

There are two types of Preparing for Assessment activity.

> Applying your knowledge: Look at a familiar scientific concept in a new context.

> Evaluating and Analysing evidence: Build your skills in drawing conclusions from evidence.

## Practical work and exam skills

A section at the end of the book guides you through your practical work, your Controlled Assessment tasks and your exam, with advice on: planning, carrying out and evaluating an experiment; using maths to analyse data; the language used in exam questions; and how best to approach your written exam.

## Bad Science for Schools

Based on Bad Science by Ben Goldacre, these activities give you the chance to be a 'science detective' and evaluate the scientific claims that you hear every day in the media.

## Glossary

Check on the meaning of scientific vocabulary that you come across.

# B1 You and your genes

## What you should already know...

### Cells are the building blocks of plants and animals

They contain specialised structures, including a nucleus. This photo shows an animal cell.

Most plants and animals are multicellular organisms. The cells are organised into tissues and organs.

 What are the differences between plant and animal cells?

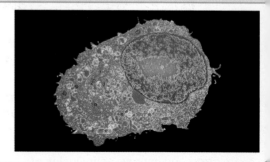

### In multicellular organisms some cells are specialised

Specialised cells have special jobs.

For example, red blood cells carry oxygen to other cells.

Nerve cells allow the brain to communicate with other parts of the body.

 Name two other specialised animal cells.

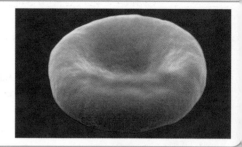

### Animals and most plants reproduce by sexual reproduction

Sexual reproduction needs a male and a female.

In animals, the male produces sperm cells and the female produces egg cells.

When one sperm fuses with one egg, fertilisation takes place. The fertilised egg develops into a new individual.

 In plants, what are the male and female sex cells called?

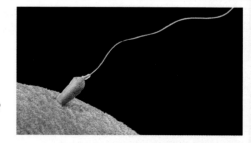

### Individuals show variation

Some differences between individuals are due to the environment. For example, boxers train to build up their muscles.

Other differences, such as eye colour, are inherited from parents.

 Give some examples of inherited features.

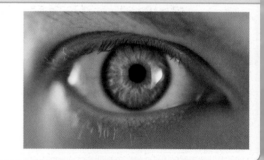

# In B1 you will find out about...

> chromosomes in the nucleus of a cell, which contain the instructions for how an organism functions

> deoxyribonucleic acid or DNA, the long molecules that make up the chromosomes

> genes, which are sections of the DNA providing particular instructions

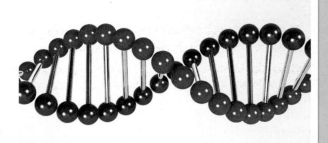

> chromosomes existing in pairs, with the same genes in the same place

> different versions of the same gene called alleles

> how genes are passed from parent to offspring

> features of an individual depending on the combination of alleles inherited

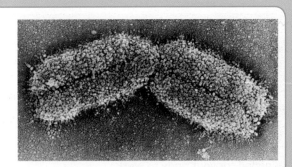

> diseases caused by a faulty allele

> the inheritance of these genetic disorders

> the implications for individuals and for society of testing for genetic disorders

> asexual reproduction to form clones

> stem cells, which are unspecialised cells with the potential to become specialised, and their medical uses

# What your genes do

**We are learning to:**
> describe what a gene is and what it does
> explain differences between structural and functional proteins
> discuss ethical issues raised by genetic research

## How alike are you and a chimpanzee?

Chimpanzees and humans may be more alike than you think. Our characteristics are determined by DNA, the substance that makes up genes. Research has shown that about 96% of the DNA sequences found in humans are also found in chimpanzees.

FIGURE 1

## What genes are and what they do

**Genes** make us who we are. They are the instructions that control how organisms develop and function.

You began as a single cell at conception, when a sperm fertilised the egg. Genes within the cell instructed your body how to grow and become the functioning human being that you are today. Genes are found in the nucleus of cells. They instruct the cells to make **proteins** needed for your body to work.

### QUESTIONS

1 Where are genes found?
2 What do genes do?

## Genes as codes for proteins

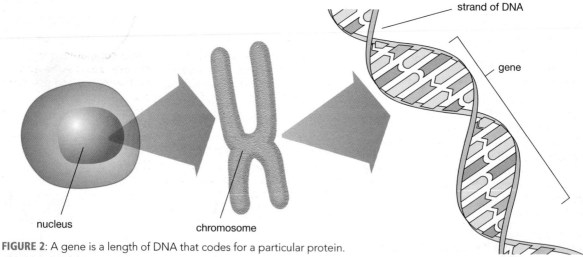

FIGURE 2: A gene is a length of DNA that codes for a particular protein.

A gene is a section of a long chemical called **DNA** (deoxyribonucleic acid). Each gene provides the code for the production of a protein. The genetic material is found within the cell nucleus. The DNA is in long strands that are coiled and packed into structures called **chromosomes**.

The strands of DNA are made up of four chemicals that we call 'bases', along with phosphate and sugar molecules. It is the order of these four different bases along each piece of the DNA strand that determines the protein code.

### Did you know?

If you stretched out the DNA from the nucleus of one of your body cells, it would be about two metres long.

🔍 gene GCSE

## Structural and functional proteins

Some proteins give an organism rigidity and strength. These are called **structural proteins**. Collagen is a structural protein that gives the skin its elasticity. It is also a component of connective tissue like ligaments and cartilage. Keratin is the structural protein that helps to make hair and nails in humans, and hooves or feathers in some animals. In plants, structural proteins help to strengthen the cell wall.

**Functional proteins** enable or speed up chemical reactions in the organism. Examples of functional proteins are the **enzymes** needed for digestion in animals, such as lipase which digests fats, and amylase which digests starch.

FIGURE 3: Human hair contains the structural protein keratin.

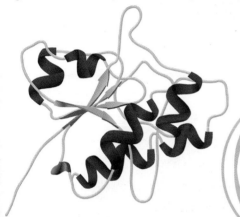

FIGURE 4: Proteins often have complex folded or coiled shapes.

*Watch out!* Make sure you know the difference between chromosomes, DNA and genes.

### QUESTIONS

**3** How do different lengths of DNA code for different proteins?

**4** Put these in order of size, largest first: cell, DNA, chromosome, gene, base, nucleus.

## DNA sequencing

Scientists can now identify the sequence of bases in DNA and determine which genes are located where on the chromosomes. The Human Genome Project identified all the 20–25 000 genes in human DNA.

The ability to 'read' a person's genome (gene set) brings ethical considerations. There have been considerable benefits to society from the Human Genome Project, for example the identification of particular sequences of genes that can lead to or protect against certain diseases. This allows individuals to be screened for genetic disorders.

Some drug companies have identified genes linked to particular illnesses, such as breast cancer and ovarian cancer. They have tried to 'patent' (claim ownership of) these genes in order to make a profit by developing drugs that target these genes. This would mean that other companies have to pay to use that gene for research. There is debate about whether this should be allowed. Some of the arguments for and against patenting genes are outlined in the table.

| FOR patenting a gene | AGAINST patenting a gene |
|---|---|
| Drugs are more likely to be developed to prevent disease caused by that gene, as there is likely to be more profit | Other companies are restricted in their research on that gene, and this could slow the search for prevention or cure |
| The patent is time-limited, so after a certain amount of time other companies are able to make cheaper versions of the treatment that has been developed | A company should not be allowed to 'own' a part of the human genome |

### QUESTIONS

**5** What are the benefits to a person of being identified as at risk of a particular disease?

**6** Discuss the arguments for and against patenting human genes. Can you add other arguments for and against?

# You're different

## Bad hair day?

You inherit the tendency for curly or straight hair from your parents. If one parent has straight hair and one curly, then your hair usually will be a combination of the two.

**We are learning to:**
> understand that characteristics may be determined by genes and by the environment
> understand that some characteristics are determined by several genes
> consider whether genes or upbringing is more important in determining characteristics

**FIGURE 1**: You can blame your genes for your wild hair.

## Why are we all different?

Some differences between individuals are due to genes. An example of this in humans is whether or not you have dimples.

There are other differences that people are not born with, and are entirely due to the **environment**. For example, somebody may have a scar from an injury, or may have dyed their hair. For humans, the 'environment' in this context doesn't just mean the physical surroundings, but also the social situation that they live in.

Some differences are due to a combination of genetics and environmental factors working together. If your parents are overweight, you may inherit the genes that make it more likely that you will be overweight, but you would only become overweight if you ate too much food and did too little exercise.

**FIGURE 2**: You can't 'develop' dimples, you are either born with them, or without.

### QUESTIONS

1 Give your own example of a difference that is due to environmental factors.

2 Do you think skin colour is due to environmental factors, genetic factors or both? Explain your answer.

**FIGURE 3**: Piercings are an example of environmental variation.

genetic and environmental variation GCSE

B1 You and your genes

# Continuous variation

In some cases, several genes work together to determine a feature. A good example of this is eye colour. Human eye colour does not follow a simple pattern of inheritance, because it is determined by a number of different genes. As a result, there is a wide range of eye colours in humans. This is an example of **continuous variation**, where there is a wide range of possible outcomes.

Some continuously varying genetic outcomes may also be affected by environmental factors. Milk yield in cows, for example, is determined by several genes, but is also influenced by the contentedness of the cow and the quality of grazing.

## QUESTIONS

**3** Use the internet to find other examples of traits that are controlled by several genes and result in continuous variation.

**4** Explain, using ideas about continuous variation, why the average human male height in the UK has increased in the last century from 166 cm to 175 cm.

# Genes versus environment (Higher tier only)

The genetic makeup of an individual organism is called the **genotype**. The characteristics that an individual displays are called the **phenotype**.

**FIGURE 4:** Identical twins (left) have exactly the same genes. Fraternal twins (right) are only as alike as siblings.

The genotype strongly influences the phenotype, but the environment also has an important role to play in determining the physical characteristics of an organism. It is thought that intelligence has a genetic component, but the full potential of an individual can only be reached if they are provided with the stimuli they need by the environment that surrounds them. Similarly, an individual's genotype might enable them to be a successful athlete, but they will only reach that goal if they train well.

Investigations into the relative importance of environmental and genetic factors are often carried out using 'twin studies'. These compare identical and non-identical twins. Identical twins are produced when one zygote splits into two, and each of the cells produced develops into a baby. Identical twins therefore have identical genotypes. Non-identical or 'fraternal' twins grow from two separate eggs each fertilised by different sperm, so have different genotypes just like other siblings. Twin studies allow scientists to more accurately discover whether 'nature' (the genotype) or 'nurture' (the environment or upbringing) is more important in the development of a characteristic.

### Did you know?

In 2009, identical twin brothers were allowed to go free from court in Germany, despite police knowing that one of them had committed a multi-million-Euro jewellery theft. Because of their identical DNA, it could not be proved which of the brothers had been present at the scene of the crime.

## QUESTIONS

**5** Explain why identical twins share all their genetic material but fraternal twins do not.

**6** Explain why individuals sharing the same genotype might have different phenotypes.

continuous variation    twin studies nature nurture

# How genes work together

## What do turkeys have 82 of, but humans only have 46?

A turkey has nearly twice as many chromosomes (82) as a human (46). The range of chromosome numbers in living things is huge. Some ant species only have 2 chromosomes, a kangaroo has 12 and a maize plant has 20 chromosomes.

**We are learning to:**
> describe how chromosomes and genes are organised in cells
> explain that offspring receive genes from both their parents
> understand what alleles are
> understand that there are some questions that science cannot answer

**FIGURE 1**

## How genes and chromosomes are organised

Genes provide the code that tells a cell what proteins to build. The genes we have determine many of our characteristics, or traits, such as hair colour and eye colour. The genes are connected in long molecules of DNA. These would take up a lot of space if they were stretched out. Instead, they are twisted and folded over into structures called **chromosomes** for storage. Each chromosome contains hundreds of genes.

Chromosomes occur in pairs. Human cells have 23 pairs, so there are 46 chromosomes in total in most human cells. The chromosomes are found in the nucleus of the cell. The sex cells – sperm and egg cells – are different from the other cells in the body, as they have only 23 chromosomes each, one from each pair. This is so that when a sperm and egg cell join together at **fertilisation**, they form a new cell called a **zygote** that has 46 chromosomes, 23 from each parent. The zygote can then develop into a human baby with genes from both its parents.

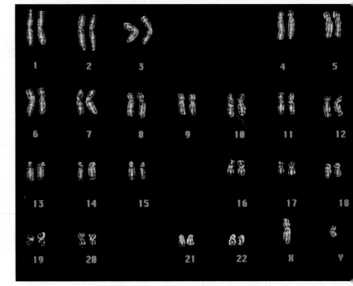

**FIGURE 2**: A **karyotype** is a picture of all the chromosomes in an individual's cells. The chromosomes are stained with a dye to make them easier to see and are arranged in order from 1 to 22 with the sex chromosomes last.

### Alleles

Pairs of chromosomes usually contain the genes for the same trait in the same position. The alternative forms of the genes are called **alleles**. Offspring get one of each pair of chromosomes from their mother and one from their father. This means they have two alleles for each gene, one from each parent.

### QUESTIONS

1. What are genes and why are they important?
2. Explain how an individual can have different alleles for the same gene.

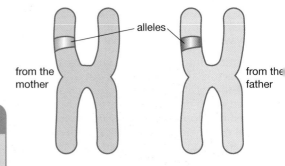

**FIGURE 3**: Alleles are alternative forms of the same gene. They are found in the same position on each chromosome in a pair.

# Chromosomal disorders

Mistakes or **mutations** sometimes occur in the process of creating sex cells or during fertilisation. This can lead to individuals who have different numbers of chromosomes from the usual 46. If a chromosome abnormality is detected during pregnancy, parents face the difficult decision of whether to continue or terminate the pregnancy. In one of the most common of these disorders, Down's syndrome, a baby receives an extra copy of chromosome 21, resulting in 47 chromosomes in total instead of 46. Around one in every thousand babies born is affected, and all will have some degree of learning disability and health problems. Doctors can advise what the likely outcomes may be for a child with a chromosomal disorder, but cannot make the decision about whether the pregnancy should continue; only the parents can make that decision.

**FIGURE 4**: People born with chromosomal disorders such as Down's syndrome can live full and rewarding lives.

## QUESTIONS

**3** Explain how a child has a combination of characteristics from each of its parents.

**4** Find out more about a chromosomal disorder and explain what causes it and what effects it has on an individual.

# Different combinations of alleles (Higher tier only)

It is possible that both alleles of a gene are the same, in which case the individual is said to be **homozygous** for that trait. The root of the word, homo-, is derived from the Greek word which means 'the same'. If the alleles for a gene in an individual are different, the individual is said to be **heterozygous** for the trait. Hetero- is from the Greek for 'different'.

**Watch out!**
Chromosomes can be represented as 'sticks', or as 'X's as here. Under a microscope they appear as 'X's. This is because they have replicated prior to cell division. The arms of the 'X' are copies of one another.

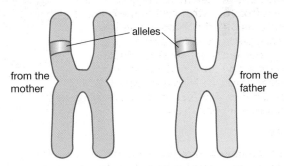

**FIGURE 5**: A pair of chromosomes with a pair of alleles. Here the alleles on each of the pair are the same, so this individual is homozygous. In Figure 3 the individual is heterozygous.

## QUESTIONS

**5** Explain the difference between an individual who is homozygous and one who is heterozygous for a trait.

**6** Explain why nearly all animals have an even number of chromosomes in their body cells.

# Variation in families

**We are learning to:**
> describe the processes that lead to variation in offspring
> explain why offspring of the same parents can be different from each other

## How alike are you and your siblings?

It is likely that you share some characteristics with your siblings, even though it is possible for siblings to look very different from each other. Perhaps you have the same fiery temper or love of music. Some of the characteristics you share will be due to genetics, and some to the environment you share.

FIGURE 1

## Why you look like your parents

In most families, it is easy to see some similarity between parents and children. Parents and children are similar because some of the parents' genes are passed on to their offspring through **sexual reproduction**. Half the genes come from the mother and half from the father. The offspring are not identical but show **variation**. This is because each egg (ovum) and sperm cell contains a different combination of genes from the parent, so each child in a family will receive slightly different genes, even though they have the same parents.

**Watch out!**
Only identical twins, triplets, and so on can have identical chromosomes to each other. All other types of siblings, no matter how similar they look, have an **assortment** of chromosomes from each of their parents.

FIGURE 2: Can you spot the family likenesses in this family?

### QUESTIONS

1 Explain why siblings look similar to each other, but not identical.
2 Can you think of an example of a trait that has been passed on in your family?

## Pairs of alleles

Chromosomes occur in pairs in the nucleus of a cell. Each pair of chromosomes has genes with instructions for the same trait on it in the same locations. These corresponding genes are called alleles. The pair of alleles can have different versions of the genes on them.

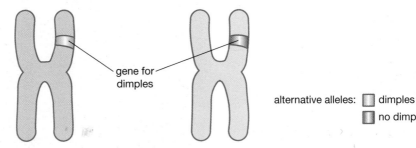

FIGURE 3: The alleles for having dimples or not are located on the same part of a pair of chromosomes. This individual has one of each allele.

For example, the gene in a particular location on a chromosome might code for whether or not a person has dimples. The alternative alleles will be for 'dimples' or 'no dimples'. An individual could have any combination of these alleles. Both alleles could be 'dimples', there could be one allele for 'dimples' and one for 'no dimples', or both alleles could be 'no dimples'.

Offspring receive one of each pair of chromosomes, and therefore one of each pair of alleles, from each parent. Which alleles they receive determines their physical characteristics.

### QUESTIONS

**3** Identical twins have exactly the same genes. They are formed when the zygote splits into two shortly after fertilisation and each of the new cells goes on to develop into a baby. Explain why, even though they have the same genes, identical twins can look different from each other.

**4** Give an example of a physical characteristic that can be affected by both genes and the environment.

## Variation in offspring

A mother and father have 23 pairs of chromosomes in each of their body cells. Each pair of chromosomes contains hundreds of pairs of alleles, and each allele has a number of possible forms. When an egg or a sperm is formed, one of each the 23 pairs of chromosomes (with one of two possible sets of alleles on it) is split off into the ovum or sperm cell. An ovum and a sperm cell only contain 23 chromosomes each. At fertilisation, this unique set of 23 chromosomes from the father (in the sperm) joins with the unique set of 23 chromosomes from the mother (in the ovum) to form a set of 23 pairs of chromosomes that is the gene set of the offspring.

Siblings will receive some of the same chromosomes from each of their parents, but they are likely to share only half their chromosomes, due to the random assortment of the pairs of chromosomes in the formation of the ovum and sperm. This mix of chromosomes, and hence mix of alleles, is one of the factors that leads to variation in the offspring.

### QUESTIONS

**5** Each chromosome in each of the 23 pairs from the father has a one in two chance of making it into a particular sperm cell. Using a probability calculation, what is the chance of two sperm cells containing the same combination of 23 chromosomes?

**6** Some alleles have more possible forms than others. What would you expect the relationship to be between the amount of variation shown in a trait and the number of possible alleles for that trait? Explain your answer.

# Genetic crosses

## Which genes dominate in your family tree?

It might be your glowing auburn locks or your hairy big toe, but most families have a trait that seems to be passed from generation to generation. Can you identify the dominant traits that run through your family?

**We are learning to:**
> explain how dominant and recessive genes interact
> interpret and draw Punnett squares and family trees
> consider ethical issues that genetic counselling might raise

**FIGURE 1**: Freckles may be a trait that dominates in your family.

## How traits are determined

Traits are passed on from parents to their offspring through genes on chromosomes. Genes for particular traits are located on the same place in each of a pair of chromosomes. These alternative versions of the same gene are called **alleles** (see page 14).

A child gets one chromosome in each of the 23 pairs from its mother and one from its father. So for any trait, a child gets an allele for that trait from its mother and one from its father. Sometimes the alleles are the same from both parents, and sometimes they are different.

Alleles can be **dominant** or **recessive**. When the alleles in a pair are different from each other, the trait shown in the offspring is that of the dominant allele. The only way to get the recessive trait is to have both recessive alleles. Look at Figure 2. Dominant alleles are written with capital letters in genetics diagrams, and recessive genes are written with lower-case letters.

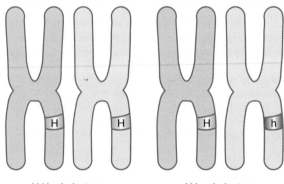 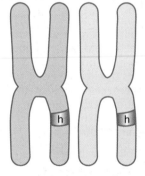

H H = hairy toes     H h = hairy toes     h h = hairless toes

**FIGURE 2**: Hairy toes is a dominant trait. The dominant gene for hairy toes is represented by H, and the recessive gene for hairless toes is represented by h. The only way to get the recessive trait (hairless toes) is to have two of the recessive alleles (hh).

### QUESTIONS

1. Do you have the same alleles as your siblings? Explain your answer.
2. a  What is a dominant trait?
   b  Using the letters F for freckles (the dominant trait) and f for no freckles (the recessive trait), what are the possible combinations of alleles? Which of the combinations would result in an individual with freckles?

dominant and recessive traits

# Genetic diagrams

In genetics, a diagram called a **Punnett square** is used to show clearly all the possible outcomes for a particular trait, and how likely the outcomes are for a particular pair of alleles. Using the 'hairy toes' example, a Punnett square of a genetic cross between a hairless-toed female and a hairy-toed male with the alleles Hh would look like Figure 3.

Another way to show genetic inheritance is using a **family tree diagram**. These diagrams have a key to help you to interpret them, and show how a trait is passed on through the family.

Huntington's disease is a serious genetic disorder caused by a faulty dominant allele. The family tree diagram in Figure 4 clearly shows the likelihood of offspring having the disorder when one of the parents has one dominant allele.

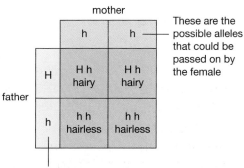

**FIGURE 3**: The Punnett Square above shows the probability of outcomes for hairy and hairless toes.

**Watch out!**
Dominant alleles are always represented by a capital letter, recessive by a lower-case letter. When you draw Punnett squares or label family tree diagrams, make sure you write the letters representing the alleles very clearly, especially for letters such as 'c'.

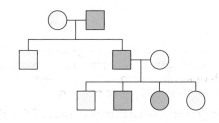

Key
○ female
□ male
● ■ person with Huntington's disease

**FIGURE 4**: The family tree above shows the inheritance of the dominant gene for Huntington's disease.

## QUESTIONS

**3 a** Draw a Punnett square for a cross between a freckle-free male (ff) and a female with freckles and the alleles FF. What proportion of the offspring would have freckles?

**b** Draw a family tree diagram showing this couple's children and grandchildren, indicating which might have freckles.

# Genotype and phenotype (Higher tier only)

> The **genotype** of an organism is the combination of alleles that the organism has.

> The **phenotype** of an organism is its observable physical characteristics.

**Watch out!**
Make sure you are clear about the difference between genotype and phenotype. A good way to remember is by thinking of the sounds of the start of these words: 'phenotype' and 'physical'.

### Did you know?
The Punnett square was first created and used by Reginald Crundall Punnett (1875–1967), a geneticist who studied Natural Sciences at Cambridge University. He had a particular interest in genetics, and studied inheritance through breeding chickens and looking at inheritance patterns in pea plants.

## QUESTIONS

**4** For question 3b above, add the genotypes of all the family members to your diagram.

**5** Family tree genetic diagrams are used by doctors to provide genetic counselling for couples with genetic disorders prior to pregnancy. Explain why they are useful and what ethical considerations such counselling might raise.

Punnett square    patterns of inheritance

# Sex determination

**We are learning to:**
> explain how chromosomes determine the sex of a human baby

## Wanted a sister instead of a brother? Blame your dad!

When an egg is fertilised by a sperm, there is a 50% chance of the resulting baby being female or male. But did you know it is always the sperm that is responsible for determining the sex of the baby?

FIGURE 1

### How the sex of a baby is determined

One of the 23 pairs of chromosomes found in human cells is the pair called the 'sex chromosomes'. In a karyotype, or picture of human chromosomes, the sex chromosomes are always shown as the last pair (see Figure 2 on page 14). Sex chromosomes come in two types, X and Y. An individual can have the sex chromosomes XX, in which case she is female, or XY, in which case he is male.

Whenever fertilisation occurs, there is the same chance that the baby will be male or female.

**XX = female**
**XY = male**

**Watch out!**
X and Y for the sex chromosomes are always written in capital letters and always with X first.

**QUESTIONS**

1. What sex chromosomes do males and females have?
2. Why do you think it is useful to arrange chromosomes into a 'karyotype'? (See page 14 for more about karyotypes.)

# Sex chromosomes

A female has the sex chromosomes XX, so each egg cell will have the sex chromosome X. A male has both X and Y sex chromosomes, so half of the sperm cells will have the X sex chromosome and half will have the Y chromosome.

The **Punnett square** shown here shows how the sex chromosomes can be combined in the offspring. Two possible combinations give a male baby and two combinations give a female baby. Remember the X chromosome is always listed first when writing combinations of sex chromosomes.

|  |  | mother | |
|---|---|---|---|
|  |  | X | X |
| father | X | XX female | XX female |
|  | Y | XY male | XY male |

**FIGURE 2**: This Punnett square shows the possible outcomes for the sex of a baby. You need to be able to draw a Punnett square like this one.

## QUESTIONS

**3** What are the chances of having a male or female baby?

**4** Explain why, for humans, it is always the male that determines the sex of the baby.

# The sex-determining gene (Higher tier only)

It is not the presence of the Y chromosome in itself, but the presence of a gene on the Y chromosome that determines whether an embryo will develop male characteristics or not. The gene is known as the **sex-determining gene**. It is thought that the sex-determining gene on the Y chromosome triggers the development of testes in the growing embryo. If this gene is not present, then ovaries develop and the embryo is a female.

If you look at the pair of sex chromosomes, X and Y, you can see that there are parts of the X chromosome in a male that have no 'matching' alleles on the Y chromosome. Normally, alleles exist as a 'pair' so each can have an alternative allele that 'dominates'. Genes that appear in the region of the X chromosome where there are no alleles on the Y chromosome are shown as traits in the individual even if they are recessive. These genes are said to be sex-linked.

Examples of genes found in this section of the X chromosome are the gene for haemophilia (a blood-clotting disorder) and red–green colour blindness. Because there is no 'matching' allele on the Y chromosome to 'dominate' these genes, males in particular suffer from these conditions.

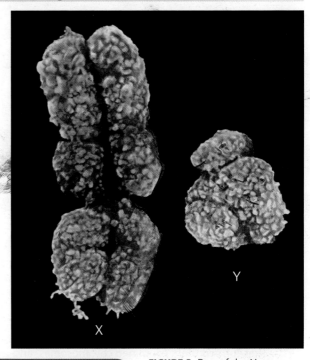

**FIGURE 3**: Part of the X chromosome has no 'matching' part on the Y chromosome, so the traits coded for by the alleles found here always appear in the individual.

## QUESTIONS

**5** Klinefelter syndrome affects about 1 in 1000 male babies. They have the genotype XXY due to receiving an extra X chromosome at fertilisation. Explain why someone with the genotype XXY is male.

**6** Fragile X syndrome is a disorder related to a faulty X chromosome. It causes mental and physical problems for those who are affected. Females tend to suffer from much milder symptoms than male sufferers. Using ideas about sex-linked traits, explain a possible reason for this.

sex-linked traits    haemophilia    Klinefelter syndrome

# Preparing for assessment: Applying your knowledge

*To achieve a good grade in science, you not only have to know and understand scientific ideas, but you need to be able to apply them to other situations. These tasks will support you in developing these skills.*

## She has her grandfather's eyes

Will (14) has a brother, Andy (11), and a sister, Ellie (16). All three are blonde, though Ellie has brown eyes, unlike her brothers' blue eyes. All three are slim, though Will has a rather stockier build.

They all inherited characteristics from their parents, of course. The nuclei of the sperm and egg cells that formed them contained chromosomes, and each chromosome contained genes, which determined their characteristics.

Some of these characteristics are easy to see. Their mother has brown eyes and their father has blue eyes (not that sons inherit from fathers and daughters from mothers). However, brown eyes are dominant over blue, so why haven't all three children got brown eyes? This is because we all carry two sets of genetic data for each characteristic. We display the one that is dominant, but still carry the other. We pass one of these on to our children, and it's just as likely to be the recessive one as the dominant one. The mother was carrying a 'blue eye gene' (from her mother) as well as a 'brown eye gene' (from her father) and it was the first of those that she happened to pass on to the boys, and the second to Ellie.

With other features patterns of inheritance are not quite so obvious. Brown hair is dominant over blonde, so it's perhaps rather surprising that all three children are blonde, since their mother has brown hair and their father is fair. However, a look in the family photo album reveals that the mother was fair until her teens, and one of her grandfathers was fair, so she may be carrying a 'blonde hair gene'.

B1 Applying your knowledge

## Task 1

Read the information opposite to see how the children compare with their parents.
> Why is it that certain features of the children are similar to those of their parents?
> Explain how the genetic information travels from parent to child.
> Should the mother be surprised to have three blonde children?

## Task 2

> Was it a certainty that at least some of the children would have blonde hair?
> Explain the reasoning behind your answer. See if you can use a Punnett square to help.

## Task 3

Will is rather stockier in build than the other two. He's like both his grandfathers in this respect. The other two children are slimmer and slightly taller.
> Suggest two reasons why Will might be different from his brother and sister in this way.

## Task 4

The complex system of inheritance means that humans vary (as do all organisms that reproduce sexually). This variation rarely amounts to anything significant – blondes are no better (or worse) at surviving than people with any other hair colour.
> Why do living things vary?

## Maximise your grade

These sentences show what you need to include in your work to achieve each grade. Use them to improve your work and be more successful.

**E**
For grade E, your answers should show that you can:
> describe how genes determine some characteristics of an organism
> describe how genes produce variation in offspring
> explain what is meant by dominant and recessive characteristics
> explain why offspring have some similarities to their parents

**C**
For grades D, C, in addition show that you can:
> explain why offspring from the same parents can differ from one another
> explain the inheritance of single gene characteristics with a dominant and a recessive allele using a Punnett Square
> describe how alleles in a pair may be the same or different, and explain the significance of different alleles
> suggest, with reasons, why organisms vary

**A**
For grades B, A, in addition show that you can:
> explain why a person will not show the characteristic of their recessive allele
> understand and use the terms homozygous and heterozygous

# Disorders caused by a single gene

**We are learning to:**
> understand how recessive and dominant single-gene disorders are inherited
> describe the symptoms of cystic fibrosis and Huntington's disease
> interpret and explain the inheritance of cystic fibrosis
> discuss the ethical implications of genetic testing

## Why is knowing about gene disorders important?

Whether or not you have a gene disorder, you need a good understanding of how they are passed on through families. A person can carry a gene for a disorder without knowing it. This becomes important if they meet and want to start a family with another 'carrier' of the disorder.

**FIGURE 1**

## Disorders caused by a single dominant gene

Some disorders are caused by a single faulty allele. When a single faulty allele causes a disease in this way, we say it is a **dominant** disorder. This means that anyone with one allele for the disease (remember – alleles come in pairs) will get the disease. An example of a single-gene disorder that is dominant is Huntington's disease.

Symptoms of Huntington's disease usually begin when people are aged between 30 and 50 years. These include:

> tremor, which is an uncontrollable shaking of legs, arms and head
> clumsiness
> memory loss
> inability to concentrate
> mood changes.

Because the symptoms do not appear for so long, many people who are not screened do not know they are sufferers until they have already had children.

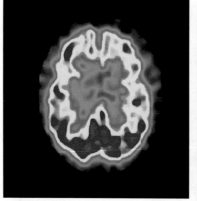 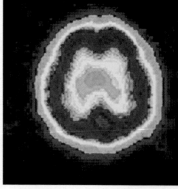

**FIGURE 2**: The photo on the left is a scan of a brain of someone with Huntington's disease. The enlarged green areas (compared with the healthy brain on the right) show regions of nerve cell loss, caused by the disease.

### Did you know?

Huntington's disease was first recognised by George Huntington in 1872. It is caused by a mistake on chromosome 4.

### QUESTIONS

1. What is a dominant disorder?
2. Which alleles could a sufferer of Huntington's disease have?

# B1 You and your genes

## Disorders caused by recessive genes

Disorders can also be caused by recessive alleles. The only way to get a **recessive** trait is to have both recessive alleles. Disorders caused by a recessive gene include cystic fibrosis. Sufferers have alleles cc. This serious genetic disorder begins in early childhood and is caused by a faulty protein in the cell membranes. Sufferers of cystic fibrosis have the following symptoms:

> thick, gluey mucus, particularly affecting their lungs

> difficulty breathing

> tendency to chest infections

> difficulty in digesting food.

The inheritance of cystic fibrosis can be shown in a family tree diagram. It is likely that the parents in the first generation shown in Figure 4 both had the alleles Cc. This is called being a 'carrier' for the disorder (see page 26). When two carriers have children together, they have a one in four chance of having a child with cystic fibrosis.

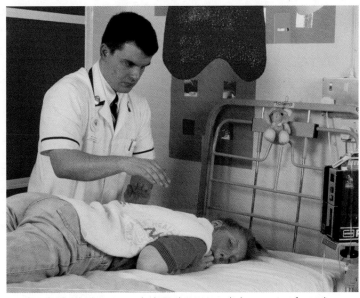

**FIGURE 3**: Physiotherapy to help to loosen and clear mucus from the lungs is an important part of treatment for some sufferers of cystic fibrosis.

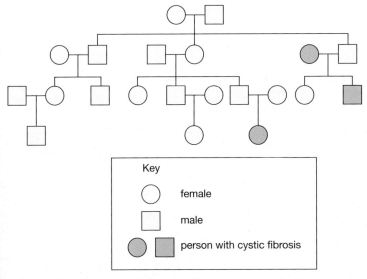

**FIGURE 4**: The inheritance of the cystic fibrosis gene in a family.

### QUESTIONS

**3** Explain why cystic fibrosis sufferers are prone to chest infections.

**4** Use a Punnett square to show the probabilities of different outcomes for offspring from a male with cystic fibrosis (cc) and a female without cystic fibrosis (CC). What proportion of the offspring would have the disease?

## Genetic testing

**Genetic testing** can be carried out for a particular condition if a person is likely to be at greater risk of the disorder because it occurs in their family. It is possible to test people to find out if they have the gene for Huntington's disease or cystic fibrosis. This allows people to get the treatment that they need and provides them with information that might help them to plan for their future, for example whether they plan to have a family.

### QUESTIONS

**5** In a small group, discuss ethical issues that might be raised by genetic testing. Draw up a list of guidelines to help address some of these ethical considerations.

🔍 recessive disorder    cystic fibrosis

# Carrying genes and passing them on

**We are learning to:**
> discuss the ethics of embryo screening
> understand how and why work with embryos is regulated

## What genes are you carrying?

All of us are carrying thousands of genes in our cells that we know nothing about. These are often recessive alleles that do not show in our physical appearance because they are 'masked' by other, dominant alleles. A good example of a gene like this is the gene for cystic fibrosis.

FIGURE 1

## Carrying a genetic disorder

Cystic fibrosis is a **recessive** condition. This means that you only have the disorder if you have two faulty alleles, one from each parent. If one parent has cystic fibrosis then they know that any child they have is at higher risk of suffering from the disorder, but it is also quite possible for two parents who have no sign of cystic fibrosis to give birth to a child who has the disorder. This is because people can have the gene for the disorder without being aware of it. This is called being a **carrier** for the condition. A carrier for cystic fibrosis has one faulty allele.

**Watch out!**
With recessive disorders, a 'carrier' has one recessive gene for the disorder, but usually suffers no symptoms. They do *not* have a mild case of the disorder. The individual only suffers from the disorder if they have *two* copies of the recessive gene.

### QUESTIONS

1. What is a recessive condition?
2. Which genes does a carrier for cystic fibrosis have?

## In vitro fertilisation and embryo screening

In vitro fertilisation (IVF) is a technique that can help infertile couples to conceive a baby. The method involves taking control of the mother's hormones, and causing her to overproduce ripe eggs in her ovaries. These eggs are then collected and mixed with a sample of her partner's sperm. Some of the eggs are fertilised and begin to develop into human **embryos**. A number (usually no more than three) of well-developed embryos are then implanted into the mother's uterus via a hollow tube through her cervix. Some of these embryos may implant in the uterus wall and develop into **fetuses**.

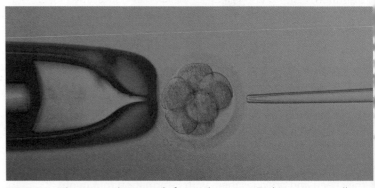

**FIGURE 2**: A human embryo ready for implantation. Embryos are usually implanted back into the mother two or three days after the eggs have been collected. There are usually between four and eight cells at that stage.

cystic fibrosis carrier

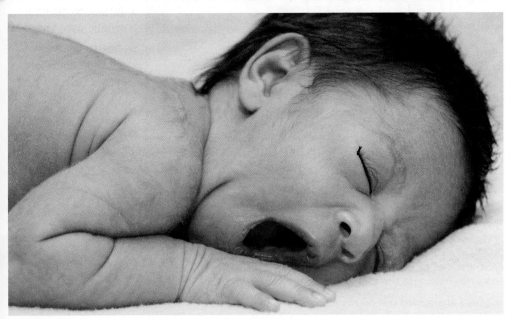

FIGURE 3: Couples with a history of a genetic disorder in their family can improve their chances of having a healthy child by having their embryos screened.

IVF is very useful as it allows doctors to investigate the genetic make-up of the embryos prior to implantation in the mother in a process known as 'embryo screening'. In families where there is a known history of a genetic disorder such as cystic fibrosis, the embryos can be screened to find out if they have inherited the faulty genes. This screening allows doctors to remove any embryos that would suffer from the disorder and only implant genetically normal embryos back into the mother. Embryos that are not required can sometimes be frozen for future use, donated for research or used by infertile couples through embryo donation. Unwanted embryos are destroyed.

## QUESTIONS

**3** What combination of alleles would doctors be looking for in embryos that would be suitable to implant into a mother with a family history of cystic fibrosis?

**4** Some people think that it is unethical to discard an 'unwanted' embryo, even if it only consists of between four and eight cells. What reasons might they give for their view?

## Regulating work with embryos (Higher tier only)

Screening embryos prior to implantation and only using healthy embryos is called **pre-implantation genetic diagnosis (PGD)**. Procedures like PGD and fertility treatments, along with embryo research, are carefully monitored in the UK. Guidelines are provided to clinics and research centres about how embryos are used, making sure ethical and moral considerations are taken into account.

When decisions are controversial about whether treatment such as PGD should take place, often an 'ethics committee' will discuss and take the decision. This committee usually consists of a number of health care professionals, but may also include other people not involved in medicine, to help to provide the most balanced view possible.

## QUESTIONS

**5** Why are fertility treatment centres and embryo research facilities monitored?

**6** What are the advantages of using pre-implantation genetic diagnosis?

### Did you know?

In 2007, 42 babies were born in the UK following pre-implantation genetic diagnosis.

# More about genetic testing

**We are learning to:**
> describe reasons for genetic screening
> understand some of the implications of genetic testing
> understand that people may have to make difficult choices about genetic testing

## Who should make decisions about genetic testing?

You may think it is obvious who should make the decision – the person being tested. But what if the person who will be tested is still an embryo? Rules are in place that cover situations like this to try to make sure the best interests of the maximum number of people are met.

### Genetic screening

Genetic screening can be carried out for a particular condition even when there is no family history of a disorder. Genetic screening of large numbers of individuals is sometimes used to identify sufferers of genetic disorders as early as possible, in order to minimise the damage such disorders can cause.

A number of screening tests are carried out on newborn babies to make sure they are not suffering from rare genetic disorders. Almost all babies in the UK are checked using a 'blood spot test' where a drop of blood is taken from the baby's heel and tested for disorders such as MCADD (a disorder where fats cannot be broken down properly) and phenylketonuria. These conditions are very unusual, but they can cause permanent disability and brain damage to the few babies that do suffer from them if they are not detected very early.

There are many other reasons why people are screened. For example, a patient may be tested before a doctor prescribes a certain drug, to see how effective that drug would be for that individual, or to check for an allergy to the drug.

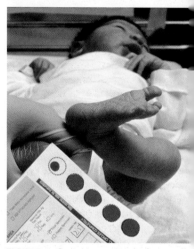

**FIGURE 1**: Genetic testing – almost all babies have a blood spot test to screen for a number of disorders, even though these disorders are very rare.

> **QUESTION**
>
> 1  Why is the newborn screening programme important, even though the conditions it tests for are rare?

### Genetic testing in pregnancy

Some genetic tests are carried out during pregnancy. These involve **cell sampling** – collecting cells from the fetus while it is developing inside the mother's uterus.

A long needle can be used to take a sample of the amniotic fluid from around the baby. The fluid contains some of the baby's cells, which can be tested for the alleles for specific genetic diseases or for chromosome abnormalities. This process is called amniocentesis.

Another test used in pregnancy is chorionic villus sampling. In this test a sample of cells from the placenta is taken through the mother's cervix or through her abdomen.

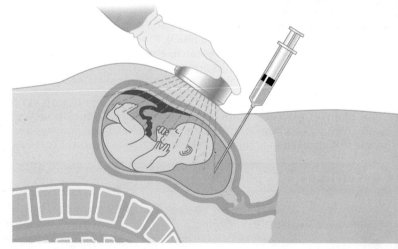

**FIGURE 2**: An ultrasound probe is used to monitor the procedure when amniocentesis is carried out to make sure the needle does not touch the fetus.

newborn screening    antenatal testing

B1 You and your genes

Both these tests carry a risk. One or two mothers out of 100 will have a miscarriage as a result of the procedure. For many, this risk is worth taking to find out whether their baby is suffering from a serious disorder.

### QUESTIONS

**2** Karyotyping is used to determine whether the baby has a chromosomal abnormality. What is karyotyping and how is it useful? (You may need to look back at page 14.)

**3** Why might parents want to know whether their unborn baby has a chromosomal or other genetic disorder?

**4** What sort of information would be important to parents who found through an amniocentesis that their child has a genetic abnormality?

## Implications of genetic testing

The possibility of genetic testing raises a number of important ethical considerations. For example, if an individual is diagnosed with a disorder such as Huntington's disease, should they have to tell other people, such as their place of work or their health insurance provider? Should they be compelled to share the information with their family?

Occasionally, a genetic test can give a 'false positive' result, where the individual *does not* have a disorder but the test had a positive result, or a 'false negative' result, where the test was negative but the individual *does* have the disorder. Different tests have different levels of accuracy. Amniocentesis is the most accurate test, giving the correct result over 99% of the time.

The results of genetic tests on fetuses often raise very difficult decisions for parents. If they have received news that their child is suffering from a genetic abnormality, they must decide whether to continue with the pregnancy or terminate it. Parents need a lot of support and information at this time. Some couples who are aware that they are likely to pass on a genetic abnormality to a child may decide not to have a family at all. Others decide that they will not have testing and will cope with a child with a disorder if they have one. Genetic disorders have a very wide range of effects, and some people live productive, happy lives despite having a genetic disorder.

**FIGURE 3**: Parents who have babies with genetic disorders usually need time to come to terms emotionally with the new situation. Finding out the news during pregnancy can give parents valuable time to prepare.

**Watch out!**
You need to be able to discuss all the ethical issues raised by genetic testing of embryos, children and adults. You should present a reasoned argument, outlining the evidence you have considered and justifying why and how you have reached your conclusions.

### QUESTIONS

**5** Which is worse, a false positive or a false negative? Explain your answer.

**6** It is possible for insurance companies to screen blood to test whether those who are seeking insurance are likely to suffer from particular genetic disorders. People would have to pay more if they were likely to get ill, but others would pay less. Should this be allowed? Justify your answer.

ethical issues antenatal testing    ethics genetic screening

# Cloning

**We are learning to:**
- explain what a clone is and how a clone is made
- explore some of the issues for society relating to cloning of humans

## Will there ever be another one of you?

You may have heard of Dolly, the cloned sheep, and perhaps stories about wealthy people who have had their favourite pet cloned. But how easy is it to clone an organism?

**FIGURE 1**: Snuppy is a clone. He has identical genes to his parent. Of the 1095 eggs created, two puppies were born and only Snuppy survived.

### Cloning in nature

**Clones** are individuals with identical genes. Bacteria, plants and even some animals can reproduce asexually to form clones. In **asexual reproduction**, only one parent is involved, so the offspring has identical DNA to the parent.

Plants can reproduce asexually in a variety of ways. Some plants, such as brambles, send out runners, which are long shoots that grow along the surface or just below the surface of the ground. The runners send out roots and grow into an identical plant at a short distance from the parent plant. Plants can also produce bulbs that grow into offspring with identical DNA to the parent plant.

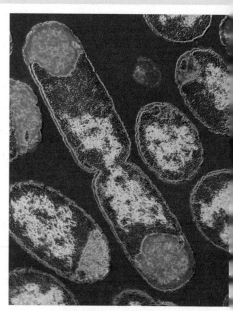

**FIGURE 2**: The bacterium in the centre is dividing to make two new individuals.

**FIGURE 3**: Daffodils, tulips and crocuses all produce bulbs as a method of asexual reproduction.

Some animals can reproduce asexually, although this is quite unusual. An animal that can reproduce in this way is the hydra. When a hydra reproduces asexually, the offspring 'buds' out from the body of the parent. Echinoderms, such as starfish, can grow back from a small piece of tissue.

> **QUESTIONS**
>
> 1  Explain the difference between sexual and asexual reproduction.
>
> 2  Bacteria reproduce by dividing in two. The number of bacteria can double every 20 minutes. If there are 10 bacteria included when you make a sandwich, how many will there be one hour later, when you eat it?

**FIGURE 4**: This hydra is reproducing by budding.

asexual reproduction    cloning

## The environment and clones

As clones have identical DNA, any differences between clones and their parents are likely to be due to the environment they are in, rather than due to their genes.

There are advantages to cloning as a method of reproduction. If the parent has been successful in their habitat, then genetically identical offspring are likely to be successful too. Asexual reproduction also allows the individual to focus their energy on growing and dividing, rather than looking for a partner for sexual reproduction. Plants and animals in very isolated and stable habitats often reproduce asexually. However, the disadvantage is that, as the offspring do not show any genetic variation, any change in conditions or a disease that affects the parent could affect the whole population of clones.

### QUESTIONS

**3** Explain what sort of differences you might see between cloned plants and their parent.

**4** Explain why plants in very stable and isolated habitats often reproduce asexually.

## Animal and human clones

Apart from the few animals that can reproduce asexually, there are other ways that animal clones can occur. Identical siblings have identical DNA, so they are clones of each other, rather than of their parents. Identical siblings, most commonly twins, are produced when the fertilised egg splits, producing two or more genetically identical individuals. Like cloned plants, cloned animals can be different from each other because of the different environments that they encounter despite having identical DNA.

### Artificial animal cloning (Higher tier only)

Clones of animals can also be created artificially. This process is complex, and is the method by which Dolly the sheep and Snuppy the dog were created.

The nucleus containing the genetic material is extracted from an adult body cell and inserted into an empty egg cell. This gives the egg cell a full set of genes, without being fertilised. If the egg begins to develop into an embryo, it can be implanted into a suitable female – a surrogate mother. The cell develops normally to become an individual with identical DNA to the 'donor'.

It is illegal to create clones of human beings in many countries of the world including the UK. Although scientists can create embryos, these cannot be implanted into surrogate mothers. In 2005, British scientists created cloned human embryos as part of research into what is known as 'therapeutic cloning', but these embryos were used only for research. Some scientists from countries where human cloning is not illegal have claimed to have cloned humans, but their claims have been shown, through DNA testing, to be false.

### QUESTIONS

**5** Find out what 'therapeutic cloning' is and how it might be used.

**6** Why do you think human cloning is illegal in many countries?

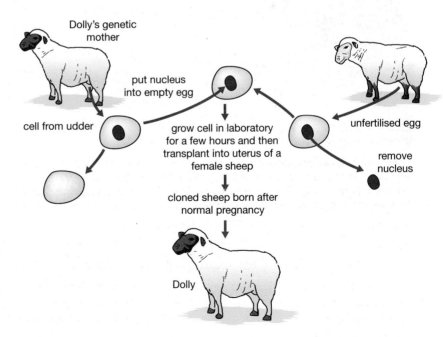

**FIGURE 5**: How Dolly the sheep was cloned.

Snuppy the dog    human cloning

# Multipurpose stem cells

**We are learning to:**
- understand the properties of stem cells
- explore the potential use of stem cells in medicine
- justify ethical decisions made involving stem cell research

## What will stem cells do for us?

Stem cell technology could transform medicine. For example, stem cells have already been used to grow skin tissue. Technology like this could be used to grow features to help people with severe facial disfigurement, such as those with serious burns.

### Stem cells

A human embryo develops from a single cell into a baby with all the different organs and cells that make up its body. This happens because the unspecialised **embryonic stem cells** that form in the first few days of life can develop into *any* type of cell. Embryonic stem cells can be removed from the embryo at around a week old when it has only about 100 cells and is a ball smaller than the full stop at the end of this sentence.

**Adult stem cells** are only found in certain parts of the body. These cells can repair or replace cells by becoming a different type of cell. For example, the bone marrow contains adult stem cells that are able to form several different types of blood cell. But adult stem cells can only develop into *some* other types of cell.

Once cells have become a particular type, or specialised, they cannot change. This process is called **differentiation**, and occurs before 12 weeks in a human pregnancy.

**Watch out!** Adult and embryonic stem cells are different. Adult stem cells can only make a limited number of other cell types. Embryonic stem cells can make *any* type of cell.

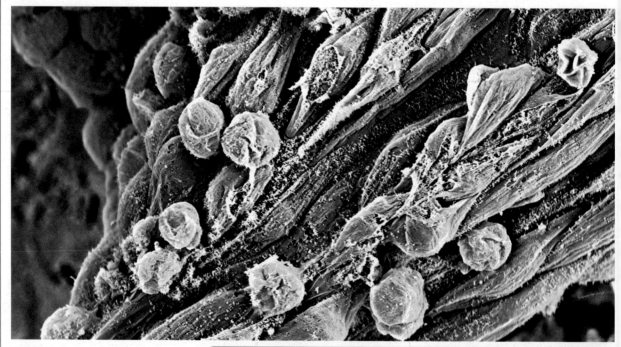

**FIGURE 1:** Stem cells like these can be grown in a lab to form millions of cells, known as a 'cell line'. The 'cell line' can then be differentiated into whatever types of cells are needed.

### QUESTIONS

1. How old are the embryos when the stem cells are removed?
2. What is the difference between adult and embryonic stem cells?

adult stem cells   embryonic stem cells

# The potential of stem cells

Stem cells are extremely useful to scientists as they can be used to produce any other type of cell. This is an exciting area of research, as it might allow treatments or even cures to be developed for many diseases. However, using embryonic stem cells is controversial. This is because the stem cells are generally taken from living human embryos, usually those that are surplus to requirements after fertility treatments. Scientists are working on ways of creating stem cells that can become any human cell by 'reprogramming' adult body cells, so that embryos do not need to be destroyed.

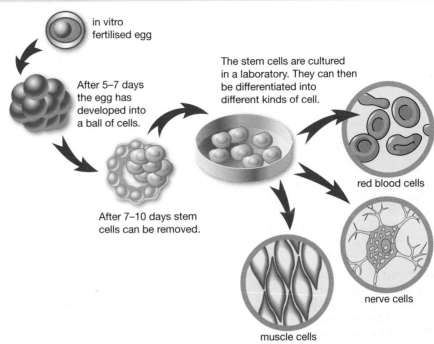

**FIGURE 2**: Stem cells from a human embryo can be used to make any type of specialised cell.

## Did you know?

Some stem cells are found in umbilical cord blood. It is already possible to harvest the cord blood when a baby is born and store it in case the child becomes ill and needs treatment with stem cells later in life, even though these treatments have yet to be developed.

## QUESTIONS

**3** Why is it useful to be able to develop cells to experiment on?

**4** What types of fertility treatments might generate surplus embryos?

# How stem cells could cure disease

Human stem cells have many potential uses. Stem cells that have been differentiated into particular specific body cells could be used to test drugs for their effectiveness when treating disease. Embryonic stem cells are the most useful as they are able to become any other type of body cell. Scientists know that differentiation of stem cells involves the switching on and off of particular genes. For example, in a heart cell, the genes for making heart muscle would be switched 'on' and many other genes that were not needed by the heart cell would be switched 'off'.

A very important application of stem cell research is the possibility of creating cells and tissues. Currently, organ donation is used but organs are in short supply. Stem cell therapies could provide a renewable source of cells to treat disorders such as spinal injury, Alzheimer's disease and heart disease. Intensive research is being carried out on how stem cells could be used in this way. Current studies into the use of stem cells to treat Parkinson's disease and spinal injury have shown very promising results.

## QUESTIONS

**5** Find out more about switching genes 'on' and 'off'. How is differentiation thought to be controlled?

**6** Much of the research into stem cells is being carried out on rodents. Is such research justified? Explain your answer.

# B1 Checklist

## To achieve your forecast grade in the exam you'll need to revise

Use this checklist to see what you can do now. Refer back to pages 10–33 if you're not sure.

Look across the rows to see how you could progress – ***bold italic*** means Higher tier only.

Remember you'll need to be able to use these ideas in various ways, such as:
> interpreting pictures, diagrams and graphs
> applying ideas to new situations
> explaining ethical implications
> suggesting some benefits and risks to society
> drawing conclusions from evidence you've been given.

Look at pages 312–318 for more information about exams and how you'll be assessed.

| To aim for a grade E | To aim for a grade C | To aim for a grade A |
|---|---|---|
| understand that genes determine some characteristics of an organism | recall what genes are and understand how they determine characteristics | understand what genes do, and recall the difference between structural and functional proteins |
| recall how many chromosomes there are in body cells and in sex cells | recall how many chromosomes there are in body and sex cells and how many alleles for each gene an individual usually has | |
| recall that an individual usually has a pair of alleles for each gene | understand that alleles in a pair may be the same or different, and understand the significance of different alleles | ***understand that having both alleles the same in a pair is called homozygous, and having different alleles in a pair is called heterozygous*** |
| understand that genes produce variation in offspring | understand how genes combine and produce variation in offspring, why offspring have some similarities to their parents and why different offspring from the same parents can differ | |
| understand what is meant by dominant and recessive characteristics | understand the meaning of dominant and recessive alleles, and use family trees and Punnett squares to show the inheritance of single gene characteristics with a dominant and a recessive allele | understand why a person will not show the characteristic of their recessive allele; ***understand the terms genotype and phenotype*** |
| recall the difference between the sex chromosomes in human males and females | | ***understand how the presence or absence of the Y chromosome triggers the development of testes or ovaries*** |

| To aim for a grade E | To aim for a grade C | To aim for a grade A |
|---|---|---|
| recall some symptoms of Huntington's disease and cystic fibrosis | recall the symptoms of Huntington's disease and cystic fibrosis and understand how these are caused, and use family trees and Punnett squares to show the inheritance of a single gene disorder | understand the roles of dominant and recessive alleles in Huntington's disease and cystic fibrosis, and understand that a person with one recessive allele will not show the symptoms of the disorder but may pass the recessive allele to their children |
| describe uses of genetic testing for screening adults, children and embryos | understand the implications of testing adults and fetuses for alleles that cause genetic disease | *understand the implications of testing embryos for selection prior to implantation in IVF, and understand the implications of the use of genetic testing by non-medical organisations* |
| understand that bacteria, plants and some animals can form clones naturally | understand how clones occur naturally and the likely cause of differences between clones | *understand how clones of plants and animals can be produced artificially* |
| understand what stem cells are, and when the majority of cells in multicellular organisms become specialised | understand the difference between adult and embryonic stem cells and how they can be used to treat some illnesses | |

# Exam-style questions

## Foundation level

**1** These graphs show death rates from heart disease in the USA by ethnic group.

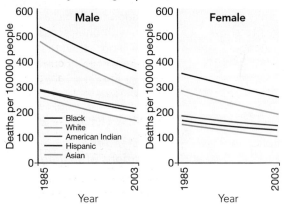

AO2 **a** How do the patterns of male and female death rates compare? [1]

AO3 **b** Which ethnic group has the lowest risk of death from heart disease? Suggest a reason why this group has the lowest death rate. [2]

AO3 **c** Suggest how this data could be useful. [2]

[Total: 5]

## Foundation/Higher level

AO2 **2** Birds have a different sex determination system from humans. In birds, the female has the chromosomes YZ while the male has the chromosomes YY.

Which of the following statements is/are true and which is/are false? [4]

A In birds, the sex chromosome in the female egg determines the sex off the offspring.

B It is likely that sex-linked disorders in birds affect the female more than the male.

C Regardless of chromosomes, it is always the male who determines the sex of the offspring.

D In humans, the male sex chromosome is responsible for determining the sex of the offspring.

[Total: 4]

AO2 **3** Dave has Marfan syndrome. It causes disorders of the connective tissue as well as heart problems. Marfan syndrome is a dominant disorder. Alison is Dave's partner. She is pregnant. The allele for Marfan syndrome is represented by the letter M. The recessive allele for not having the syndrome is m.

**a** What alleles *could* Dave have? [2]

**b** Dave has genetic testing and doctors tell him he has only a 50% chance of passing Marfan syndrome onto his baby. What alleles *does* Dave have? [1]

**c** Dave wants the baby tested for Marfan syndrome. Alison doesn't, as she would want to continue with the pregnancy whatever the result. Explain the advantages and disadvantages of testing the baby before birth to find out whether it has Marfan syndrome. [4]

[Total: 7]

## Higher level

AO1 **4** Look at the diagram below, which shows alleles on pairs of chromosomes from different individuals.

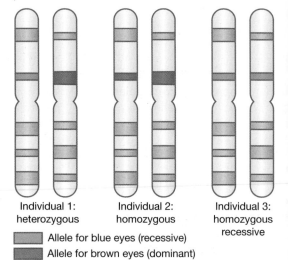

Individual 1: heterozygous
Individual 2: homozygous
Individual 3: homozygous recessive

☐ Allele for blue eyes (recessive)
■ Allele for brown eyes (dominant)

**a** Explain what colour eyes each of the individuals 1, 2 and 3 will have, and what their genotypes are. Use B for the dominant brown-eyed gene and b for the recessive blue-eyed gene. [3]

**b** Draw a Punnett square to show the possible offspring if individual 1 had children with individual 2. Identify both genotypes and phenotypes of the offspring. [4]

[Total: 7]

AO1 **5** Describe what *pre-implantation genetic diagnosis* is and explain why it is useful. Give an example of a situation in which it might be used.

The quality of your written communication will be assessed in your answer to this question. [6]

[Total: 6]

# Worked example

This graph shows the number of laboratories that offer genetic testing and the number of diseases that can be tested for.

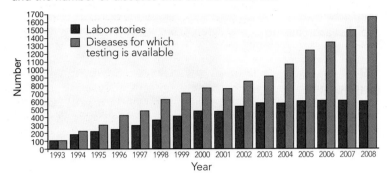

**AO2 a** Describe the pattern shown on the graph. [2]

*The number of laboratories that offer genetic testing has increased from 1993 to 2008, and the number of diseases that can be tested for has also increased.* ✔

This answer would only receive 1 mark, as, although a simple pattern has been stated, the answer gives insufficient detail. For 2 marks, the student should have included dates and figures in the answer to illustrate the point. For example: The number of genetic disorders that can be tested for has increased from 100 diseases in 1993 to nearly 1700 diseases in 2008, with a rapid increase since 2003.

**AO3 b** Suggest why the number of diseases it is possible to test for has increased so rapidly since 2003. [2]

*Because scientists know more about human genes now and have found out the genes which cause a lot of the diseases. When they know which genes cause the diseases, it is possible to test people for them to see if they will get the disease or to see whether they have the genes for the disease.* ✔ ✔

This is a good thorough answer. It would score the full 2 marks.

**AO3 c** Although it is possible to test adults to find out whether they have inherited a genetic disease that runs in their family, some adults choose not to be tested. Explain why. [3]

*Some people might have seen other people in the family have the disease and be ill or die from it and might not want to know if they were going to have the same disease. Some people might prefer to live their lives not knowing what they might die from. Also, it might be less important to someone to know if they will get a disease if they are not going to have a family. Someone might change their mind if they were going to try for a family as they might not want to pass on a genetic disease to their children.* ✔ ✔

## How to raise your grade

Take note of the comments from examiners – these will help you to improve your grade.

This is another good answer, although the end of the answer is slightly less relevant to the question asked. The points about not wishing to know whether they would suffer from the disease and also being less concerned if they did not plan on having children is a good one. The answer would score 2 marks out of 3. It does not take into consideration any discrimination the individual might suffer if they were found to have the gene for the disease, for example from an employer or insurance company; a comment on this would have gained the third mark.

# B2 Keeping healthy

## What you should already know...

### Your body has defences against disease

The body has built-in defences to stop disease-causing organisms from getting in.

The skin is waterproof and self-healing.

The blood clots quickly to form scabs if the skin is damaged.

Acid in the stomach ensures that most bacteria in the food we eat are killed before they can cause us harm.

Even our tears have antibiotic properties to help prevent eye infections.

All these defences help to keep infections out and to keep us healthy.

- Why is it important that scabs form quickly if skin is damaged?

### The body is made of cells, tissues and organs

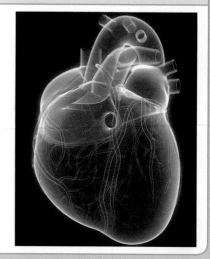

The cells, tissues and organs work together to keep us healthy.

The heart is the organ that pumps blood around the body in the circulatory system.

The heart is made of muscle cells.

The blood takes oxygen and nutrients to the cells and removes waste products.

- Name two other organ systems in the body.

### Our body systems keep us the same, whatever our environment

We are warm-blooded. This means that we maintain our temperature at 37 °C, whatever the temperature is outside our body.

Our nervous system detects changes around us and causes us to respond.

- Give an example of another warm-blooded species.

# In B2 you will find out about...

> microorganisms that can reproduce rapidly inside the human body and cause symptoms of infectious disease

> how the immune system defends the body against infection by microorganisms: the roles of the white blood cells and antibodies

> how vaccinations work

> the use of antimicrobial substances and the development of resistance to these

> how drugs and vaccines are tested in clinical trials, and the ethical issues around using humans in trials

> why the heart is called a 'double pump'

> how to minimise your risk of heart disease

> the link between lifestyle factors and disease in different countries

> what we mean by a correlation

> heart rate and blood pressure and how they are measured

> what homeostasis is and how automatic control systems in the body keep us alive

> how the amount of water in our bodies is controlled, and how alcohol and Ecstasy interfere with this system

> the role of the kidneys in controlling the water level in the body

# Microbes and your body

**We are learning to:**
- explain why some bacteria and viruses cause illness
- appreciate that bacteria and viruses reproduce rapidly in the human body
- interpret data about the rate of reproduction of microorganisms

## How are you?

Most of us have suffered from a bad cold or flu, a chest infection or a sore throat. These illnesses are caused by a variety of microorganisms which grow very well inside our bodies. In fact, our bodies are home to millions of harmless microorganisms that we are not aware of. It is only when harmful microorganisms make an appearance that the body fights back.

FIGURE 1: Illnesses like the flu can make you feel dreadful; here you will find out why.

## What causes disease?

Many diseases that humans suffer from are caused by **microorganisms** such as bacteria and viruses. Harmful microorganisms are called **pathogens**. It is the effects of these pathogens on our bodies that cause us to feel ill.

### QUESTIONS

1. What is the name given to a disease-causing organism?
2. Name two microorganisms that cause disease in humans.

## How microorganisms can harm our bodies

Some bacteria release **toxins** (poisons) that have effects on cells and tissues. Diseases caused by bacteria include

> meningitis, which is an infection that attacks the membrane around the brain

> tetanus, which causes muscle spasms and eventual death if untreated

> salmonella, which causes severe stomach upset and dehydration

> tuberculosis, which infects and damages the lungs, causing breathing problems.

### Did you know?

Many bacteria, such as the *Escherichia coli* (*E. coli*) food-poisoning bacterium, have a tail-like structure called a flagellum, which they use for propulsion.

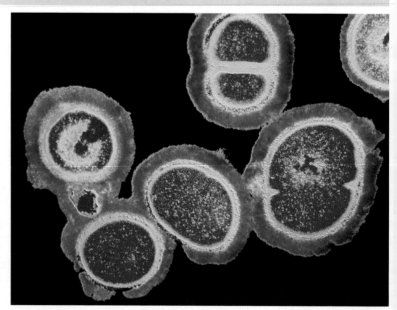

FIGURE 2: One type of the *Staphylococcus aureus* bacterium, which causes a variety of infections including boils and food poisoning, acts by producing a toxin that breaks down the skin layers.

pathogen virus GCSE    diseases caused bacteria / viruses

Viruses cause illness in a different way from bacteria. Once a virus enters a body cell, it takes over the cell and uses it to make copies of itself. These copies can then infect other cells. Sometimes, there is so much of the virus in a cell that it explodes. This damages or destroys the cell and also releases thousands of copies of the virus to infect other cells or be spread to other people. Viruses cause diseases such as flu, the common cold, measles and chickenpox.

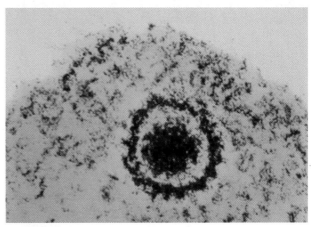

**FIGURE 3**: Flu virus within a body cell. The virus uses the cell's mechanism for making protein to make copies of itself.

## QUESTIONS

**3** Give two examples of diseases caused by bacteria.

**4** Explain how bacteria and viruses cause the person infected by them to feel ill.

## Reproduction of microorganisms

Bacteria and viruses can reproduce very rapidly in ideal conditions. In warm, moist environments with plenty of food (such as inside the human body), bacteria can reproduce every 20 minutes. Bacteria reproduce by a process called **binary fission**, which is a type of asexual reproduction. This means that after 20 minutes one bacterium becomes two, after another 20 minutes those two become four, after another 20 minutes those four become eight, and so on. This type of growth is known as 'exponential' growth.

Exponential growth can only continue as long as there are no resources in short supply. Generally bacteria run out of food or space and the growth rate begins to slow down after the initial phase of exponential growth. A graph of bacterial growth over time shows distinct phases (see Figure 5).

Viruses cannot reproduce without a 'host' cell. They hijack the host cell's own mechanism for making proteins and use it to replicate (make copies of themselves). When a virus has taken over a host cell, it can reproduce very rapidly. Once one cell has been infected and used to replicate the virus, many thousands of viruses are released into the body, each of which will be able to make thousands more once it enters another host cell. Viruses are also very effective at spreading beyond the body of the original host to infect other people. Some symptoms of diseases such as the common cold cause the host to spread the virus to those around them.

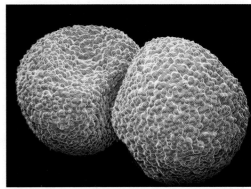

**FIGURE 4**: A bacterial cell undergoing binary fission.

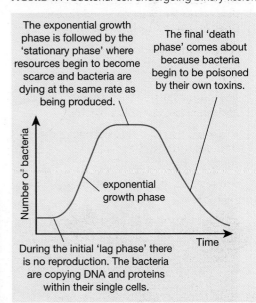

**FIGURE 5**: The growth rate of bacteria in a Petri dish with a thin layer of nutrient agar.

## QUESTIONS

**5** Bacteria reproduce every 20 minutes. If 10 bacteria are present on your tonsil at the start of the school day at 8.30 am, how many will there be at lunchtime (1.30 pm)?

**6** Explain some of the symptoms of the common cold and how these symptoms may help to spread the virus.

how does a virus replicate

# Defending against disease

**We are learning to:**
- describe how the immune system responds to pathogens
- explain the roles of white blood cells in the immune system
- understand how the body responds to pathogens it has encountered before

## Have I met you somewhere before?

Our body's defence system is so sophisticated that it can recognise pathogens that it has already met. These pathogens are in for a rough ride as the body destroys them rapidly and effectively.

### How our bodies fight back against disease

Our bodies have a range of ways to defend us against the invasion of pathogens. The first line of defence is external: our skin with its ability to heal itself, our saliva and our tears. These body fluids, and also stomach acid, either wash away invading microorganisms or destroy them before they can do any damage.

If a pathogen gets past the external defences and into our bodies, then the internal system of defence, called the **immune system**, starts to work. This system attacks pathogens by using several different types of **white blood cell**. These white blood cells are made in the bone marrow. There are two different mechanisms used to destroy pathogens that enter the blood.

> Some white blood cells engulf microorganisms and digest them.

> Another type of white blood cell produces **antibodies** that recognise and destroy particular microorganisms.

### QUESTIONS

1 Outline the defences our bodies have against disease.

2 Explain how two types of white blood cells work in the immune system.

**Watch out!**
White blood cells 'engulf' and 'digest' invading microorganisms, but they don't 'eat' them.

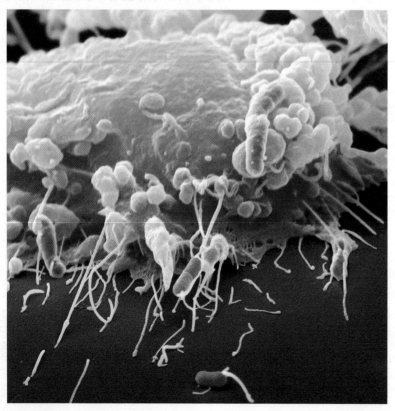

**FIGURE 1:** White blood cells are often large so that they can engulf pathogens that enter the body. This cell is engulfing *E. coli* bacteria (pink rods).

immune system GCSE

# Responses to antigens

The white blood cells that engulf and digest microorganisms attack any invader.

The response of the white blood cells that produce antibodies is targeted at particular pathogens.

Pathogens have proteins in their cell membranes that our immune systems recognise as foreign. These proteins are called **antigens**. Specific antibodies are made for each antigen by the white blood cells.

Antibodies attach to the antigens on the surface of the microorganisms. This inactivates the microorganisms and makes them clump together. They can then be engulfed by other white blood cells or destroyed.

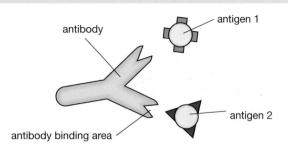

**FIGURE 2**: You can see that the antibody only matches antigen 2. A different antibody is needed to respond to antigen 1

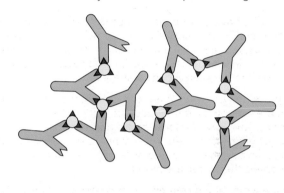

**FIGURE 3**: Antibodies cause the pathogens to clump together.

## QUESTIONS

**3** Draw the antibody that would be needed to bind to antigen 1 in the diagram.

**4** Why do you think the antibody–antigen interaction is sometimes called the 'lock and key' mechanism?

# Recognising antigens

If the immune system has encountered a particular pathogen before, it can respond very quickly to the threat.

The immune system recognises pathogens that it has encountered because the antigens in the pathogen's cell membrane stay the same.

If an antigen is recognised, cells of the immune system called **memory cells** are able to rapidly produce large numbers of antibodies to the pathogen and disable or destroy it before it is able to take hold in the body. When this occurs, the person is said to be **immune** to that pathogen.

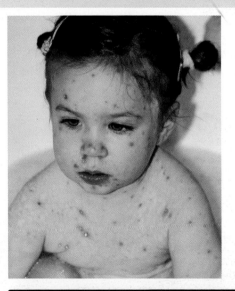

**FIGURE 4**: Chickenpox is an illness that generally only strikes once in a lifetime, as the body develops immunity after the first bout of illness.

## QUESTIONS

**5** What advantage does having memory cells in the immune system give to the body?

**6** Explain what is meant when someone is said to be immune to a disease. Why do you think it is possible to have immunity against some illnesses and not others?

### Did you know?

Parents sometimes hold 'chickenpox parties'. They invite a child who is ill with chickenpox to come to play with their children and other young friends. This is so that the children catch chickenpox when they are young and become immune to it.

# Vaccination

**We are learning to:**
> describe how vaccinations keep us healthy
> explain why it is important for society that children are vaccinated

## Are you part of the herd?

FIGURE 1

Immunologists sometimes refer to groups of people as a 'herd' when they are talking about vaccination. 'Herd immunity theory' claims that it is more difficult for an infection to spread when large numbers of a population are immune. If a significant proportion of the population is vaccinated, the disease stands much less chance of becoming widespread.

## How vaccination works

The immune system is one of the ways that the body defends itself against disease. One of the parts of the immune system is the white blood cells that make antibodies to the proteins on the surface of pathogens. These proteins are called **antigens**. Special cells called **memory cells** are able to produce vast numbers of appropriate antibodies when they find an antigen the body has met before. This is a huge advantage because the disease can be stopped before it has damaged the body, and often before the person even knows that they have contracted it.

A **vaccine** contains a safe form of the microorganism that causes a disease. Once the vaccine is in the body, the immune system attacks the vaccine and develops memory cells against the antigens it carries. These memory cells respond very quickly if they meet the real disease-causing pathogens.

## QUESTIONS

1. What does a memory cell do?
2. How do vaccines work?

# B2 Keeping healthy

## Vaccination programmes

Babies and children undergo a course of **vaccinations** in their first year and as they prepare to go to school. Vaccines usually contain a weak or non-living form of the pathogen to ensure that the vaccination itself does not make the patient ill.

Some pathogens are very stable, with antigens that do not change much from year to year. The vaccines against these pathogens can stay the same for a long period of time. Other pathogens, such as the flu virus, change rapidly and new vaccines must be developed every year to make sure the body will recognise the altered antigens on the pathogen. A vaccination from the previous year would not protect someone from the new strain of flu that appeared the following winter.

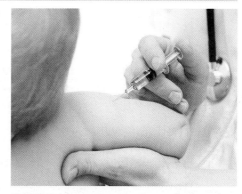

**FIGURE 2**: Vaccinations are generally given by injection under the skin.

### QUESTIONS

**3** Why do babies in the UK have a course of vaccinations in their first year?

**4** Why do vaccines often use a non-active form of the disease?

### Did you know?

Polio is a disease that causes disability and death, particularly in children under 5 years old. In 2002, all European countries were certified 'polio free', thanks to the routine vaccination of young children against this disease.

## Preventing epidemics (Higher tier only)

Epidemics can occur when diseases spread rapidly through populations. The flu epidemic of 1918 killed an estimated 50 million people. Because many diseases have become much less common as a result of widespread vaccination, some parents are now less concerned about whether or not their own children are vaccinated. In addition, people worry about how safe vaccines are. A scare that was begun by a very small study of 12 children published in *The Lancet* medical journal stirred up media interest in vaccine safety. The study and the doctor concerned have since been discredited, but many people stopped having their children vaccinated.

The danger of fewer children being vaccinated is that, if the disease were to begin to spread through the population, the unvaccinated children would be much more likely to contract the illness and to pass it on to other unvaccinated or otherwise susceptible people. To avoid an epidemic it is necessary for a high percentage of the population to be vaccinated. The population then has 'herd immunity'.

Childhood diseases like measles, mumps and rubella can cause particular problems for children, pregnant women and the elderly and can cause infertility, deafness and, in rare cases, even death. The World Health Organisation states that measles is one of the leading causes of death among young children, even though a safe and cost-effective vaccine is available. The aim is to eradicate these diseases through global vaccination programmes.

**FIGURE 3**: The masks worn by these soldiers could not protect them from flu.

### QUESTIONS

**5** There are many more thousands of deaths and long-term health problems from diseases that are preventable by vaccination than there are from reactions to the vaccination itself. Why, then, do you think some people choose not to vaccinate their children? Discuss this and decide what you think.

**6** Write an e-mail to a family member explaining the importance of vaccination for her new baby.

🔍 vaccination GCSE

# Making vaccination safe

## What risks do you take?

Almost every activity we take part in has some degree of risk attached to it. Whether it is walking to school or motocross, we risk accident or injury every day of our lives. We often do not think of the risks, particularly if the activity is something we enjoy.

**We are learning to:**
- explain that some risks are believed to be greater than they actually are
- make links between the use of antimicrobials and the development of resistance to them
- explain why scientists have more confidence in large data sets

**FIGURE 1**: Most activities carry risk but we are unlikely to consider these risks much, particularly if the activity is one we enjoy.

## How safe is safe?

Scientists go to great lengths to make sure that vaccines and prescription drugs are as safe as they can be. Prescription drugs and vaccinations, if used in the right way, are very safe.

Hardly anything is without some degree of risk, even if that risk is very small. However, the *perception* of risk of some prescription drugs and vaccinations is higher than the actual risk. There are many reasons for this. Research has shown that people tend to be more concerned about risks from technology, such as vaccination or nuclear power, than they are about 'natural' risks, like heart disease. Heart disease is the number one cause of death worldwide, yet millions of people do not take actions such as giving up smoking or maintaining a healthy weight to prevent it.

When a new medicine or vaccine is being developed, it is tested very carefully to make sure it is safe and to find out if there are any **side effects**. These are effects that are not wanted and may be harmful or unpleasant. Some individuals will have more side effects from prescription drugs than others, due to genetic variation.

**FIGURE 2**: Obesity, which can lead to heart disease, is a far greater risk to health than most prescription drugs, but it is a risk many people readily accept.

### Did you know?
It can take 10–15 years for a drug to get through all the trials needed to make sure it is safe.

### QUESTIONS

1. Why do you think people are happier to accept risk from activities they enjoy than from vaccination?

2. Why is it important to test drugs thoroughly and on many people before they become widely available?

## Reactions to vaccination

Vaccinations are extremely safe and hundreds of millions of children have benefited from the immunity from disease that they provide. However, in a number of cases, a child may suffer a minor adverse reaction from the vaccination, such as a rash or mild fever. In very rare cases the reaction may be more serious.

Any adverse reactions are carefully documented and followed up. Because of the huge numbers of children that have been vaccinated (for example, it is estimated that over 260 million children have been vaccinated against measles), it is possible for scientists to be very confident of the safety of each vaccine. Vaccines and drugs that, despite clinical testing, seem to be generating an unusual number of adverse reactions, are quickly withdrawn.

### Did you know?

The first ever vaccination was against smallpox and was carried out by Dr Edward Jenner in 1796. Smallpox has now been eradicated worldwide, saving tens of thousands of lives a year.

### QUESTIONS

**3** Why can scientists be confident that vaccines are very safe?

**4** Why is it important that any adverse reactions are reported and followed up?

## Other ways of reducing infection

**Antimicrobials** are a group of substances that are used to kill microorganisms or inhibit (slow) their growth. They are effective against bacteria, viruses and fungi. **Antibiotics**, which are effective against bacteria but not against viruses, are a type of antimicrobial. Antibiotics transformed medicine when they first became available, allowing doctors to treat a variety of common and serious illnesses caused by bacteria, such as tuberculosis.

There has been concern about microorganisms developing **resistance** to these chemicals. Resistance is thought to have come about due to the very wide use of antimicrobials, not just in medicine, but also in the intensive farming industry, where antimicrobials are added in low doses to animal feed as they increase growth rates. Resistance to antimicrobials has already led to some strains of bacteria that are very difficult to control – most types of antimicrobial are ineffective against them. Antimicrobial-resistant microbes are a particular problem where antimicrobials are used frequently, such as in hospitals.

**FIGURE 3:** The use of antimicrobials to increase growth rates of intensively farmed animals has increased the rate at which resistance has developed.

### QUESTIONS

**5** Explain how antimicrobial resistance comes about.

**6** Explain why regular use of antimicrobials, for example in hospitals, can lead to resistance.

antimicrobial resistance GCSE

# Preparing for assessment: Applying your knowledge

*To achieve a good grade in science, you not only have to know and understand scientific ideas, but you need to be able to apply them to other situations. These tasks will support you in developing these skills.*

## New polio vaccine to save thousands of lives worldwide

It is the 1950s. Abi is 6 years old and lives in Chicago, America. She is going to be vaccinated against polio. Polio is a disease that can be prevented but not cured. At this time, teams of scientists have just recently developed vaccines and polio is being virtually eradicated in North America and Europe.

At first Abi doesn't understand why she is at the doctors as she doesn't feel ill. Her mother explains though that polio is a horrible disease, leaving victims crippled, but that it is easy to prevent with a vaccine. A nurse puts a few drops of the vaccine on a sugar lump, which Abi then eats. She doesn't see how eating a sugar lump will stop her catching anything!

What happens is that the sugar lump dissolves in the mouth and stomach, and the polio vaccine, containing a harmless version of the polio virus, enters the bloodstream. The body's white blood cells respond. They start producing antibodies to attack the virus. Because this version of the virus cannot cause the disease, there is no risk to Abi. However, now she has antibodies ready and waiting if the harmful polio virus infects her in the future.

Having the vaccination meant that it was almost impossible for Abi to catch polio, even if the virus entered her body. Discuss in your group how the vaccination works.

# B2 Applying your knowledge

## Task 1

> If Abi caught polio, why might it spread through her body?
> Why is it important to stop people catching polio?

## Task 2

> Explain why, prior to vaccination, the body is unable to use its existing antibodies to deal with the polio virus.

## Task 3

> Vaccination involves putting a live but weakened virus into the body. How does the body recognise a virus?
> What would happen if the polio virus got into Abi's body after she had been vaccinated?

## Task 4

> Describe how immunity to polio comes from prior infection.
> What does the term *immunisation* mean?

## Task 5

When Abi goes back to school and talks to her friends about going to the doctors, she finds out that most of them have also been vaccinated.

> Explain how Abi and her friends' experiences relate to the concept of herd immunity.
> Why is this concept important in the control of infectious diseases?

## Maximise your grade

These sentences show what you need to include in your work to achieve each grade. Use them to improve your work and be more successful.

**E**
For grade E, your answers should show that you can:
> understand that microorganisms can produce substances that damage the body
> understand that our bodies provide ideal conditions for microorganisms to grow and multiply

**C**
For grades D, C, in addition, show that you can:
> explain how our immune systems, including white blood cells, defend us against disease
> explain how antibodies protect us from pathogens and also how they recognise and respond to organisms that have been encountered before, using memory cells
> understand what vaccines are

**A**
For grades B, A, in addition, show that you can:
> understand what vaccinations are and that they work by triggering an immune response to a dead or inactive disease
> explain what is meant by herd immunity and how it can protect a species

# Mutation and resistance

**We are learning to:**
- describe the problems that antibiotic resistance can cause
- understand guidelines that reduce the risk of resistance developing
- explain why a correlation does not necessarily mean that one thing is caused by another

## Superbugs at home?

It is not just in hospitals and doctors' surgeries that superbugs are developing. Many household cleaning products contain antimicrobial agents that could encourage resistance to develop in the microorganisms in your home, making them more difficult to get rid of.

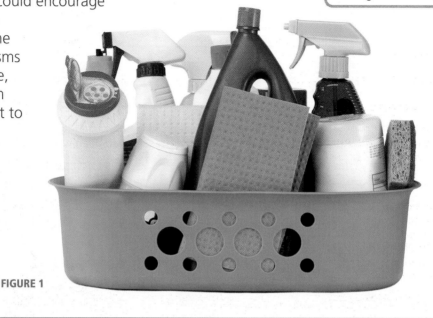

FIGURE 1

## The use of antimicrobials

The word **antimicrobial** refers to a group of chemicals that are effective in destroying microorganisms. There are many different types of antimicrobial, and not all of them are effective against all organisms. Antibiotics are effective against bacteria, antifungals against fungi, and antivirals against viruses.

**Antibiotics** are effective against bacterial infection, but not against viruses. This means they are of no use against common viruses such as colds and flu, unless a bacterial infection (such as a chest infection) develops. This is why doctors do not prescribe antibiotics for the flu. The drugs would have absolutely no effect.

### QUESTIONS

1 Name three types of antimicrobial, and state which microorganisms they are effective against.

2 Explain why antibiotics are not effective against the common cold.

## Resistance to antimicrobials

Bacteria and fungi may develop **resistance** to antimicrobials. Resistant microorganisms are not easily killed by antimicrobial substances. The normal population of microorganisms contains some that are resistant to the antimicrobial. When we use the antimicrobial, all those that are susceptible will be destroyed, leaving the microorganisms that are more resistant to antimicrobials. These resistant microorganisms then reproduce, resulting in a more resistant population (see Figure 2). The more the antimicrobial is applied, the more and more resistant the remaining microorganisms are. These extremely resistant microbes are known as **superbugs**.

antimicrobial   antibiotic resistance

B2 Keeping healthy

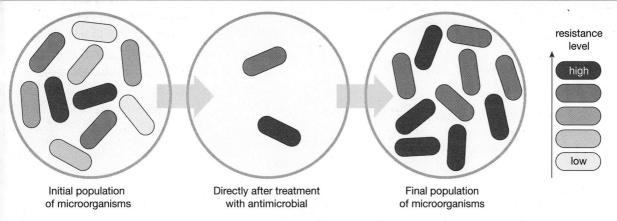

FIGURE 2: This diagram demonstrates how the use of antimicrobials can lead to a more resistant population of microbes.

## How to take antibiotics

Microbial resistance is not only caused by the use of antimicrobials for cleaning, but also when antibiotic drugs are not used correctly by patients. A patient who has an infection that has become resistant to antibiotics is in a very serious position and such infections are sometimes fatal.

There are a number of measures that can be taken to reduce the rate at which antibiotic resistance develops. Antibiotics should only be prescribed if they are really needed, so people should never take antibiotics that have been prescribed for someone else. The course of antibiotics should be taken as directed, and patients should take all the tablets they are told to.

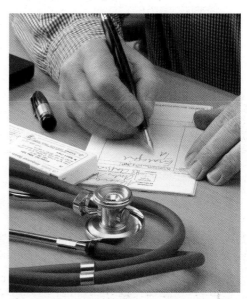

FIGURE 3: Doctors are very careful about how they prescribe antibiotics, in order to reduce the rate of development of resistance.

### QUESTIONS

**3** Summarise in a flow chart how antibiotic resistance develops.

**4** Write a leaflet for patients describing what antibiotic resistance is, how it comes about and how they can reduce the risk of the problem.

## Mutations for resistance (Higher tier only)

Bacterial resistance to antibiotics is a particular problem that scientists are working hard to resolve. As the number of types of antibiotic used has increased, so the incidence of antibiotic resistance has risen. Scientists call this a **correlation**. Further work has to be done to determine whether there is actually a 'cause and effect' relationship when a correlation like this is found. In the case of antibiotic resistance, studies have found that there *is* a cause and effect relationship to this correlation. The use of antibiotics, particularly if they are not used correctly, actually causes antibiotic resistance to develop.

Antibiotic resistance has led to some strains of bacteria that are extremely difficult to eradicate. An example of a bacterium like this is MRSA, which can infect wounds and is particularly found in hospitals. Hospitals are regularly cleaned using antimicrobials to reduce the risk of infection, but this also provides the opportunity for antimicrobial resistance to develop. Random changes (**mutations**) in the genes of microorganisms can allow them to survive the initial application of the antimicrobial. These microorganisms are able to reproduce, thus spreading this mutation for resistance.

### QUESTIONS

**5** Why is antibiotic resistance a problem?

**6** How can some microbes survive the application of an antimicrobial substance?

MRSA    taking antibiotics

# Human guinea pigs

**We are learning to:**
> describe how medicines and vaccines are tested
> consider the ethical implications of clinical trials
> know about different types of clinical trial

## Would you be a human guinea pig?

Human volunteers are, at this very moment, being used to test the safety and effectiveness of new drugs, vaccines and medicines. As well as earning money, these volunteers risk their health to help to ensure that drugs are safe to use.

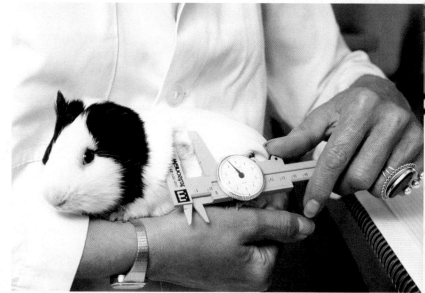

**FIGURE 1**

## Making medicines and vaccinations safe

Before new medicines, treatments and vaccinations are made available to everyone, they are very carefully tested. The initial stages of the testing involve trying out the substance on animals and on samples of human cells to see what the effects of the drug are and to check for toxicity (whether it has any poisonous effect). Medicines and vaccinations that are likely to cause severe side effects do not make it past this stage of testing.

If the drugs seem safe at this stage, they are then tested on humans in **clinical trials**. There are two main types of clinical trial. One type tests the drug on healthy participants and closely monitors them for side effects to ensure that the drug is safe. The participants in clinical trials, particularly this first type, are usually paid for taking part. The other type of trial uses participants who are suffering from a condition and have consented to being part of a trial. The trial will test a new drug or treatment for how well it treats the disease as well as safety.

Clinical trials can be carried out in a single location or in clinics in several locations. For each trial there are very detailed instructions that explain the method that must be followed. This allows the results of trials in several different clinics to be combined.

**FIGURE 2**: Vaccinations are tested for safety in the same way as new medicines.

clinical trials GCSE    how new drugs are tested

## QUESTIONS

1 Participants in some clinical trials are paid volunteers, but occasionally the effects of the drug can be very damaging, even fatal, to the test participants. Do you think it is reasonable to test drugs on humans even though they may be harmed?

2 Why is it important that the methods used are identical if the trial is carried out in more than one clinic?

**FIGURE 3:** Some clinical trials involve actual patients. The trials may be designed to compare the effectiveness of a new treatment for disease with existing treatments.

## Use of placebos

When drug tests are carried out to compare an existing treatment with a new one, some participants receive the existing treatment – they are the **control group**. The other participants receive the new treatment. Sometimes, drug tests are carried out that compare the effects of a drug against a **placebo**. A placebo is a tablet or solution made to look just like the drug being tested, but without the active ingredient. In this case, the control group would be given the placebo.

There are ethical issues to consider when placebos are used in clinical trials. For example, should a placebo ever be used in a clinical trial on patients who are suffering from a disease when an effective treatment for that disease exists? There are guidelines in place to ensure patients are not disadvantaged by taking part in a clinical trial.

### Did you know?

It is thought that the first controlled clinical trial was performed in 1898. This study showed that serum treatment (blood from animals that had been immunised against the disease) was effective against the disease diphtheria. At that time this was a very deadly disease, killing around 50 000 children a year in Germany alone.

## QUESTIONS

3 Why is it important to have a 'control' group?

4 What would be the problem with giving a patient a placebo if an existing treatment existed?

## Trial methods (Higher tier only)

In an 'open-label' trial, both researchers and patients know which drug the patient is receiving. A 'blind' study is one in which the patients do not know which drug they are receiving, but the researchers do know. In a 'double-blind' study, neither patients nor researchers know who is receiving the trial drug and who is receiving an existing treatment or a placebo.

Some trials look at the effects of a drug taken over a long period of time. This is important as some of the side effects may increase or appear over time, or the drug may become less effective as the body deals with it more efficiently.

## QUESTIONS

5 Why is a 'double-blind' study considered to generate the most reliable results?

6 Suggest how long-term monitoring of a drug might be carried out.

# Your amazing heart

## Can you die from a broken heart?

'Broken heart syndrome' is a real medical disorder. Following a shock or period of great emotional stress, sometimes an individual can suffer symptoms very like a heart attack, brought on by physical changes to the heart. Fortunately, with the right care, most people survive.

**We are learning to:**
- describe how the heart works
- link the structure and function of blood vessels
- explain what causes a heart attack

### Two pumps in one

The cells in the body receive the nutrients and oxygen they need from the blood. The blood also removes waste products from the cells. The blood is pumped around the body through a system of specialised vessels by the heart. The heart and the blood vessels make up the **circulatory system**. The heart is a specialised pump called a 'double pump'.

**Watch out!** The heart is labelled left and right as though it was inside a body. When you look at it on paper, the left is on the right!

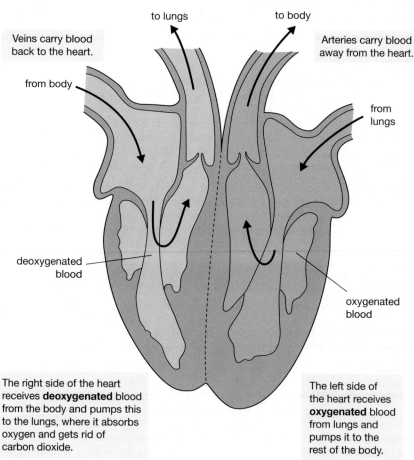

Blood is pumped independently to the lungs and the rest of the body.

Veins carry blood back to the heart.

Arteries carry blood away from the heart.

from body

from lungs

deoxygenated blood

oxygenated blood

The right side of the heart receives **deoxygenated** blood from the body and pumps this to the lungs, where it absorbs oxygen and gets rid of carbon dioxide.

The left side of the heart receives **oxygenated** blood from lungs and pumps it to the rest of the body.

**FIGURE 1**: The heart acts as a double pump.

### QUESTIONS

1. How are human cells provided with nutrients and oxygen?
2. Why is the heart called a double pump?

**Did you know?**

A heart is estimated to pump over 2.5 billion times in a lifetime.

# Blood vessels

There are three types of blood vessels in the body: **arteries**, **veins** and **capillaries**. Because the pressure of blood being forced from the heart as it pumps is high, arteries have very thick, elastic muscular walls to withstand the pressure.

Gradually, the arteries branch into smaller and smaller vessels until blood cells can only just fit through. These tiny vessels are called capillaries. Because they are so narrow, the blood cells travel slowly and because they are squashed against the capillary walls, which are very thin, maximum transfer of substances such as oxygen can occur across the capillary walls.

Capillaries join together again into larger vessels now containing blood with the oxygen removed and with a higher amount of carbon dioxide and other waste products. These larger vessels carry blood back to the heart and are called veins. Because the blood is at low pressure, veins do not need thick walls like arteries. Veins contain valves to prevent the slow-moving blood going backwards or 'pooling' in the lower parts of the body. As the large muscles of the body are moved, blood is squeezed through the veins and back to the heart.

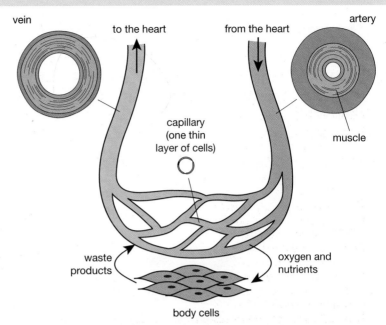

**FIGURE 2**: Arteries, veins and capillaries have different structures due to their different functions.

## QUESTIONS

**3** Explain why artery walls are thick and elastic.

**4** Why do veins have valves in them?

# Supplying the heart with blood

Heart muscle cells can be seen to twitch in very tiny embryos, even before the shape of the heart itself can be seen. Heart muscle works hard and needs a good blood supply. Blood is provided to the heart muscle itself by the **coronary arteries**. These arteries are located over the surface of the heart muscle and provide the oxygen and nutrients it needs, while coronary veins remove waste products such as carbon dioxide.

Like all blood vessels, the coronary arteries around the heart can be partially or fully blocked by fatty deposits. These deposits can slow down or stop the blood flow to part of the heart, which can lead to a heart attack. A heart attack occurs if parts of the heart muscle are not getting sufficient blood supply and stop contracting (beating) in the way they should. How serious a heart attack is depends on where the blockage has occurred.

## QUESTIONS

**5** Suggest why eating a healthy diet might reduce the risk of a heart attack.

**6** Explain how the structure of blood vessels is related to their function.

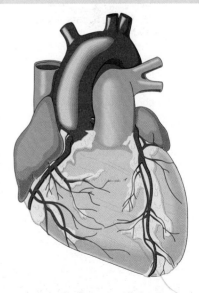

**FIGURE 3**: The heart muscle is supplied with blood by the coronary arteries, shown here in red.

# Keeping your heart healthy

**We are learning to:**
> explain how lifestyle affects the risk of heart disease
> correctly use the term 'correlation'

## How healthy is your heart?

Every day you make choices that affect the health of your heart. Whether you choose to walk the dog or watch the television, eat pie and chips or salmon and vegetables for your dinner, your choices affect your heart health. Read on to discover how to keep your heart as healthy as it can be.

###  Reducing the risk

**FIGURE 1**: Fast foods and convenience foods tend to contain a lot of saturated fat, which can increase the chance of suffering from heart disease.

**Coronary heart disease** is caused by the build-up of fatty substances in the coronary arteries – the blood vessels that supply the heart with blood. Heart disease causes more deaths in the UK than any other disease. It is not caused by a microorganism, but mostly by the choices we make about how we live our lives.

There are many things that you can do to keep your heart healthy, such as avoiding the main risk factors. Smoking is one of the main risk factors for heart disease. The carbon monoxide and nicotine in cigarette smoke causes hardening of the arteries, which can lead to them becoming blocked.

Other factors that increase your risk of suffering from heart disease are a poor diet that contains a lot of **saturated fat**, excessive alcohol consumption, and a lifestyle that causes stress. So you are more likely to suffer from heart disease if you smoke, drink alcohol and have a high-fat diet, but not everyone who has these factors gets heart disease.

**Did you know?**

In the UK, 10 people an hour die from a heart attack.

**Watch out!**

A heart attack is not the same as heart disease. A heart attack occurs when heart disease is so severe that one of the arteries becomes blocked, starving part of the heart muscle of oxygen and causing it to die.

### QUESTIONS

1 Why is heart disease such an important health issue?

2 State four lifestyle factors that increase the chances of heart disease.

coronary heart disease    saturated fat

# B2 Keeping healthy

## Keep on moving

When factors are associated with increased risk of disease, we say there is a **correlation** between these factors and the disease. Smoking, a high-stress lifestyle, and a poor diet are all correlated with heart disease. Members of families that have a history of heart disease are also more likely to suffer from it themselves.

One of the lifestyle choices that can help to prevent heart disease is regular exercise. It is thought that regular exercise strengthens the heart muscle and enables it to pump blood with less effort. Exercise can also help people to maintain a healthy weight and can reduce stress.

One of the reasons saturated fat causes heart disease is because of the **cholesterol** it contains. There are different types of cholesterol, and not all are correlated with heart disease; it is thought that exercise increases the levels of 'good' cholesterol and decreases the level of 'bad' cholesterol, thus reducing risk.

### QUESTIONS

**3** Explain why exercise reduces the risk of heart disease.

**4** Heart disease rates in the UK have fallen recently. Suggest why this might be.

**FIGURE 2**: Any type of exercise will help to reduce your risk of heart disease. If you enjoy the exercise you choose, you are more likely to continue it as an adult.

## A disease of the wealthy?

There are differences in the incidence of heart disease across the world due to different lifestyle factors. For example, in industrialised countries such as the UK and the USA, heart disease rates are high compared to less industrialised nations such as China and India. Heart disease rates in less industrialised countries are rising as these countries become wealthier. It is predicted that as citizens of these nations adopt a more 'Western' lifestyle, with diets higher in saturated fats, a less active lifestyle and more of them take up smoking, heart disease rates in these countries will rise dramatically.

Japan, although highly industrialised, has one of the lowest rates of heart disease in the world. The diet in Japan includes more rice and fish and less meat and dairy products than the diet in countries with higher heart disease risk. Research has shown, however, that when Japanese citizens move to North America and change their lifestyle and diet, their heart disease rates begin to increase. Eventually they are nearly as high as the heart disease rates of people who were born in North America.

### QUESTIONS

**5** What type of diet would lead to lower rates of heart disease?

**6** What is the evidence that the lower risk of heart disease amongst the Japanese in Japan is not due to genetics?

**FIGURE 3**: The lifestyle of many people in the UK puts them at risk from heart disease.

# Get your pulse racing

**We are learning to:**
> explain what the 'pulse rate' is
> understand what a blood pressure measurement tells us
> carry out calculations using experimental data

## How fit are you?

Cardiovascular fitness can be measured not only by the resting heart rate, but also by how long it takes the heart rate to return to normal after strenuous exercise. The fitter you are, the more rapidly your heart rate returns to its 'resting' value after exercise.

## Feel the beat

In some places, blood vessels which carry the blood throughout your body are close to the skin. In these locations, you can measure the **heart rate** by counting the 'pulses' of the blood through the blood vessels; we call this 'taking your pulse'. The **pulse rate** is measured in beats per minute (bpm). The main places that the pulse can be felt are the wrists, the neck and either side of the groin.

The **resting heart rate** is that measured when a person is relaxed. Generally, a low resting heart rate is an indicator of high fitness. A resting heart rate of around 70–100 for teenagers is normal, and a resting heart rate of 50–70 can indicate good fitness. When a person has high fitness, their heart beats more efficiently, pumping more blood around the body per beat.

When you exercise, the muscles need more oxygen and nutrients, and create more waste products to take away. In order to provide the muscles with what they need, the heart beats faster and so pumps the blood more quickly around the body when exercise takes place.

**FIGURE 1**: When sprinting, an athlete's heart rate can rise to over 200 bpm, but then quickly returns to normal.

### Did you know?

A hormone, adrenaline, also increases the heart rate, increases the breathing rate and makes the pupils of the eyes dilate. Known as the 'fight, flight or frolic' hormone, in times of high stress, anger or arousal, this hormone prepares the body for action.

### QUESTIONS

1. How can you measure the heart rate through the skin?

2. What units are used to measure heart rate?

pulse rate    athlete    heart rate    aerobic exercise

# Blood pressure and health

As well as heart rate, **blood pressure** is an important indication of health. The blood pressure measurement records the pressure of the blood on the walls of the artery and is taken with a special device called a sphygmometer. Blood pressure monitors that can be used in the home are also available and they can be useful for those with high or low blood pressure to monitor their condition. Blood pressure is such an important indicator of health that insurance companies will sometimes refuse health insurance cover to those who have high blood pressure.

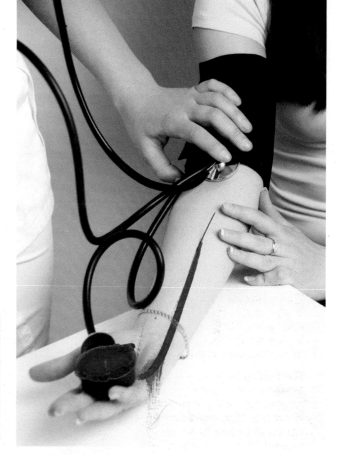

**FIGURE 2**: The blood pressure measurement is a standard check that doctors carry out to check the general health of a patient.

## QUESTIONS

**3** Explain why arteries are under higher pressure from the blood than veins.

**4** Why might insurance companies be reluctant to insure someone who has high blood pressure?

# Too high or too low?

The blood pressure reading is measured in millimetres of mercury (mmHg) and is given as two figures, for example 110/80 (spoken as '110 over 80'). The first (higher) figure is the pressure when the heart is contracting and the second (lower) figure is the pressure when the heart is relaxed.

There is a range of normal blood pressure readings, but very low or very high readings can be a cause for concern. **High blood pressure** often causes no symptoms, but a number of health conditions such as stroke and heart attack are more likely if a person has high blood pressure. **Low blood pressure** can cause dizziness and fainting.

## QUESTIONS

**5** Take your own resting pulse rate and collect the data of the whole class so that you have class data set. Find the mean resting pulse rate and the range of resting pulse rates for the class.

**6** Explain why a narrowing of the arteries, such as is present in coronary heart disease, would lead to an increase in blood pressure.

blood pressure GCSE

# Blood pressure and lifestyle factors

> **We are learning to:**
> - interpret blood pressure readings and explain the risk of high blood pressure
> - understand that a range of heart rates and blood pressures are normal
> - describe how scientists carry out epidemiological studies
> - understand some of the risks of cannabis and Ecstasy

## Would you be part of a medical study?

One day your doctor might ask you whether you would agree to be part of a medical study. This is not a clinical trial, as you will be treated the same if you agree to the study or not, but your response to treatment is recorded and added to the results. Studies like these give scientists useful information on what causes and prevents disease.

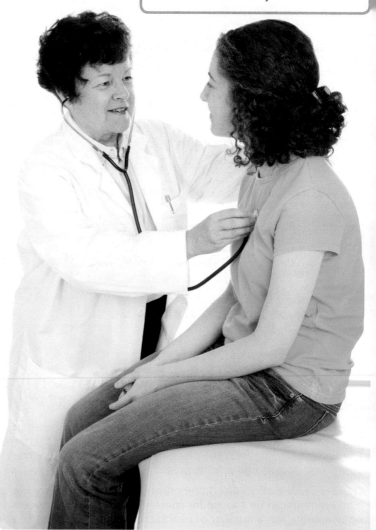

**FIGURE 1**

## High blood pressure

**Blood pressure** measurements record the pressure of the blood on the walls of the arteries. People with consistently **high blood pressure** have an increased risk of heart disease. This is true even in individuals who do not smoke, drink little alcohol and have a healthy diet. This is because high blood pressure can damage the blood vessels and cause them to narrow or harden, reducing the blood flow and making the heart work harder. High blood pressure can also damage the organs, particularly the kidneys, heart and the brain, making a heart attack or stroke more likely.

### QUESTIONS

1. State a health problem caused by high blood pressure.
2. How does high blood pressure cause this health problem?

NHS high blood pressure

## A range of normal readings

As with heart rate, blood pressure varies from person to person, and even in an individual it can differ depending on what they are doing and whether they are relaxed or anxious. The NHS states that 'the ideal blood pressure for a young, healthy adult is 120/80 or less. If you have a reading of 140/90, or more, you have high blood pressure'.

The range of 'normal' readings is because what is healthy and normal for each individual may be different. It is important for adults to have their blood pressure monitored occasionally as there are sometimes no symptoms of high blood pressure.

FIGURE 2: A blood pressure reading is always given as two numbers. The first (higher) figure is the pressure when the heart is contracting. The second (lower) figure is the pressure when the heart is relaxed.

### QUESTIONS

3 High blood pressure is sometimes known as 'the silent killer'. Explain why.

4 Explain why blood pressure is given as two figures (e.g. 120/80).

## Epidemiological studies

In order to investigate whether a lifestyle factor (such as a high-fat diet or smoking) increases the chance of a person getting a disease, scientists compare samples of individuals who are 'matched' on as many factors as possible, but differ only in the factor being investigated. Alternatively, individuals can be chosen at random so that other factors are just as likely in both samples. Studies such as these tend to involve large numbers of individuals and are known as **epidemiological studies**.

For example, doctors trying to find out whether smoking was a factor in the development of heart disease would study two large groups of people, those who already smoked and those who did not smoke, to see whether those who smoked were more likely to develop heart disease. In studies like this, people are chosen who already smoke; it would be against the ethical code for such experiments to ask subjects to take up smoking in order to find out whether it affected the chances of getting heart disease. Some large-scale studies also look at the genes carried by individuals and whether this affects their risk of suffering from particular health problems.

FIGURE 3: Epidemiological studies have found out that the misuse of non-prescription drugs (for example cannabis, Ecstasy, nicotine and alcohol) can increase heart rate and blood pressure, increasing the risk of a heart attack.

### Did you know?

In 2010, half a million adults were recruited into the largest ever epidemiological study in the UK. The study will take DNA samples from the subjects, and then track them for 30 years to find out which genes are associated with cancer and heart disease.

### QUESTIONS

5 Explain why, in epidemiological studies, groups of individuals have to be 'matched' on as many factors as possible, and differ only on the factor being tested.

6 Why would it be unethical to ask an experimental subject to take up smoking in order to investigate its effects on heart disease?

# Homeostasis

## Are you working hard?

Even if it does not look like it, your body is constantly working to keep its internal environment constant, no matter what you do. Whether you biked to school or walked through snow, had nothing for breakfast or a huge fry-up, your body will have kept its temperature and blood sugar almost constant.

> **We are learning to:**
> - explain what homeostasis is and why it is necessary
> - describe the roles of the parts of the homeostatic control system
> - understand the control mechanism

### A constant internal environment

For the body to function as it should, the inside of the body must be a stable environment, even though the environment outside changes regularly. For example, a person might be in a sauna or a cold swimming pool, but the body temperature and the amount of water in the body have to stay almost exactly the same in both environments. Our cells contain enzymes that keep us alive. These function best at a certain temperature and pH, and also at particular levels of water, salt and blood sugar. The maintenance of a steady internal state is called **homeostasis**.

If the body's environment does not stay within a very narrow range of values, the consequences can be catastrophic. If the body temperature drops below 35 °C, for example, as in hypothermia, the body systems begin to shut down and death can occur very quickly.

**FIGURE 1**: Hypothermia is particularly dangerous because, as the condition worsens, the person's behaviour is affected. They can begin to feel that they are warm and start to shed their clothes, hastening their death.

> **QUESTIONS**
>
> 1. What is homeostasis?
> 2. Give two examples of factors in the body that need to be kept stable.

### How the control systems work

The system that maintains homeostasis is in three parts:

> the **receptors** that detect any change in the environment

> the **processing centres** that receive the information and determine how the body systems respond

> the **effectors** that produce the response.

Using a model for the control system can help us to imagine how it works. One example of such a model is a tropical fish tank.

**FIGURE 2**: A fish tank is a good model for temperature regulation in the body.

homeostasis GCSE

The fish tank contains a fluid that needs to be kept at a constant temperature. In this model, the receptor is the thermometer in the fish tank that measures the temperature of the water. The processing unit is the computer that controls the water heater, and the effector is the water heater.

## QUESTIONS

**3** Describe the three parts of the control system that maintains a steady environment in the body.

**4** In the fish tank example, explain what would happen if the water became too cold. Use the words receptor, processing centre and effector.

## Negative feedback (Higher tier only)

An important part of the homeostatic control system is the reversal of changes that have put the body system off balance. This reversal helps the body back to the steady state it should be in. It is called **negative feedback**.

The body's response to temperature changes is a good example of negative feedback. If the body temperature rises over 37 °C, the receptors detect this and send a message to the processing centre in the brain. The processing centre then directs the effectors, in this case the blood vessels, to **vasodilate** (become slightly wider and closer to the skin surface) so that heat can be lost through the skin and the body temperature drops. Once the temperature has dropped back to normal, the receptor would detect this and pass the message to the processing centre, which would direct the effectors (the blood vessels) to return to their normal diameter.

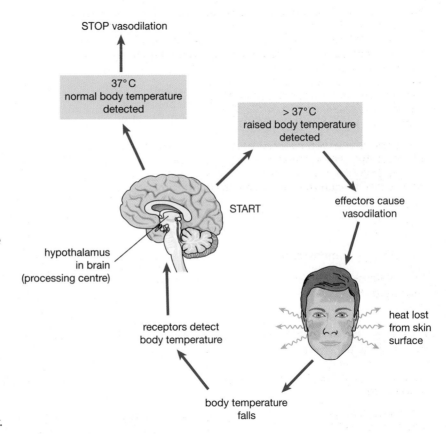

**FIGURE 3**: Negative feedback reverses a change in body temperature.

**Watch out!**
It is important to remember to include the part of negative feedback where the steady state has been reached and the effector is directed to stop.

## QUESTIONS

**5** What is negative feedback?

**6** When the body is too cold, the blood vessels vasoconstrict, conserving heat, and the muscles contract and relax quickly (shivering) to generate heat. Explain, using the terms receptor, processing centre and effector, what happens when the temperature of the body drops below 37 °C.

# Water in, water out

**We are learning to:**
- describe how humans take in and lose water
- understand the importance of a balanced water level
- understand how the kidneys maintain water balance

## How much water is inside you?

Approximately 60–75% of the body is made up of water. Some parts of the body are more watery than others. The blood contains a lot of water (over 90%), whereas the brain contains only about 70% water. Bones are only about 20% water.

## Keeping the right amount of water in our bodies

The amount of water in our bodies, like blood sugar, temperature and level of salts, is controlled by the mechanism of homeostasis. Homeostasis keeps the environment inside our bodies constant, no matter what our external environment is like.

We take water into our bodies through eating and drinking, and water is also made through the process of respiration, when energy is released from food. Water is lost from our bodies when we sweat, when we breathe out (think of the water vapour you can see on a cold day), in our faeces and in the excretion of urine. Sometimes, such as when we exercise a lot, we need to drink fluid to replace the water we have lost through sweating and breathing. Doctors sometimes need to measure the fluids taken in and check the urine excreted by patients, to make sure that their water balance is being maintained.

All the cells in the body are bathed with a fluid called **blood plasma**. It is important that the blood plasma remains at the correct concentration, as the enzymes within the cells work best at a particular concentration of water and salts. If the concentration of water is too high, then the body cells will absorb water. The concentration of the blood plasma can vary according to how much salt we have consumed in our diet, as well as how much fluid we have drunk.

**Watch out!** When you are listing the ways water *enters* our bodies, do not forget the water produced by the cells when they respire.

**FIGURE 1**: The amount of urine excreted affects the body's water balance.

### QUESTIONS

1. How is the amount of water in our bodies controlled?
2. Why do we need to drink after strenuous exercise?

### Did you know?

Seawater and human blood plasma have a similar composition of salt and minerals.

water balance GCSE

# B2 Keeping healthy

## Why water level is important

The body's cells are affected if the concentration of the blood plasma varies. If the blood plasma is too dilute, the cells will absorb water until they burst. If the blood plasma is too concentrated, the cells will lose water until they become dehydrated.

Plant cells can be used to illustrate this, although the plant cells will not burst as, unlike animal cells, plant cells have a cell wall. If you look at red onion cells under the microscope and add a drop of distilled water, you can see the cells swell as they absorb the water. If you then replace the distilled water with concentrated sugar or salt solution, the water leaves the onion cells and the cells become small and dehydrated.

### QUESTIONS

**3** Explain what would happen to body cells if the concentration of water around them was too high.

**4** Why don't we swell up and burst when we are in the bath?

## The kidneys maintain optimum water level

The **kidneys** are responsible for maintaining the levels of water, **urea**, salts and other chemicals in the blood. Urea is produced when proteins are broken down in the liver. It is poisonous to the body and so must be excreted. The kidneys are the effectors in the control system that maintains the optimum amount of water in the body.

The collecting ducts of the kidney pass through a region of the kidney that has very high concentrations of salt. How much water can move through the walls of the collecting ducts varies, and is under the control of a hormone. That means that when the collecting ducts are permeable to water, most of the water passing through them will be re-absorbed into the body rather than passing into the bladder. When the collecting ducts are impermeable to water, all the water will pass into the bladder and be **excreted** by the body. As the amount of urea and other waste materials stays relatively constant despite the amount of water lost, when we have too little water and the body is conserving water, our urine will be very dark and concentrated. Conversely, if we need to lose a lot of water, our urine will be very pale and dilute.

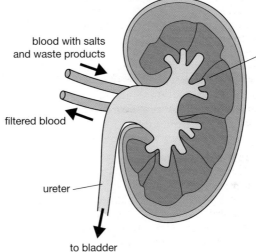

FIGURE 2: The structure of the kidney allows it to re-absorb water into the body if that is needed, or release water into the ureter and eventually the bladder if there is enough water in the body.

### QUESTIONS

**5** Explain where urea comes from and why it must be excreted.

**6** Explain why our urine may be a different colour, depending on how hot the weather is.

urea GCSE    osmosis and blood

# Your kidneys

**We are learning to:**
> describe how the kidneys respond to changes in blood plasma concentration
> explain how alcohol and Ecstasy can disturb the water balance of the body

## Are you looking after yourself?

Your body is equipped with a complex and delicately balanced system that makes sure it always stays in the optimum condition for functioning. Abuse of alcohol and drugs like Ecstasy can upset this delicate balance and lead to unpleasant effects, and even death.

**FIGURE 1**: A teenager died in 1995 after taking Ecstasy and drinking to much water because her blood plasma became so diluted that it caused her brain to swell.

## The kidneys

It is vital that the amount of water in the body, and so the concentration of the **blood plasma**, is kept within certain limits. This is to make sure that the body cells have exactly the environment that they need to function. Too little water, and the cells would become dehydrated. Too much water, and the cells could begin to absorb water and burst.

The concentration of the blood plasma is monitored by receptors in the body. The blood plasma becomes more concentrated if a person consumes a lot of salt, does not take in enough fluid, exercises hard or is in a hot environment. The **kidneys** respond to changes in the blood plasma by changing the concentration of the urine that is excreted.

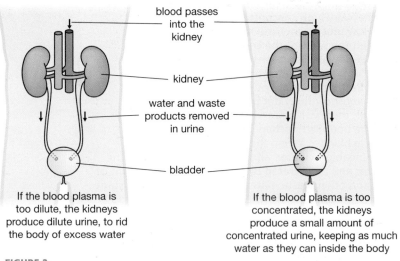

FIGURE 2

### QUESTIONS

1 What would happen to the body's cells if the blood plasma became too concentrated?
2 What situation would give very dilute blood plasma?

how kidneys work

# B2 Keeping healthy

## Drugs that affect water balance

Some recreational drugs can affect how the body responds to a change in the concentration of blood plasma. When alcohol is consumed, the body produces a lot of dilute urine, even when the blood plasma is concentrated. This can lead to dehydration, which is one of the causes of the after-effect of excessive alcohol intake, the 'hangover'. Dehydration can be damaging to health.

The drug Ecstasy has the opposite effect to alcohol. Even if the blood plasma is dilute, only small amounts of concentrated urine are produced. Taking Ecstasy combined with drinking large quantities of water can lead to the blood plasma becoming much too dilute, and the body's cells swelling up with water. If the tissues of the brain swell, this can cause seizures and bleeding into the brain which can be fatal.

### QUESTIONS

**3** Explain what causes a 'hangover' after drinking excessive alcohol.

**4** Why is it dangerous if the blood plasma becomes too dilute?

## The kidneys and ADH (Higher tier only)

Hormones are chemicals that control many of the systems of the body. **Anti-diuretic hormone (ADH)** is released into the bloodstream by the pituitary gland in the brain in response to changes in the concentration of the blood plasma. ADH acts upon the kidneys and causes them to reduce the amount of water lost in urine.

The effects seen when alcohol or Ecstasy are taken are due to the effects of these substances upon ADH.

> Alcohol suppresses the action of ADH, leading to a greater volume of more dilute urine, even when the blood plasma has become too concentrated.

> Ecstasy increases ADH production, leading to a smaller volume of less dilute urine, even when the blood plasma has become too dilute.

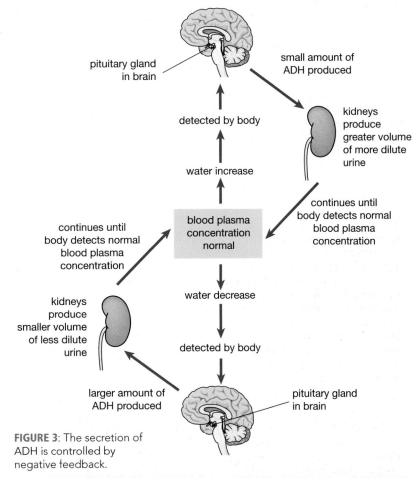

**FIGURE 3**: The secretion of ADH is controlled by negative feedback.

**Watch out!**
A 'diuretic' is a drug that increases the amount of urine produced. Anti-diuretic hormone, therefore, reduces the amount of urine produced.

### QUESTIONS

**5** Explain the role of ADH in the regulation of the concentration of blood plasma.

**6** Explain why excessive drinking of alcohol can lead to a hangover the following day.

# B2 Checklist

## To achieve your forecast grade in the exam you'll need to revise

Use this checklist to see what you can do *now*. Refer back to pages 40–67 if you're not sure.

Look across the rows to see how you could progress – **bold italic** means Higher tier only.

Remember you'll need to be able to use these ideas in various ways, such as:
> interpreting pictures, diagrams and graphs
> applying ideas to new situations
> explaining ethical implications
> suggesting some benefits and risks to society
> drawing conclusions from evidence you've been given.

Look at pages 312–318 for more information about exams and how you'll be assessed.

| To aim for a grade E | To aim for a grade C | To aim for a grade A |
|---|---|---|
| understand that microorganisms can produce substances that harm the body | explain how our immune system, including white blood cells, defends us against disease | understand what vaccinations are and that they work by triggering an immune response to a dead or inactive disease |
| understand that our bodies provide ideal conditions for microorganisms to grow and multiply | explain how antibodies protect us from pathogens and also how they recognise and respond to organisms that have been encountered before, using memory cells; understand what vaccines are and how they are tested | *understand why, to prevent epidemics of infectious diseases, it is necessary to vaccinate a high percentage of a population* |
| recall that chemicals called antimicrobials can be used to kill bacteria, fungi and viruses, and understand that over time bacteria and fungi may become resistant to antimicrobials | understand the ways to reduce resistance to antimicrobials and antibiotics | *understand that random changes (mutations) in the genes of microorganisms sometimes lead to varieties which are less affected by antimicrobials* |
| recall that new drugs and vaccines are tested for safety and effectiveness using animals and human cells grown in the laboratory and then using human trials | recall that human trials may be carried out on groups of healthy volunteers to test the safety of a drug and also on people with illness to see how effective the drug is; understand the ethical issues related to using placebos in human trials | *describe and explain the different types of drug trial, including 'open-label', 'blind' and 'double-blind' human trials, in the testing of a new medical treatment; understand the importance of long-term human trials* |

| To aim for a grade E | To aim for a grade C | To aim for a grade A |
|---|---|---|
| describe the role of the heart as a double pump in the circulatory system and understand why heart muscle cells need their own blood supply | understand how the structure of arteries, veins and capillaries is related to their function, and how lifestyle factors can lead to fatty deposits in the blood vessels that supply the heart muscle leading to a heart attack | *describe actions that could be taken to reduce the risk of heart disease when provided with lifestyle and genetic data* |
| understand what heart rate and blood pressure are and how they can be measured | understand that there is a range of normal measurements for heart rate and blood pressure; that high blood pressure increases heart disease risk, and how large-scale studies are used to identify factors that put people at risk of heart disease | |
| understand that the body keeps the environment inside it the same whatever the external conditions, and this is called homeostasis | understand that the systems that maintain the body's environment have detectors, processing centres and effectors | *understand what negative feedback is and how it reverses changes to a system* |
| understand how the body takes in and loses water, and why it is important for the amount of water in the body to remain stable | understand that the kidneys are important in controlling the level of water, waste and other chemicals in the blood, and that the kidneys change the concentration of the urine according to whether the body needs to retain or lose water; understand that alcohol and ecstasy disrupt this process and can lead to too much or not enough water being lost by the body | *understand that the concentration of urine is under the control of a hormone called ADH, which is released into the bloodstream by the pituitary gland, and understand that ADH release is controlled by negative feedback* |

# Exam-style questions

## Foundation level

**AO1 1** Copy and complete this table to summarise the structure and function of the three types of blood vessel in the human body.

| Blood vessels | Walls | Oxygenated or deoxygenated blood? | Link the wall structure to the function |
|---|---|---|---|
| Artery | Thick and elastic | | | [2] |
| Vein | Thin with valves | | | [2] |
| | Single cell thick | Blood changes from oxygenated to deoxygenated as it travels through | | [2] |

[Total: 6]

## Foundation/Higher level

**2** Look at this graph showing the resistance of a bacterium, *S. pneumoniae*, to three different types of antibiotic.

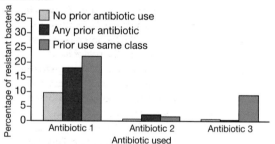

**AO2 a** Describe the pattern seen for antibiotic 1. [3]

**AO3 b** Which antibiotic does there seem to be least resistance to, even when it has been used before? Can you explain why this might be the case? [2]

**AO1 c** How can antibiotic resistance be reduced? [3]

[Total: 8]

**3** This graph shows the heart rate in response to exercise of a fit and unfit person.

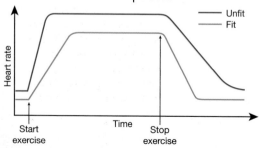

**AO2 a** Describe and explain the differences in heart rate between the fit and unfit person:
(i) as exercise begins [2]
(ii) when exercise has stopped. [2]

**AO3 b** What does this suggest about the impact of fitness on heart rate? [2]

[Total: 6]

**4** Jake has been swimming in the cold sea for over an hour. His mother makes him come out of the sea as she is worried he will become too cold and get hypothermia. Jake is shivering when he comes out of the sea and looks rather grey. Within a few minutes he has stopped shivering and looks pink again.

**AO1 a** What is hypothermia is and why is it dangerous? [2]

**AO2 b** Using the correct terms for the parts of the homeostatic system, explain how Jake's body acted to control his temperature:
(i) while he was in the water [5]
(ii) once he was out of the water in the sunshine. [4]

[Total: 11]

## Higher level

**5** This graph shows deaths from measles and also the number of cases of the disease notified to the health authorities (notification of the disease was not compulsory prior to the 1940s).

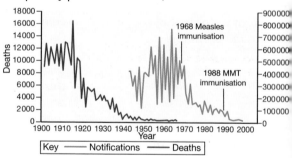

**AO2 a** The death rate from measles clearly starts to decline before the vaccination is introduced. Can you explain what could have brought about the reduction in death rate at this time? [2]

**AO2 b** Since the first measles vaccine was introduced, there have been very few deaths from measles, and the number of cases of measles has also reduced dramatically. Can you explain this? [2]

**AO3 c** The UK recently had its first death from measles for more than 14 years. The 13-year-old boy who died had not had the MMR vaccination. Use the graph to explain why some people choose not to vaccinate their children. [2]

[Total: 6]

AO1 recall the science   AO2 apply your knowledge   AO3 evaluate and analyse the evidence

B2 Keeping healthy

# Worked examples

**1** Hamish is a 45-year-old lorry driver. His family is short of money and he works long hours, including stressful city routes that he admits he does not enjoy. He has gone to the doctor's surgery because he has been feeling breathless and has been suffering from chest pain. His father died from a heart attack at the age of 50. Hamish is overweight and eats mainly fast food from roadside cafes due to his job. He does not exercise. Hamish gave up smoking two years ago.

AO1 **a** What symptoms of heart disease does Hamish have? [2]

*He is breathless and has chest pain.* ✔ ✔

AO1 **b** State and explain three risk factors that Hamish has for coronary heart disease. The quality of written communication will be assessed in your answer. [6]

*Hamish has a stressful job and he feels breathless. He also does not do much exercise and he eats a lot of fast food.*

AO2 **c** What suggestions would you make to Hamish to reduce his chance of suffering a heart attack? [2]

*Hamish could do more exercise as that would make his heart stronger.* ✔

AO1 **2** Choose the correct word to complete the following statements about clinical trials.

**a** A substance used in a clinical trial that looks the same as the test drug but does not contain the active ingredient is called a

control / (placebo) ✔ / subject [1]

**b** Patients do not normally know whether they are being tested with the real drug or not. This is important so that the clinical trial is

safe / (ethical) / reliable [1]

## How to raise your grade

Take note of the comments from examiners – these will help you to improve your grade.

Two symptoms are correctly identified: 2 marks.

'He feels breathless' is a symptom, not a risk factor. Several factors are identified, but not explained.

Spelling and sentence construction are appropriate. The answer meets the criteria for 2 marks (see the example mark scheme on page 317). Some of the criteria for 4 marks have been met but not all, so the answer would get 3 marks.

To get more marks an explanation of the risk factors is needed.

A good answer might be: 'He has a stressful job; stress can raise the blood pressure and increase the risk of heart disease due to damage to the blood vessels. He does no exercise and eats a lot of food high in saturated fat and this can build up in artery walls, making a clot more likely. His father had heart disease, and it runs in families.'

This suggests what the change would involve and why it would help Hamish, but only offers one idea: 1 mark only.

Part a is correct: 1 mark. Part b is incorrect; the correct answer is 'reliable'. The reason for not telling patients whether they are taking the trial drug is to avoid effects caused by the patient imagining an effect from the drug (the 'placebo effect').

# B3 Life on Earth

## What you should already know…

### Living things can be classified into groups

Animals with backbones are classified as vertebrates.

The five groups of vertebrates are amphibians, birds, mammals, fish and reptiles.

Each of the groups of vertebrates has particular characteristics. For example, mammals give birth to live young, suckle their young, have hair and are warm blooded.

 What are the characteristics of the amphibians?

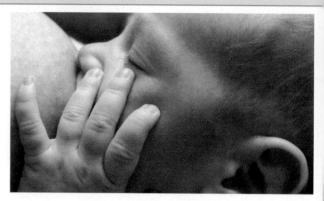

### Feeding relationships can be represented by food chains

Plants can make their own food through photosynthesis.

Photosynthesis is a process that takes place in all green plants.

Green plants are known as producers because they produce their own food.

All food chains begin with a producer, usually a green plant.

The arrows in food chains represent the flow of energy.

 Give an example of a food chain from woodland.

### Animals and plants are adapted to survive in their habitats.

An example of adaptation in animals is the thick white fur on a polar bear that keeps it warm and camouflages it from its prey.

Another example is the large stomach of a camel that allows it to store a lot of water.

A habitat is a place where animals and plants live, for example woodland, a rock pool, a hedgerow or a garden pond.

 Name a plant that is adapted to its habitat and explain how it is adapted.

# In B3 you will find out about...

> the classification system that scientists around the world use to make sure they are all using the same name for the same organism

> evidence that life on Earth began about 3500 million years ago

> fossils and DNA evidence, which can help us to understand how living things are related to each other

> mutation of genes and other variation, which meant that some organisms survived better than others and became more common, and how this led to evolution

> the scientist Charles Darwin who began to form ideas about evolution because of his observations

> how environmental change can cause species to change or go extinct

> how food chains can be linked together to form a food web

> the effect on a whole food web if one species of organism is removed

> how to calculate the efficiency of energy transfer in food chains

> how vital nutrients in the environment are recycled so that they can be used again

> what biodiversity is and why we should maintain it for future generations

> how we can live more sustainable lives

# Species and adaptation

**We are learning to:**
> explain how species are adapted to their environments and give examples
> explain that species compete for resources
> predict the effect on species of disturbance to a food web

## Are you a survivor?

It is thought that some of the **adaptations** of humans have allowed us to dominate the species on this planet. Scientists have proposed that when we adapted to walking on two legs, this freed up our hands to use tools. This gave us a huge advantage over other species.

FIGURE 1: Early humans not only use tools, but also made tools in advance for future use.

## How species are adapted to their habitat

The term **species** is used to mean a group of organisms that can breed together to produce fertile offspring; that is, offspring that are able to produce offspring themselves. Species are adapted to ensure that they survive in the environment in which they are living. We call the environment where an organism lives its **habitat**.

Being adapted to a habitat means that an animal or plant has features that help it to survive in that habitat. These might be physical features (like a crocodile's muscular tail for strong swimming) or ways of behaving (such as desert kangaroo rats sleeping during the hottest part of the day to keep cool).

Organisms are adapted to live in their habitats in a huge variety of ways. Two examples of well adapted organisms are shown in Figure 2 below.

**Habitat:**
Hot dry conditions, e.g. desert

**Adaptations:**
> Thick, waxy cuticle to prevent water loss
> Spines instead of leaves to reduce water loss and to discourage animals from eating it
> Many cactuses have a stumpy shape which gives a small surface area : volume ratio, which reduces the amount of water lost
> Very shallow and wide root systems to gather the maximum water when it rains

**Habitat:**
Water, e.g. coral reef

**Adaptations:**
> Fins help to keep the fish upright and steady when swimming
> Gills are the gaseous exchange surface; water passes over them and oxygen and carbon dioxide are exchanged
> Different fish have different mouth positions depending on how they feed
> Many bony fish have a swim bladder which helps them maintain their position in the water. Gases from the blood are added or removed to allow them to swim at different depths

FIGURE 2: Adaptations of the cactus and the fish.

### QUESTIONS

1. What does the term 'species' mean?
2. Describe how a fish is adapted to its habitat.

# Competition

It is not just the environment that affects the chances of an individual surviving. Organisms are also dependent on other species for their survival. They may depend on other species for food or for their habitat. Many species of insect, for example have habitats within or on particular plants.

In a particular habitat there will be **competition** between species for resources. Animals may compete for breeding sites or territories, food, mates or living space. Plants may compete for space, light, nutrients and water.

A **food web** shows the feeding interactions of some of the species within a habitat.

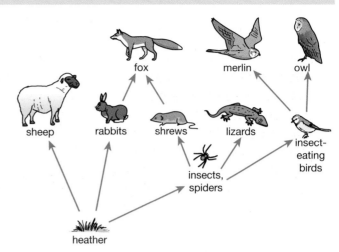

FIGURE 3: Most species will have feeding links with several other animals and plants within the habitat. These interactions can be shown using a food web.

**Watch out!**
Food webs are not the same as food chains. Food webs contain a greater number of the plants and animals in a habitat and food chains show only one route through a food web.

## QUESTIONS

3 Describe what plants growing on a woodland floor might compete for.

4 Using the food web in Figure 3, describe what could happen if the insect population suddenly decreased.

# Interdependence (Higher-tier only)

Because species depend on each other in many ways, any impact on one species within a food web will have a knock-on effect on many of the other species that are part of the same food web. Food webs are extremely useful because they give us a visual representation of how species in a habitat depend on each other for survival. We call this **interdependence**.

## QUESTIONS

5 Draw an example of a food web in a marine (sea) or other habitat.

6 Explain, with reference to your food web, what is meant by 'interdependence' of organisms. What would happen if the number of one species dropped or increased rapidly?

FIGURE 4: Interdependence means that all species in a food web have an influence on one another.

# Changes and challenges

> **We are learning to:**
> - understand reasons why species can become extinct
> - explain why scientists must think carefully before introducing non-native species into habitats

## Have I introduced you?

Native species are those that have evolved in a particular habitat. Non-native or 'introduced' species are from elsewhere and are placed, either deliberately or accidentally, in other habitats. This often has disastrous consequences for the native wildlife in those habitats.

## When species cannot survive change

The Siberian tiger and the Mountain gorilla have both suffered from the loss of or damage to their habitat and are at grave risk of **extinction**. When environments or habitats change, species that were previously well adapted have to make rapid changes to the way they live or they may become extinct.

Many species are able to cope with small changes in their environment, such as perhaps breeding earlier as the weather becomes warmer earlier in the season. Some species are not able to cope with such changes. A large number of species are currently at risk of extinction because they are not able to respond quickly enough to changes in their habitat.

### QUESTIONS

1. What does the word 'extinction' mean?
2. Give an example of an animal under threat of extinction, and the reasons for this.

**FIGURE 1**: The Adélie penguins of Antarctica are under threat as their habitat has become smaller and warmer.

## When the competition is too great

Species may become under threat if another species that competes with them, preys on them or causes disease is introduced into their habitat. A good example of such a threat is the Signal crayfish. It is estimated that 95% of the UK native crayfish population has been lost due to the introduction of this species from North America.

**FIGURE 2**: The Signal crayfish was introduced into the UK in the 1970s and 1980s. As it rapidly spread through the UK it passed on diseases and outcompeted the native British crayfish for food.

Another example of a species that was introduced into a habitat and subsequently had a devastating effect on the **native species** already present is the Cane toad. This **non-native species** was introduced into Australia deliberately to try to control the Cane beetles on sugar cane. The Cane toad is a voracious predator, eating insects, small snakes and sometimes small rodents. Cane toads are also poisonous and have few natural predators, even in their native America. Once in Australia, they competed with native frogs for breeding and egg-laying sites. The toads soon reproduced and spread, reducing the populations of their prey as well as affecting other organisms in the food web.

**FIGURE 3**: Native wildlife has been killed by eating the poisonous Cane toads, although some species have learned to avoid the toads.

## QUESTIONS

**3** Explain why species might become extinct due to the introduction of another species into the habitat.

**4** Why have Cane toads been so successful in Australia?

### Did you know?

An adult Cane toad can reach 25 cm in length and weigh over 2 kg.

## Effects of species extinction

The extinction of one organism in a habitat will have effects on all the other organisms in the food web. This is the case whether it is an animal, a plant or a microorganism that becomes extinct.

For example, in this food web, if the squirrels became extinct, this would have effects on the mouse, rabbit and bird populations. Although there might be more plants available for food, these remaining small animals would be predated upon more heavily by the foxes, hawks and owls.

In chalk grassland, the Large Blue butterfly had a very specialised life cycle that depended on ants. The larvae of the butterfly were taken into ants' nests during the winter because they secreted a sugary solution that the ants fed from. This kept the larvae alive over the winter.

When the habitat was not grazed sufficiently, the grass became too tall for the ants to live beneath as it shaded the ground and reduced the temperature. The ants disappeared from the habitat and so the Large Blue butterfly larvae were not able to overwinter. The Large Blue butterfly became extinct from the UK but conservationists are now trying to re-introduce them.

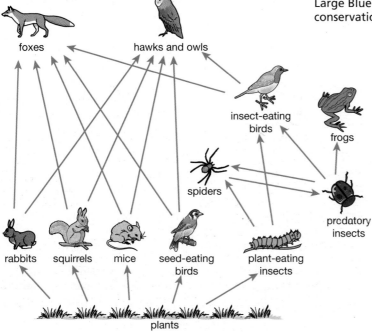

FIGURE 4: Complex relationships in a food web.

## QUESTIONS

**5** Choose two organisms in the food web in Figure 4 and explain how they might be affected by the extinction of the squirrels.

**6** Use ideas from these pages to explain why it is so important that introductions of species are handled extremely carefully.

Signal crayfish  Cane toad

# Preparing for assessment: Evaluating and analysing evidence

To achieve a good grade in science, you not only have to know and understand scientific ideas, but you need to be able to apply them to other situations and to analyse evidence. These tasks will support you in developing these skills.

## Visitors cause mayhem in paradise

Guam is an island in the Pacific Ocean. It was used by American troops during and after the Second World War as a staging post for supplies and it is likely that, during this period, the Brown tree snake was inadvertently introduced. It may have arrived in the holds of cargo ships from the South Pacific Islands. It is native to Australia, but hadn't previously been seen on Guam. When it arrived, it found itself in an environment with few predators – and plenty of prey.

The Brown tree snake is nocturnal, lives in the trees and feeds on a range of animal life, including birds. The effect of the arrival of the snake was devastating; 10 of Guam's 12 native bird species are now extinct on the island and the other two have dwindled to dangerously low numbers. The birds played a key role in the island's ecosystem, being pollinators, spreading seeds and preying on insects that feed on plants.

The Brown tree snake is both poisonous and a constrictor. However, its fangs are at the back of its mouth and it's not capable of delivering a fatal bite to adult humans. In fact, many people on the island have never seen one. Brown tree snakes normally grow to between 1 and 2 m in length, but those on Guam now sometimes reach over 3 m long.

A range of strategies has been tried to control the snake population – and to prevent it from spreading to other islands where it might have similar effects. Measures have included traps, trained dogs and night-time patrols on areas such as the edge of air strips. However, one of the latest techniques is to use dead mice treated with a drug called Tylenol. This is used effectively as a medicine for humans, but is fatal to the snakes if they eat it.

In order to develop this approach, scientists had to study the behaviour of the snakes. Most snakes won't eat anything they haven't killed themselves, but Brown tree snakes are scavengers and will eat dead meat. However, they live in a fairly densely forested area so the dead mice have to be placed by being dropped from aircraft. The snakes live in trees though; the mouse bodies have to be attached to strips of card and paper so they get caught in the trees instead of falling to the ground.

The scientists are developing a way of finding out whether this approach works. It isn't easy in dense jungle to search either for uneaten mouse bodies or dead snakes. What they are planning to do is to insert small radio transmitters into a small proportion of the mice. They can then track the position of the mouse bodies to see if they have been moved.

The long-term effects of the rapidly rising snake population are still being investigated. It has been established that trees with seeds that were distributed by birds have been affected; young trees are now only found very close to older ones. With few birds left on Guam, it is the lizard population that is now coming under pressure from the snakes. However, the spider population seems to be significantly higher than on other similar islands.

# B3 Evaluating and analysing evidence

## Task 1

> With reference to the Brown tree snake, explain the difference between predators and prey.
> Why are the snakes growing to sizes not commonly seen elsewhere?

## Task 2

> Construct a food chain that includes a Brown tree snake.
> Why will the Brown tree snake population not continue to rise without limitation?

## Task 3

> What problems might be associated with dropping dead mice containing drugs into the jungle?
> How are scientists trying to find out whether the poisoning works?
> What assumptions do they have to make?
> Wouldn't it be better to place transmitters in all the mice?

## Task 4

A snake that has eaten a poisoned mouse takes around 60 hours to die. Some people think that the culling of the snake population is essential, whereas others have argued that it is inhumane. Do you think this approach is justified?

## Task 5

Construct a food web that enables you to explain the impact of the Brown tree snake on birds, spiders and lizards.

## Maximise your grade

These sentences show what you need to include in your work to achieve each grade. Use them to improve your work and be more successful.

**E**
For grade E, your answers should show that you can:
> understand what a species is, and how species are adapted to their environments

**C**
For grades D, C, in addition, show that you can:
> understand what a species is, and that organisms that share a habitat compete for resources
> understand that when the environment changes, or a new predator or disease arrives in the habitat, some species will go extinct because they cannot adapt quickly enough
> explain how a population change in one organism can impact on other organisms in the same food web

**A**
For grades B, A, in addition, show that you can:
> explain the interdependence of living organisms using food webs

# Chains of life

**We are learning to:**
- describe how energy passes out of food chains
- calculate the efficiency of energy transfer in food chains

## Could you live in the dark?

You may think you could live without the Sun's light. But it is not just the convenience of daylight we have to thank the Sun for. The energy it radiates powers the most important process on Earth – photosynthesis.

### Energy in sunlight

The Sun is important to all of us. Nearly all organisms on Earth are dependent on the energy from the Sun for their survival. Without the Sun, there would be very little of life as we know it on Earth. We are all able to benefit from the energy of the Sun because of plants. Plants are able to absorb some of the energy in sunlight by the process of **photosynthesis**.

Plants then store this energy in the carbon-based chemicals that make up their cells and tissues. Other organisms can get energy by eating plants. Almost every food chain begins with a plant absorbing energy from the Sun.

**QUESTIONS**

1. Why is the Sun important to life on Earth?
2. By what process are plants able to absorb and store some of the energy from sunlight?

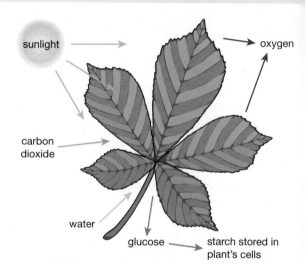

**FIGURE 1**: The process of photosynthesis can be represented as: carbon dioxide + water → glucose + oxygen.

### Energy transfer through food chains

The energy plants absorb is used to power the process of photosynthesis; the reaction that uses water and carbon dioxide to make oxygen and glucose. The oxygen is released into the atmosphere, but the sugars can be stored by the plant as starch, a molecule made of many joined glucose molecules. When animals eat plants, this energy that has been absorbed by the plant and used to create glucose passes into the animal.

Not all the energy that plants absorb from the Sun can be stored. The plant needs to use some of the energy to reproduce and respire. Of the energy that is stored by the plant, not all is passed on to the animal that eats it. For example, there are parts of the plant that cannot be digested.

**FIGURE 2**: Energy passes out of a food chain when there are parts of organisms that cannot be eaten. This energy reduction within the food chain at each stage limits the length of food chains.

food chain efficiency GCSE

At *each* part of a food chain, as energy is transferred from one organism to the next, some of that energy passes out of the food chain through:

> heat as organisms use energy to respire, move and keep themselves warm

> waste products from excretion

> uneaten parts such as stalks in plants and hair and claws in animals.

This energy cannot be used by the next organism in the food chain.

The **efficiency** of the transfer of energy at each level of a food chain is the ratio of the energy available in the organism's tissues (as food) to the total amount of energy the organism has ingested (eaten).

$$\text{percentage efficiency} = \frac{\text{energy in tissues}}{\text{energy in food eaten}} \times 100\%$$

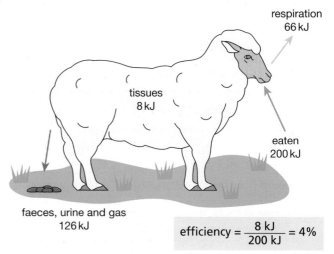

$$\text{efficiency} = \frac{8\,\text{kJ}}{200\,\text{kJ}} = 4\%$$

**FIGURE 3**: This sheep has transferred a lot of the energy from the food it has eaten into heat energy that has been released into the environment. Only a small proportion of the energy is available to the next level of the food chain.

### QUESTIONS

**3** Explain what happens in the process of photosynthesis.

**4** Why isn't all the energy passed from one level of a food chain to another?

**5** Calculate the efficiency of energy transfer for a salmon that has ingested 80 kJ of food and has 10 kJ of energy in its tissues.

### Did you know?

Shorter food chains are more efficient as less energy passes out of the chain. A vegetarian or vegan diet is much more efficient in terms of energy than a carnivorous one.

## The role of decomposers

Another important transfer of energy goes on after organisms have died. There are a very large number of organisms and microorganisms that feed on dead and decaying matter. Most of these **decomposers** are bacteria and fungi, which release enzymes to break down organic matter before they digest it. Decomposers play an important role in recycling the nutrients in the ecosystem.

### Detritivores (Higher tier only)

Organisms called **detritivores** begin the decay process by breaking dead matter into smaller pieces by feeding on it. Examples of detritivores are earthworms and woodlice.

### QUESTIONS

**6** Cold-blooded animals such as fish have a higher efficiency of energy transfer than warm-blooded animals. Explain why.

**7** Explain how organisms are broken down after they have died.

# Recycling nutrients

**We are learning to:**
> describe the carbon cycle
> describe the nitrogen cycle

## How big is your footprint?

Your 'carbon footprint' is one way to measure your impact on the environment in terms of the amount of carbon dioxide your activities generate. A person's carbon footprint is influenced by the transport they use, the way their home is heated, the things they eat and the products they buy.

FIGURE 1

## The carbon cycle

**Carbon** is the basis of the molecules we are made up of. It is continuously recycled in the environment, in the **carbon cycle**.

Carbon enters the carbon cycle as **carbon dioxide** in different ways:

> through **combustion** (burning)

> through **respiration** (when living things release the energy from food)

> through **decomposition**, where dead animals and plants are broken down by microorganisms. Their carbon molecules can then become part of the cycle again.

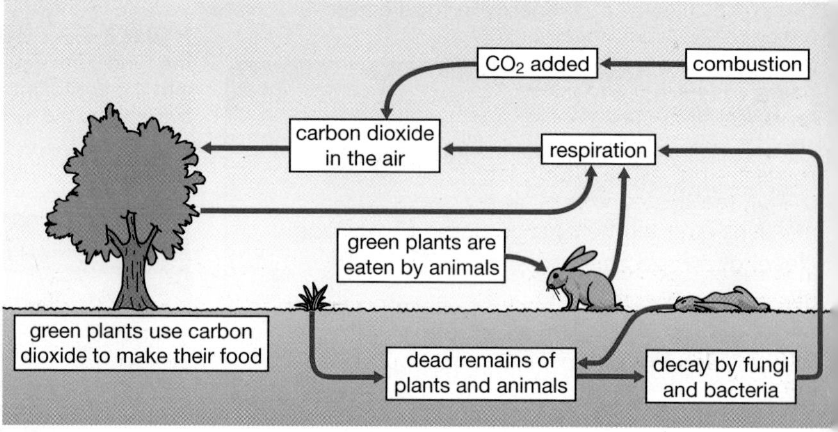

**FIGURE 2**: Because carbon is so important to so many living things, it is fortunate that it is recycled through a process called the carbon cycle, so it will never run out.

Carbon is removed from the atmosphere when plants photosynthesise. They are able to 'fix' the carbon from carbon dioxide so that it is stored in their tissues.

In the past, the amount of carbon dioxide in the atmosphere was kept in balance by this natural cycle. Since industrialisation began in the western world, the amount of carbon dioxide in the atmosphere has been increasing. This contributes to global warming because carbon dioxide is a greenhouse gas (see page 110).

### QUESTIONS

**1a** How does carbon enter the atmosphere?

**b** How is carbon removed from the atmosphere?

**2** What is causing an increase in the amount of carbon dioxide in the atmosphere?

## The nitrogen cycle

As well as carbon, nitrogen is an essential component of living things, and this too is recycled in a process called the **nitrogen cycle**.

Nitrogen is an important nutrient in animals' diets, in the form of nitrogen compounds in **proteins**. Plants use **nitrates**, which they absorb through their roots, to make proteins. The nitrogen compounds are passed along food chains when animals eat the proteins in plants, and animals eat other animals.

When animals excrete waste, nitrogen compounds are returned to the soil. When animals and plants die, the nitrates in their bodies are released in the process of decomposition, and pass back into the soil. Plants are then able to absorb these nitrates into their roots to make more proteins, and the cycle begins again.

# B3 Life on Earth

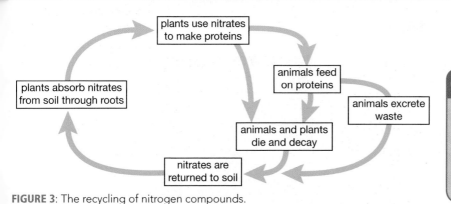

FIGURE 3: The recycling of nitrogen compounds.

> ## QUESTIONS
>
> **3** Describe how animals take in nitrates.
>
> **4** How are nitrates released from animals and plants back into the soil?

## The role of bacteria in the nitrogen cycle (Higher tier only)

Bacteria are a key part of the nitrogen cycle. Leguminous plants (peas and beans) have **nitrogen-fixing bacteria** in nodules on their roots. These bacteria, which are also present in the soil, are able to convert nitrogen gas, abundant in the air, to nitrates, which the plants can use to make proteins.

In waterlogged conditions, **denitrifying bacteria** convert nitrates in the soil back into nitrogen gas, which enters the atmosphere.

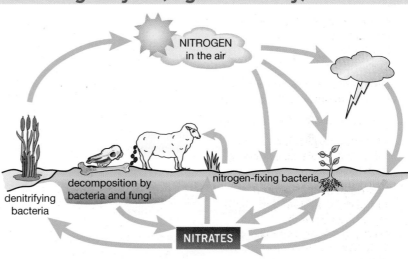

FIGURE 4: The nitrogen cycle, showing the role of bacteria.

> ## QUESTIONS
>
> **5** Explain the roles of bacteria in the nitrogen cycle.
>
> **6** Peas and beans contain more protein than other plant foods. Can you use what you know about the nitrogen cycle to explain why?

**Watch out!**
The nitrogen cycle involves different types of bacteria: nitrogen-fixing bacteria, which turn nitrogen to nitrates, denitrifying bacteria, which turn nitrates to nitrogen, and also the bacteria that cause the decay of dead animals and plants. Make sure you remember the difference.

### Did you know?
Lightning is another way that the nitrogen in the atmosphere can be converted into nitrates, which can be used by plants to make proteins.

nitrogen cycle GCSE

# Environmental indicators

## What can we learn from a mayfly?

Mayflies can tell us a lot about the environments in which they live as they are sensitive to environmental change. They are an example of a **living indicator**.

FIGURE 1: A mayfly.

**We are learning to:**
> explain how environmental change can be monitored
> interpret data to investigate environmental change

## Indicators of environmental change

An important part of the work of scientists is to monitor environments to see whether they are changing. There are a variety of ways to check whether an environment has changed, been damaged or improved over time. Both living and non-living indicators can be used to keep track of the changes in the environment.

**Non-living indicators** that scientists might monitor are the levels of carbon dioxide, the temperature, or the levels of nitrates in water. Nitrates are found in fertiliser and are a common cause of pollution in waterways.

**Living indicators** include organisms such as **phytoplankton** (microscopic plankton that can photosynthesise), mayflies and **lichens**, which are very sensitive to the environment they live in.

### QUESTIONS

1 How can scientists keep track of changes in the environment?

2 Give examples of living and non-living indicators.

## What environmental indicators can tell us

Environmental change takes place on many scales, from chemical changes in local habitats like streams and rivers to global-scale changes such as the global increase in carbon dioxide in the atmosphere. Scientists monitor environmental change on all scales.

Non-living indicators can give scientists a good idea of the changes taking place in our environment. Measurements of carbon dioxide in the air, for example, have been rising over recent years, leading to concerns about global warming.

Living indicators give us very precise information on how a habitat is being degraded or whether the habitat is improving. Phytoplankton (Figure 2) photosynthesise and are at the start of aquatic food chains, so any effect on them will have knock-on effects on the whole food chain.

Mayflies are insects that spend nearly all their lives in the water, only coming out briefly to mate and lay eggs. Mayflies need a high level of oxygen in the water and little pollution to survive.

Lichens absorb water and minerals directly from the air over their whole surface area, and so are very sensitive to levels of pollution in the air. Different lichens are sensitive to different types and levels of pollution. Air quality can be monitored by observing the communities of lichens that are able to grow.

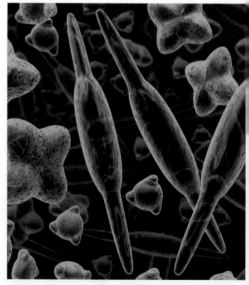

FIGURE 2: Phytoplankton, which have a variety of forms, are sensitive to changes in the temperature of the water in which they live. Changes to ocean currents will affect their populations.

B3 Life on Earth

FIGURE 3: The lichen growing on these trees indicates that the air is very clean.

**Hint:**
The worksheet (b3_05 Environmental indicators) provides an important data analysis task, based on a lichen survey. You need practice in this type of task.

## QUESTIONS

**3** Why are phytoplankton so important to aquatic ecosystems?

**4** Why do you think communities of lichens are studied instead of just a single species?

## Interpreting data on environmental change

Data from both living and non-living indicators can be used to provide information about a habitat. Living indicators are particularly useful as they give us a view of what the environmental conditions have been like in a habitat over time.

### Did you know?

Invertebrates in streams can be sampled by 'kick sampling'. The sampler holds a net downstream of their feet. They kick the bottom of the stream and catch in the net the invertebrates that have been disturbed.

Some organisms are very tolerant of some types of pollution, whereas other organisms may be very sensitive. The different sensitivities can be used to gauge the health of an aquatic environment. The organisms most sensitive to pollution are given a higher 'score' than those less sensitive. A survey of the organisms present can then give a good indication of the health of a stream, river or pond. A site where a number of organisms are found that are very sensitive to pollution, for example, is likely to get a high score on the survey, indicating that it is unpolluted.

Long-term studies of non-living indicators such as carbon dioxide levels or concentrations of chemicals can help us to see whether there are trends over time. If there are, action might be needed to protect the environment.

FIGURE 4: Testing the nitrate or phosphate levels only gives a snapshot of the conditions at the time the sample is taken. However, using non-living indicators like this is cheaper and quicker than carrying out surveys of invertebrate life.

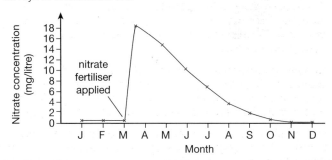

FIGURE 5: The variation in nitrate level in a stream that runs alongside farmland. The crops are fertilised in March. Nitrates are very soluble in water, and are washed out of soil into nearby waterways.

## QUESTIONS

**5** Look at the graph above (Figure 5). Describe what the data on the graph shows.

**6** Use the information on the graph and what you have learned to explain why it is important to collect data from non-living indicators such as nitrate levels over a period of time.

freshwater invertebrates

# Variation, mutation and evolution

**We are learning to:**
> understand that scientists have used clues to discover the origins of life
> understand that genetic variation between individuals can be passed on to the next generation
> understand that scientific explanations are developed from data

## Where did we come from?

There was nobody around to observe the very first life forming on Earth. Scientists have had to use the clues available to piece together how all the immense variety of life on Earth came to be here.

### The start of life on Earth

Evidence shows that life on Earth began around 3500 million years ago. Most scientists agree that the first life forms were very simple. Over an immensely long period of time, these simple living things changed to become all the species we see on Earth today. Many, many millions of species will also have become **extinct** in that time.

**Fossils** are found in rocks. They are made from the dead bodies of plants and animals, or traces such as footprints, which became buried. Over millions of years, these remains turned to rock. It is possible to date fossils accurately because of the rock they are found in. Most of the animals and plants now found as fossils are extinct, but they give us very important information about how organisms changed over time, or evolved.

**Evolution** has come about because of variation. Individuals within the same species are different (see page 12). Some of these differences between individuals are due to the environment but some are due to the genes that they carry. Variation that is caused by an individual's genes can be passed on to their offspring.

**FIGURE 1**: This fossil of a trilobite was found in Utah, USA, and is about 550 million years old.

 **QUESTIONS**

1 When did life on Earth begin, and what were the first life forms like?

2 What are fossils?

3 Look back at page 12. Describe some differences in humans that are likely to be due to genetics rather than environment.

**FIGURE 2**: Because of the shuffling of chromosomes when eggs and sperm are produced, offspring of the same parents can be very different.

### What is a mutation?

A **mutation** is a change in the genetic information in a cell, and can occur at random, sometimes during the process of copying the DNA for a new cell. Some mutations result in a change in the physical appearance or characteristics of an organism. If there is a mutation in a sex cell (an egg or sperm), this mutation can be passed on to the next generation. Such a mutation can cause variation between individuals.

when did life on Earth begin?  what is evolution?

Sometimes this variation is an obvious difference in appearance. On other occasions a mutation may affect patterns of behaviour. Sometimes the outcome of a mutation has a useful effect for an organism that gives it a competitive advantage. For example, if a mutation occurred that made an organism faster at running away from predators, that organism might be more likely to survive to reproduce. This would increase the frequency of the gene for faster running in that population.

FIGURE 3: In the Antarctic fur seal population, about one in 800 seals is pale instead of dark. The frequency of the mutation is so predictable that population surveys are carried out by counting the pale seals and multiplying by 800 to get the total population.

## QUESTION

4 Explain how parents can give birth to offspring who are different from each other.

## The theory of evolution

Scientists believe that the enormous variety of organisms we see around us today came about because of evolution. Evolution is the change in the frequency of genes in a population over enormously long periods of time. Sometimes genes change because of mutations. Some of these mutations are beneficial to the organism, but some are not. Some genes become more common over time, and some become less common. Some genes disappear from the **gene pool** altogether. As the frequencies of different genes varied over time and differently in different habitats, different species gradually emerged.

The theory of evolution is a good example of a scientific theory. It is based on data and observations, but scientists also needed creative thought and a vision of how everything might fit together. The theory of evolution is based on the things we can observe around us now, and also what we have in the **fossil record**, which is the information from all fossils collected and recorded. Although the fossil record contains thousands of pieces of evidence that support the theory of evolution, not all parts of animals and plants form fossils; for example soft animals like jellyfish do not form good fossils. This means that scientists have had to accept that they are unlikely to find examples of certain species, but instead draw conclusions about what they might have been like using their knowledge of other similar species. When new fossils are found, they fill in the gaps in our understanding of how life changed over time to become the animals and plants we see around us today.

## QUESTIONS

5 Antarctic fur seals live in habitats that are often cold and icy. They are preyed on by Elephant seals and they eat fish. Explain the possible advantages and disadvantages of being a pale fur seal in that environment.

6 Explain why fossils provide good evidence for evolution.

### Did you know?

We do not know how many species exist on Earth as so many have not yet been identified. Current estimates of the number of species on Earth range from 5 million to 30 million.

# The great competition of life

## Are you a winner in the competition of life?

Yes, you are. The fact that you are here at all shows that your ancestors were really very good at surviving and reproducing. Many, many millions of organisms have become extinct since life began on Earth but humans have survived and thrived, and are now the dominant species on Earth.

FIGURE 1

**We are learning to:**
> explain the process of natural selection
> distinguish between natural selection and selective breeding
> explain how variation between individuals can eventually lead to change in species
> interpret data on changes in species caused by natural selection

## Variation and natural selection

Individuals of a species have variations that affect their physical appearance and behaviour. A difference in an individual can give it an advantage compared to others of the same species. This advantage may allow the individual to be more successful at reproducing, or more likely to survive to reproduce. If this advantage is due to a genetic difference, the individual will pass the advantageous genes on to the next generation. Less successful individuals are less likely to breed and pass their genes on to the next generation. This means that in the next generation, more individuals will have this gene and thus this useful characteristic. This is known as **natural selection**. Over time, these changes may lead to the development of a new species.

### QUESTIONS

1. Why do individuals of a species vary?
2. How can a genetic advantage be passed to the next generation?

## Natural selection and selective breeding

Humans have found ways to exploit the process shown by natural selection to their own advantage. If we want an animal with particular characteristics, we can choose two animals with characteristics nearer to these than is normal in the population and breed them together. The result will be an animal or plant with characteristics closer to those we want than in the previous generation. This is then repeated until the required characteristic is sufficiently exaggerated. A good example is the way **selective breeding** has been used is in the production of cattle. Different breeds of cattle have been selectively bred to have the characteristics we require.

**FIGURE 2**: Holstein cows (top) are used for the production of milk. They have high milk production and little muscle mass. The Hereford (bottom) is bred for meat production and has a lot of muscle.

# B3 Life on Earth

|  | Natural selection | Selective breeding |
|---|---|---|
| Needs genetic variation upon which to act | Yes | Yes |
| Favours genes that promote survival in the wild | Yes | No |
| How many generations of breeding are needed | Many, many generations | Many generations |
| Human intervention needed? | No | Yes |

Similarities and differences between natural selection and selective breeding.

## QUESTIONS

**3** What is selective breeding?

**4** Explain the differences and similarities between selective breeding and natural selection.

## Natural selection in action

Natural selection is an important part of the evolutionary process. Evolution is the change in the frequency of genes in a population over enormously long periods of time.

Imagine an animal that lives in open grassland, for example a lizard, which is preyed upon by another species. If a mutation occurred that led to one of these lizards having better camouflage in the grassland, this might have consequences.

The better camouflaged animal might not be preyed upon as the predator would find it harder to see, and it would also have to spend less energy escaping from predators. The better camouflaged lizard might also find it easier to find food, as its own prey would be less likely to see it coming. These advantages would make this lizard more likely to survive to reproduce, and it might therefore produce more offspring than other lizards. Thus its genes (which include the advantageous one for better camouflage) will occur slightly more often in the next generation than the genes of other lizards of the same species. The next generation will have more individuals with the genes for 'better camouflage'.

If this is repeated for many thousands of generations, providing the genes continue to be advantageous, nearly all the members of the species will end up with the gene for better camouflage, and that will become the 'norm' for that species.

**Watch out!**
A genetic advantage may be due to a mutation or just to normal variation between individuals.

| Year | % Melanic (dark) peppered moths | Smoke pollution (µg/m³) |
|---|---|---|
| 1980 | 76 | 164 |
| 1985 | 63 | 103 |
| 1990 | 27 | 34 |
| 1995 | 6 | 12 |
| 2000 | 3 | 7 |
| 2005 | 1 | 5 |
| 2010 | 2 | 4 |

This table shows data on the percentage of dark peppered moths found and the smoke pollution levels near one city.

**FIGURE 3**: The melanic peppered moth was camouflaged on the dark, sooty bark.

## QUESTIONS

**5** Explain how an advantageous gene or mutation might spread through a population over time.

**6** Look at the data in the table. Explain the pattern shown in the table and suggest why the peppered moth is now mostly found in its pale form.

selective breeding GCSE

# Evolution has the answers

## How much do you know about your ancestors?

The theory of evolution does not imply that humans descended from apes. We are *one* of the great apes, along with gorillas and orang-utans, which began to evolve separately from monkeys about 30 million years ago.

FIGURE 1

**We are learning to:**
> describe how the effects of mutations, natural selection, isolation and environmental change can cause new species to evolve
> explain how scientists can make predictions about what might fill 'gaps' in the fossil record

## Natural selection and evolution

Individuals of a species can vary due to **mutations**. These are changes in genes that affect the physical features or behaviour of an organism. The process of natural selection favours individuals who have the most useful traits, as these individuals are more likely to survive and pass their genes on to the next generation. Over a very long time, these genes are likely to become the norm in the population. This is how evolutionary change takes place.

The fossil record is made up of all the known and recorded fossils in the world. It is not a collection in terms of all the fossils being stored in one place (there are millions of fossils in the world), but is all the information that can be found from studying the fossils.

The fossil record provides evidence for evolutionary change over very long periods of time. If there is enough pressure on a species, and the species reproduces very quickly, then evolution can also take place quickly. An example of this is the development of bacteria that have become resistant to antibiotics.

### QUESTIONS

1. Where can evidence be found for evolution?
2. Why does natural selection favour useful features in individuals?

**Watch out!**
Natural selection and evolution are not the same thing, although natural selection allows evolution to happen. Natural selection acts upon individuals, but evolution changes how often genes occur across a whole population.

## Isolation and environmental change

A number of factors influence the rate at which evolution takes place. If organisms are **isolated** in their habitats (for example on an island, or separated from other organisms of the same species by a mountain range or rift valley), then natural selection will act independently on the two populations. Different genes are likely to become more frequent in the different populations. Over time, this will lead to different species evolving in each location, no longer able to reproduce with each other.

Environmental change can have huge effects on evolution. When the climate or habitat changes, species must adapt or die.

**FIGURE 2**: Animals that are well-adapted to their habitats, such as this brown tree hopper, may face extinction if that habitat changes.

## QUESTIONS

**3** Explain why isolation can cause different species to evolve.

**4** Why might animals and plants that are well-adapted to their habitats become extinct if that habitat changes, for example due to climate change? Use the internet to discover why bluebells are under threat from climate change.

## The tree of life

It is thought that all life on Earth evolved from simple forms of life that existed 3500 million years ago. From these simple life forms, an immense variety of new species evolved, suited to different ways of life and different habitats. As the environment changed, so did the species, and many also became extinct.

**FIGURE 3**: On the tree of life, humans and the rest of the great apes split from monkeys about 30 million years ago, while humans only split from chimpanzees 8 million years ago. We have more genes in common with the chimpanzee (centre) than the Vervet monkey (left) has.

Scientists are able to determine how closely related organisms are to each other in evolutionary terms by analysing their DNA. More closely related organisms will have more genes in common. The relationship between all organisms can be shown in a diagram often called the 'tree of life', as it is tree-shaped and has branching paths. The tree of life shows the relationship between all organisms, with those on the same branches being more closely genetically related than those further away. Such DNA analysis has allowed scientists to quite precisely pinpoint when different branching events took place.

These ideas about how life evolved allow scientists to explain all the observations that are seen in the fossil record. They also allow scientists to predict what intermediate forms of animals or plants might look like, which can then be checked when further fossils are found.

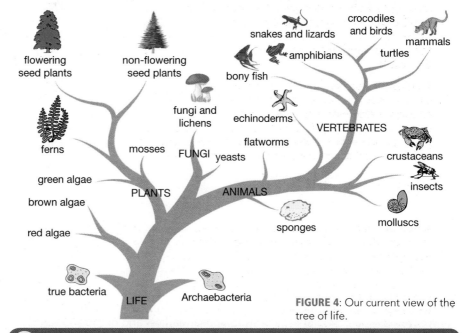

**FIGURE 4**: Our current view of the tree of life.

## QUESTIONS

**5** Explain why we are able to be much more confident in the theory of evolution than scientists were over 100 years ago, when it was first proposed.

**6** How do you think scientists predict what fossils might be found to 'fill the gaps' in the fossil record?

tree of life interactive

# Evidence from fossils and from DNA

**We are learning to:**
- understand that a scientific idea is the result of creative thought and can be tested by observations
- explain how the theory of evolution is supported by the fossil record and DNA analysis

## Are you sorted?

Scientists sort and classify things to help us all to make better sense of the world. The classifications of living things and fossils can make it clear how different species, including our own species, *Homo sapiens*, came about, and how different species are related.

##  Classifying organisms

Organisms can be **classified** in many different ways. In the past, organisms were classified mainly by their appearance: scientists grouped organisms that looked similar together. Today, we have a much more accurate way of classifying organisms, based on their DNA sequences.

Organisms with more similar DNA are likely to be more similar species. Classifying organisms in this way helps us to see how they evolved.
A common ancestor is the most recent individual from which all organisms in a group are descended. Organisms with the most similar DNA have more recent common ancestors than those with less similar DNA.

**FIGURE 1**: Guinea pigs and mice look similar but DNA analysis shows that, though they are both rodents, they have very different DNA and have not evolved recently from a common ancestor.

###  QUESTIONS

1. How were organisms classified in the past?
2. How does DNA sequencing give us information about evolution?

### Did you know?

The oldest DNA that has successfully been extracted was from 40-million-year-old extinct stingless bee called Proplebeia.

##  The fossil record and DNA evidence for evolution

The fossil record has given scientists a huge amount of evidence for evolution, and has also provided information about how species have evolved.

The theory of evolution predicts a very clear pattern in the fossil record. We would expect simpler organisms to appear earliest and become gradually more complex. We would also expect the features of newer organisms to look like adaptations or developments of those of earlier organisms. This is exactly what we do see in the fossil record.

DNA analysis of organisms that exist today has confirmed predictions made based on the fossil record, for example predictions about when animals split off from each other to form different groups of species, and how long ago common ancestors were shared. Both the fossil record and DNA analysis of species give us strong evidence for evolution.

**Watch out!**
Remember, even organisms of the same species do not all have the same DNA. A variety of different forms of genes are possible for every individual.

## QUESTIONS

**3** Explain why the fossil record provides evidence for evolution.

**4** How has DNA analysis supported our understanding of evolution?

## Charles Darwin and his observations

The theory of the origin of species by natural selection was originally proposed by Charles Darwin in 1859. He had taken a five-year voyage on a ship called *The Beagle* and it was on this expedition that he observed the animals and plants on a group of islands off South America called the Galapagos Islands. He noticed that the animals and plants were similar to those on the South American mainland, but had slight differences. One group that proved very useful in developing his theory was the mockingbirds. He observed these and noticed that the birds were slightly different on each island. He collected specimens of the mockingbirds from different islands, which are now at the Natural History Museum in London. Darwin deduced that a species could change according to the conditions where it lived. He used his creativity to come up with an explanation for his observations, and he proposed the theory of evolution by natural selection. Other scientists later found that the finches on the Galapagos Islands also showed variation that gave good evidence for Darwin's theory.

**FIGURE 2**: Finches on the Galapagos Islands have beaks that are adapted to eat the type of food found on each island.

A previous theory for evolution, proposed by a scientist named Jean Lamarck around 50 years earlier, was quite different. Lamarck suggested that the characteristics that organisms acquire in their lifetime, through use or development, could be passed on to offspring. This would mean that a giraffe, constantly stretching its neck to reach tall branches might develop a slightly longer or more flexible neck, and that this 'long neck' characteristic would be passed on to its offspring. Lamarck did not suggest *how* acquired characteristics might be passed on, and there was no good evidence for his ideas. It was plain that human parents who had developed particular characteristics (such as muscular bodies) during their lifetime did not pass these on to their offspring.

Darwin's theory could be explained using ideas about genetics that were only just beginning to be understood in Darwin's time, and current understanding of the way genes are passed on confirms that his theory makes scientific sense.

## QUESTIONS

**5** Why was Lamarck's theory not considered such a good explanation of the changes in species as Darwin's theory?

**6** Explain how different species of finches could have evolved to have different beaks, using ideas about natural selection.

OU Darwin's theory

# We need diversity

**We are learning to:**
> understand what biodiversity is and why it is important
> explain the increasing extinction rate of species
> describe how and why species are classified

## Conservation – what's in it for you?

We do not know which of the thousands of species that become extinct every year might be important to the human race. Plants and animals with useful features are becoming extinct before we have the chance to make use of them. Conserving these species is important to all of us.

FIGURE 1: A substance extracted from a rainforest plant, the periwinkle, is a very effective drug for treating cancer. It has dramatically increased the survival rate for children suffering from leukaemia.

## Conserving biodiversity

Tropical rainforests are being destroyed or damaged at a very fast rate. These habitats and others like them are very important for large numbers of species. Some of these species may be ones we have not discovered yet, but they could be very important for us as they may be useful for medicines or for food.

**Biodiversity** is the word used to describe the variety of living things in the world. It includes the variety of different species (across all species of animals, plants, bacteria and fungi). Biodiversity also refers to the **genetic diversity** (variation) *within* each species. Habitats such as tropical rainforests have high biodiversity – they contain a lot of species. It is important that the rainforests are protected.

### QUESTIONS

1. What is biodiversity?
2. Explain why it is important to maintain as much biodiversity as possible.

FIGURE 2: Chemicals originally found in rainforest plants are used to treat many health problems such as high blood pressure, malaria, arthritis and tuberculosis.

## Rate of extinctions

Species are becoming extinct more rapidly now than at any other time in recorded history, except for the **mass extinction** events seen in the fossil record. It is thought that human activity is driving species to extinction as they are hunted, and their habitats are broken up by roads or destroyed for building and farmland. Many species have become extinct without even being identified and classified.

Climate change, which is likely to lead to rising temperatures and more extreme events such as floods, droughts and storms will also accelerate the rate of extinctions, as habitats change more quickly than species can adapt to the changes.

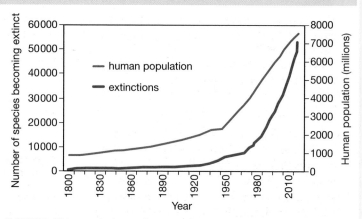

FIGURE 3: This graph shows the rate of species extinction and the growth in human population.

Center for Biological Diversity

## QUESTIONS

**3** Describe the information shown by the graph in Figure 3.

**4** Explain, using information from the paragraphs on page 94 and a specific species example from your own research, why species are becoming extinct so rapidly.

**Watch out!**
A species is a group of organisms that can breed together to produce fertile offspring.

## Classifying living things

In order to record and monitor species accurately across the globe, scientists need to make sure they are all referring to the same species by the same name. A worldwide agreed system of classification helps to do this. Species that share characteristics specific to that group are put together. Some of these characteristics are visible, such as whether or not the organism has a skeleton, or what type of flower a plant has. Other groupings may depend on more detailed study of the organism, for example looking for similarities in DNA may help scientists decide to which group a species belongs.

**FIGURE 4**: Australian magpies (left) and UK magpies (right) look similar, but are actually not the same species.

### Did you know?

We know many species by their 'common name', but this can lead to confusion. For example, a 'magpie' in Australia is not the same species as a 'magpie' in the UK.

There are different levels of classification that become more and more detailed as the levels progress from Kingdom to species. The left part of Figure 5 shows how organisms are grouped. On the right is the classification of modern humans.

## QUESTIONS

**5** Sometimes there is disagreement about whether two organisms are the same species or not. How could DNA analysis be used to settle such a disagreement?

**6** In a small group, evaluate the risks to society from loss of biodiversity. Are there any benefits to society from the loss of biodiversity? (Hint: Think carefully about why habitats and thus species are being lost.) What should be done to address this issue?

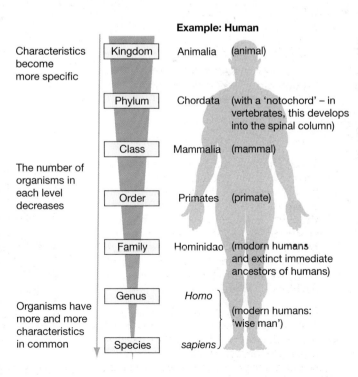

**FIGURE 5**: Explaining classification.

classification of living things GCSE

# A sustainable future

## What will you save for your grandchildren?

It has been estimated that half of existing species may become extinct by 2100. How would you feel telling your grandchildren about swans, or barn owls or badgers, and knowing they would never see one? Scientists are working hard to conserve as many species as they can.

FIGURE 1

**We are learning to:**
> define sustainability
> apply knowledge about biodiversity to explain why large-scale monoculture is unsustainable
> explain how scientists have tried to minimise the impact of habitat loss

## What we mean by sustainability

**Sustainability** is finding a way of meeting our needs now without stopping future generations from meeting their needs. For example, the way we farm the land must leave it in a state that allows it to produce the same amount of crops in the future. With so many humans now living on Earth, ways must be found of limiting our impact on the wildlife, habitats and environment that we all share and rely upon.

To increase our own contribution to sustainability we can reduce what we use, for example by choosing items with less packaging, re-using things that we have used before or that others no longer have a use for, and recycling rather than throwing away.

### QUESTIONS

1 What is sustainability?
2 How can we make our own lives more sustainable?

## The importance of maintaining ecosystems

One important aspect of sustainability is the protection of the environment. By maintaining important habitats for wildlife, we can conserve a variety of different species and maintain biodiversity. Biodiversity is important for a number of reasons. One of these reasons is that the loss or extinction of only a few species can have a big impact on the whole ecosystem, causing it to break down and stop functioning as it previously did.

Sea otters provide an example of the impact on one species on an ecosystem. In the 1800s sea otters were hunted nearly to extinction. The decline in sea otters led to a huge increase in the sea urchin population, as sea otters feed on sea urchins. The increased numbers of sea urchins fed on and destroyed the kelp beds, where spawning fish laid their eggs. Fish stocks began to decline as their breeding habitat was so reduced.

To ensure sustainability we need to maintain ecosystems. Maintaining biodiversity is an important aspect of this.

**FIGURE 2**: A sea otter.

International Year of Biodiversity

# B3 Life on Earth

### Did you know?
A species that is crucial to the functioning of an ecosystem is called a 'keystone species' after the 'keystone' at the top of a stone arch that stops the arch collapsing.

### QUESTIONS

**3** What can happen to an ecosystem if a few key species are lost?

**4** Use the example of the sea otter to explain why is it important to maintain biodiversity in ecosystems.

## Improving biodiversity on farmland

A widespread cause of the loss of biodiversity is **intensive crop production**. Intensive crop production involves the large-scale planting of one type of crop, called **monoculture**. Often hedgerows and other areas of vegetation at the sides of the fields are removed in order to increase the space for the crops. Farming in this way enables the farmer to make use of machinery available to plant, maintain and harvest the crop, and maximises yields and profits. Such farming methods have negative effects on biodiversity and are not sustainable. When a large area of only one crop is grown as a monoculture, the land will support very few other species. The application of pesticides and weed killers further reduces the numbers and diversity of other organisms that the farmland can support.

Scientists have identified the negative impacts that intensive crop production has had on biodiversity on farmland. They have worked to find ways to increase the biodiversity on farmed land, and some of these strategies have been very effective. Beetle banks at the sides and within the crop provide a habitat for beetles that **predate** on aphids that would otherwise damage the crop. As well as increasing the biodiversity, this can also lead to a reduction in the amount of pesticide that needs to be used. Farmers are also being encouraged to maintain hedgerows to support biodiversity on their farms.

**FIGURE 3**: A large-scale monoculture like this one has very low biodiversity. It is not a sustainable way to produce crops, although yields can be very high.

**FIGURE 4**: Maintaining hedgerows and wide field margins allows many species to survive in farmland and increases biodiversity. This is a more sustainable way of farming.

### Watch out!
Biodiversity is not just the *number* of species. It also refers to the genetic diversity within a species.

### QUESTIONS

**5** Explain how large-scale monoculture impacts on biodiversity.

**6** Describe steps farmers can take to maximise biodiversity on their farms.

🔍 environmental stewardship scheme   sustainability biology GCSE

# Thinking ahead

**We are learning to:**
- understand that scientific advances can make life better, but can also have unforeseen consequences
- explain the factors that determine the sustainability of a product or process

## Are we filling up the landfill?

Much of the waste we generate in our homes finds its way to landfill sites. These huge areas used for the disposal of rubbish are filling up fast. New areas of land are being turned into landfill sites to meet the increasing demands for places to dispose of waste.

FIGURE 1

 ## Reducing waste

Advances in science and technology can have unexpected effects. Before about 1950, for example, disposable nappies did not exist. Parents would use folded squares of absorbent cloth as nappies on their babies, and these cloth nappies would be soaked in disinfectant and washed daily. Modern disposable nappies with self-sticking tapes at the sides were first available in 1970s. They made life easier for millions of parents around the world and they are soft and comfortable for babies. They contain a super-absorbent polymer to make sure they do not leak.

**FIGURE 2**: Line-drying washable nappies minimises energy use as well as avoiding the waste caused by disposable nappies.

But disposable nappies have a significant disadvantage. They create huge amounts of waste that does not easily decompose. Eight million disposable nappies are thrown out every day. It has been estimated that a disposable nappy could take 500 years to break down in a landfill site.

Modern washable nappies are now available that are easy to use and comfortable for babies. But there is debate about whether they are actually better for the environment than disposable nappies, as so much energy is needed to wash and dry them in modern machines.

### Did you know?

The use of disposable nappies creates 400 000 tonnes of waste every year – the same amount as a city the size of Birmingham.

### QUESTIONS

1. Explain the disadvantages of using disposable nappies.
2. Why is it unclear whether washable nappies are better for the environment?

green living    living more sustainably

# Improving sustainability

There are a number of ways to improve sustainability when products are manufactured. The most sustainable products are produced with little energy, using locally available materials and create little pollution in their manufacture. If transporting the ingredients or the final product can be avoided, this increases the sustainability of a product. The way that products are packaged and the packaging materials used have a big impact. Minimal packaging, or the re-using of packaging, increases the sustainability of a product.

When looking at the sustainability of an item, it is important to consider the whole 'life cycle' of the product, from manufacture to disposal. Something that is very sustainable in manufacture could need a lot of energy in its disposal, which would reduce its overall sustainability. The **Life Cycle Assessment** (see pages 196–197) of a product tracks its environmental impact from sourcing the raw materials, through manufacture, transport to distribution centres, usage and then disposal.

FIGURE 3: Supermarkets now encourage customers to re-use shopping bags rather than use them once and throw them away.

## QUESTIONS

**3** Explain what is meant by a product that has been produced sustainably.

**4** Local 'farmers' markets' where residents in an area can buy produce from farmers who are local to the area are a way to increase the sustainability of food supply. Can you explain why?

# Biodegradable packaging

The use of **biodegradable** packaging has become popular for some products including plastic carrier bags. This biodegradable material, however, breaks down to release carbon dioxide. In landfill sites, where the rubbish can be packed in tightly, this decomposition process can take a long time. Also, like all packaging products, those that are biodegradable still use energy in their manufacture and in the transport needed to get them to their place of use. Cutting down or omitting packaging altogether is by far the most sustainable solution to reducing waste.

## QUESTIONS

**5** What are the likely environmental impacts of orange juice, grown in Florida and sold in the UK in biodegradable cartons? Carry out a Life Cycle Assessment, identifying the environmental impacts at each stage.

**6** Using the example above, do you think drinking Florida orange juice in the UK is a sustainable activity? Explain your answer.

reducing packaging waste

# B3 Checklist

## To achieve your forecast grade in the exam you'll need to revise.

Use this checklist to see what you can do *now*. Refer back to pages 74–99 if you're not sure.

Look across the rows to see how you could progress – **bold italic** means Higher tier only.

Remember you'll need to be able to use these ideas in various ways, such as:
> interpreting pictures, diagrams and graphs
> applying ideas to new situations
> explaining ethical implications
> suggesting some benefits and risks to society
> drawing conclusions from evidence you've been given.

Look at pages 312–318 for more information about exams and how you'll be assessed.

### To aim for a grade E | To aim for a grade C | To aim for a grade A

| To aim for a grade E | To aim for a grade C | To aim for a grade A |
|---|---|---|
| understand what a species is, and explain through examples how different species are adapted to their environments | understand that organisms that share a habitat compete for resources; understand that if the environment changes, or a new predator or disease arrives, some species will go extinct because they cannot adapt quickly enough; explain how a population change in one organism can impact on other organisms | ***explain the interdependence of living organisms using food webs*** |
| understand that the Sun provides the energy for nearly all food chains, and that plants use this energy for a process called photosynthesis | understand that the energy that plants absorb from the Sun through photosynthesis is stored in their tissues and is transferred to other organisms when the plants are eaten; understand that energy passes out of a food chain at each level via heat, waste products and inedible parts, and this means that food chains have a limited length | calculate the efficiency of energy transfer between levels of a food chain |
| understand the carbon cycle, including the processes of combustion, respiration, photosynthesis, and decomposition and the role of microorganisms | understand the carbon cycle and also the recycling of nitrogen compounds in the nitrogen cycle | ***understand the roles of nitrogen-fixing bacteria and denitrifying bacteria in the nitrogen cycle*** |

| To aim for a grade E | To aim for a grade C | To aim for a grade A |
|---|---|---|
| understand that environmental change can be measured using living and non living indicators | describe how the environment can be monitored using living and non-living indicators, and explain the advantages and drawbacks of each of these methods; give examples of living indicators | explain why it is useful for scientists to monitor long-term environmental change |
| recall that life on Earth began about 3500 million years ago and that it has evolved from very simple organisms | understand that there is variation between individuals of the same species and that some of that variation is due to genetic differences; and that these genetic differences can be passed on and this can lead to evolution | |
| understand the term natural selection | explain the process of natural selection and describe the similarities between natural selection and selective breeding; understand how Darwin developed his theory of evolution by natural selection; understand that the fossil record and DNA analysis of species provide evidence for evolution | understand the role of mutations, environmental changes, natural selection and isolation in the formation of new species; understand why Darwin's theory of evolution is a better scientific explanation than others |
| understand that biodiversity refers to the variety of life on Earth; understand that organisms are classified into groups according to similarities in characteristics | understand that there is a vast range of organisms on Earth and many of these have not yet been identified and classified; understand how the classification of organisms is organised and why it is useful | understand that classification of organisms and fossil organisms is aided by DNA analysis and shows their evolutionary relationship; understand why biodiversity should be conserved for future generations, and explain why the rate of extinctions is increasing |
| describe what is meant by sustainability | understand that maintaining biodiversity is important because it helps to improve sustainability; give examples of unsustainable farming practices and explain how these could be improved | understand that sustainability can be improved when manufacturing and packaging products, and explain ways to make these processes more sustainable |

# Exam-style questions

## Foundation level

**1** The following statements are about sustainability. Choose the correct word in each sentence. [AO1]

**a** Sustainability is about *maintaining/changing/measuring* the environment so that future generations have what they need. [1]

**b** Packaging, even if it is biodegradable, *increases/decreases/does not affect* the sustainability of a product. [1]

**c** Shopping at *supermarkets/locally/online* to reduce the distance food travels increases sustainability. [1]
[Total: 3]

**2** Look at the chart.

*Orang-utan* 48 chromosomes (24 pairs)
*Gorilla* 48 chromosomes (24 pairs)
*Chimpanzee* 48 chromosomes (24 pairs)
*Bonobo* 48 chromosomes (24 pairs)
*Human* 48 chromosomes (24 pairs)

Present
3 million years ago
6 million years ago — Extinct common ancestors of chimpanzee and bonobo
8 million years ago — Extinct common ancestors of chimpanzee (including bonobo) and human
— Extinct common ancestors of gorilla, chimpanzee and human
13 million years ago — Extinct common ancestors of orang-utan, gorilla, chimpanzee and human

**a** How long ago did orangutans, gorillas, chimpanzees and humans share an ancestor? [1] [AO2]

**b** Where does the evidence for this chart come from? [2] [AO2]

**c** What would we expect to notice if we compared humans with chimpanzees, and then compared humans with orangutans? Explain your answer. [2] [AO3]
[Total: 5]

**3** Give an example of a plant that lives in an arid environment. Describe the adaptations of the plant and the function of each adaptation. The quality of written communication will be assessed in your answer to this question. [6] [AO1]
[Total: 6]

## Foundation/Higher level

**4**

**a** Calculate the efficiency of energy transfer at each stage of this food chain. [3] [AO2]

|  | Grass | Grasshopper | Bird |
|---|---|---|---|
| Energy taken in (kJ) | 100 | 80 | 500 |
| Energy in tissues (kJ) | 50 | 8 | 15 |
| Efficiency (%) |  |  |  |

**b** What does this suggest about energy efficiencies higher up a food chain? [2] [AO3]
[Total: 5]

**5** Read this extract and answer the questions. [AO2]

**Why is your farm important for farmland birds?**
*Farmland birds need farmland and farmers. The birds on your farm are a good indicator of the overall health of biodiversity, as they sit high up the food chain. If bird populations are doing well then it indicates that the plants and insects on which they feed are thriving too. Since the mid-1970s, there has been a steep decline in the country's farmland bird populations, with many declining by over 50%. A broad range of studies have shown that these declines have been caused by the loss of breeding and year round foraging habitats, meaning that our farmland birds have fewer places to nest, raise fewer young and are less able to survive the winter.*
From 'Farming For Birds' Environmental Stewardship Scheme.

**a** Explain what is meant by the sentence, 'The birds on your farm are a good indicator of the overall health of biodiversity, as they sit high up the food chain'. [2]

**b** What reasons are given for the steep decline in farmland birds since the mid-1970s? [3]

**c** Suggest ways for farmers to improve the habitats for farmland birds, using information from this paragraph and your own ideas. [3]
[Total: 8]

**6** The Northern Spotted owl is a species under threat. The owl lives in the forests of North America and needs very large territories in which to breed and hunt. It does not migrate readily in response to habitat change. To thrive it needs old forests which have nesting sites in dead trees. These forests are in demand for logging as their wood is valuable. [AO2]

**a** Explain why the destruction or fragmentation (breaking up) of habitat would have a particularly bad effect on the Northern Spotted owl population. [3]

**b** What measures could be taken to protect the Northern Spotted owl? [3]
[Total: 6]

---

| AO1 recall the science | AO2 apply your knowledge | AO3 evaluate and analyse the evidence |

#  Worked example (Higher)

Head lice are a common parasite, particularly of children, as they are spread easily when children play. The resistance of head lice to insecticide shampoos is an example of natural selection. Permethrin is the active ingredient in a commonly used head louse treatment.

Read the following passage.

*Resistance*
*The first real evidence of head lice resistance to permethrin in the UK came in late 1994, when head lice taken from a patient who had had a number of consecutive unsuccessful treatments with permethrin were found to survive exposure to around 20 times the normal lethal dose. It was anticipated that this was not an isolated occurrence and within three months we had been contacted by health workers from several localities in the UK who shared our convictions. In each case lice were sent from patients who had received several applications of pyrethroids. All showed signs of resistance (Burgess et al., 1995).*

Taken from Brown, M.; Burgess, I: Management of insecticide resistance in head lice *Pediculus capitis*

**AO2 a** Describe the first piece of evidence that head lice were beginning to become resistant to permethrin. [1]

*In 1994, head lice taken from a patient survived exposure to permethrin at 20 times the normal lethal dose.*

**AO2 b** Explain the process by which head lice could have become resistant to permethrin. [4]

*Head lice could have become resistant to permethrin because when it is applied, some head lice will be more resistant than others.* ✓

*When the permethrin shampoo is put on, the resistant head lice won't die, but all the others will.* ✓

*This would leave just resistant head lice which then wouldn't have much competition from the other head lice. When they bred together, they would pass on their resistant genes to their offspring.* ✓

**AO2 c** There are several different types of chemical available to treat head lice. Suggest a possible approach to reduce the chance of resistance to head louse treatments in a particular area. [3]

*To stop the head lice getting resistant to the shampoos, you need to make sure all the head lice have been killed by the first treatment. Head lice that haven't been killed (these would be most likely to be the resistant ones) could be removed using a nit comb and could be squashed.* ✓

*Also, maybe different treatments could be used in turn, so head lice that are resistant to one would be killed by the other one.* ✓

## How to raise your grade
Take note of the comments from examiners — these will help you to improve your grade.

This is correct and would be awarded 1 mark.

The point about the resistant head lice not having much competition from other head lice is excellent and this is a good use of key language. The final sentence could be better if it said 'When the resistant head lice bred together'; however, coupled with the previous sentence, it earns a mark.

This is quite a good answer but is incomplete, and scores 3 out of a possible 4 marks. In order to be awarded full marks, a sentence is needed to explain that the genes for more resistance to the permethrin would spread through the population of head lice.

One more suggestion is needed to score the full 3 marks, for example making sure that the instructions on the shampoo are followed correctly.

The question asks for ideas to be suggested, so any reasonable idea can score a mark. The word 'suggest' in a question usually means that you will not have come across this particular context before, but you should apply what you know to come up with a sensible idea.

# C1 Air quality

## What you should already know...

### All materials are made up of particles

Materials are found as solids, liquids and gases.

Particles are arranged differently in solids, liquids and gases, and this is what gives the differing properties of solids, liquids and gases.

  Why can you smell a burning bonfire from a distance away from it?

### Elements are the building blocks of materials

Elements are the simplest substances known. They are each made up of one type of particle called an atom.

Elements join together in chemical reactions to form compounds.

A mixture contains two or more materials, which may be elements or compounds, that have not reacted together chemically.

Filtering separates solids from liquids or gases.

  Name an element, a compound and a mixture.

### There are patterns in the way that substances react together

Chemical reactions can be described by word equations and symbol equations.

Groups of elements and compounds react in a similar way.

Acids are compounds which behave in a similar way.

Equations of chemical reactions show these patterns of behaviour.

  How could you show a material is an acid?

### The Earth's atmosphere changes

The air is an important resource, used in many natural processes and by humans.

Changes take place in the atmosphere as a result of natural processes and human activities.

Controlling human activities that change the atmosphere is necessary to make development more sustainable.

  What human activities cause changes in the atmosphere?

# In C1 you will find out about...

- the gases that make up the air
- how the air has changed since the Earth was formed
- how humans are now changing the composition of the air
- the effects of changing air quality on health

- the chemical reaction that takes place when fuels burn
- the rearrangements of atoms in chemical reactions
- the effects of burning fuels containing sulfur
- the sources of pollutants in the air
- what happens to the pollutants when they are in the air

- how we can reduce pollutants from power stations
- ways of reducing the amount of carbon dioxide we produce
- ways of reducing the pollutants given out by vehicles

# The air around us

**We are learning to:**
- describe the composition of the air
- explain the properties of air as a gas
- explain how the true value of a quantity is estimated

## What is air?

We breathe it and burn fuels in it. Plants use it to grow. It is all around us but we hardly notice it except when the wind blows. Aircraft and birds fly in it. More than 100 kilometres above us, satellites look down through it.

**FIGURE 1**: The atmosphere forms a layer about 100 km thick over the surface of the Earth. It is made up of air, clouds and dust.

### The substances that make up the air

Air is a **gas**. It is a mixture of lots of different substances, which are all gases. The most common gases in air are nitrogen, oxygen and argon. There is also water vapour, carbon dioxide and many other substances.

**QUESTIONS**

1. Make a list of substances that are found in the air.
2. Explain why clouds and dust are part of the atmosphere but are not part of the air.

**Watch out!** Water vapour is a gas that is part of the air. Clouds are droplets of liquid water or ice.

### What we mean by a gas

A gas spreads out to take up all the space available. When a gas is inside a container, such as a balloon, it can be squeezed into a smaller volume.

The explanation for these properties is that a gas is made up of particles. The particles are very small and move around. Compared with the size of the particles, there is a lot of space between them.

In air most of the particles are small **molecules**, mainly the elements nitrogen ($N_2$), oxygen ($O_2$) and argon (Ar). There are also small molecules that are compounds, such as water ($H_2O$) and carbon dioxide ($CO_2$).

**FIGURE 2**: Most of a gas is empty space.

gas particles

# C1 Air quality

Scientific data, shown in the table, tell us that nearly all the air is nitrogen, oxygen and argon. Less than 0.1% of dry air is made up of all the other gases. These percentages are almost the same all over the Earth. Only the percentage of water vapour varies, between 0% and 1%, because the water can condense to form clouds and rain.

| Substance | % in dry air |
|---|---|
| nitrogen | 78 |
| oxygen | 21 |
| argon | 1 |

The relative amounts of gases in dry air.

## QUESTIONS

**3** What happens to the particles of gas in a balloon when a force squashes it?

**4** A classroom contains about 100 $m^3$ of air. What volume of the classroom is taken up by the oxygen in the air?

## How we know what's in the air

In the apparatus shown here, air is passed backwards and forwards from one syringe to the other while the copper in the tube is heated. The oxygen from the air joins with the copper, so the volume of the air gets smaller. The amount of oxygen in the air is the difference in volume before and after the air has been passed over the copper.

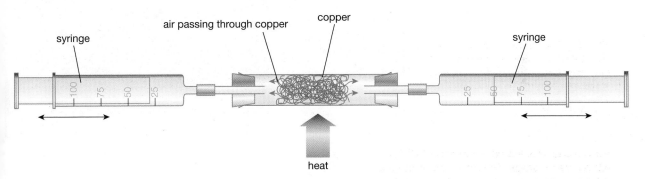

**FIGURE 3**: This apparatus can be used to find out how much oxygen there is in the air.

### Watch out!

Measurements always vary from the true value because of errors. The **range** of a set of data is from the smallest measurement to the largest. The **mean** is usually close to the true value.

## QUESTIONS

**5** An experiment using the apparatus shown to find the amount of oxygen in the air was done four times. The results for the percentage of oxygen were: 20, 21, 22, 21.

  **a** Why were several results needed?

  **b** Why were the results not all the same?

  **c** What was the range of results?

  **d** From these results, estimate the true value for the percentage of oxygen in the air. Explain why you chose your answer.

### Did you know?

About 80% of all the Earth's air is in a layer that extends to a height of about 15 kilometres above the surface.

air composition

# Changing air

## What makes the Earth's atmosphere special?

The Earth's atmosphere is just right. There's plenty of oxygen for us to breathe but not so much that fires would spread uncontrollably. The temperature over most of the Earth is suitable for life. The Earth is the only planet in the solar system which has these conditions.

> **We are learning to:**
> - explain how the Earth's atmosphere was formed
> - describe processes that have changed the atmosphere
> - explain why scientists are sceptical about new ideas until there is good evidence

### How did the Earth's atmosphere form?

Scientists look at what comes out of a volcano when it erupts. Apart from **lava** and tiny particles of dust, they find that volcanoes also give out a lot of water vapour and carbon dioxide.

By examining very old rocks, scientists have discovered that the Earth was formed about 4 billion years ago. There were many more volcanoes then, so scientists think that the Earth's early atmosphere probably consisted of carbon dioxide and water vapour, given out by the volcanoes.

#### QUESTIONS

1. Name two gases that probably formed most of the early atmosphere of the Earth.
2. How have scientists worked out which gases were in the Earth's early atmosphere?

**FIGURE 1**: The surface of the early Earth was covered with volcanoes like this.

### Life-forms change the Earth's atmosphere

Four billion years ago the Earth's atmosphere was hot and much thicker than it is now. Gradually the temperature fell and most of the water vapour condensed. It rained for thousands of years and the water became the oceans.

About 3 billion years ago life began in the form of tiny **bacteria**-like creatures. They began to use the carbon dioxide in the air and the light from the Sun to grow. This process of **photosynthesis** removed carbon dioxide from the atmosphere and produced oxygen as a waste product. Gradually the amount of oxygen in the air increased. Animals evolved that were able to breathe the oxygen.

**Watch out!** Oxygen is a waste product from plants. Animals evolved to make use of it.

**FIGURE 2**: These structures are the remains of photosynthesising bacteria that lived 2 to 3 billion years ago.

early atmosphere changes

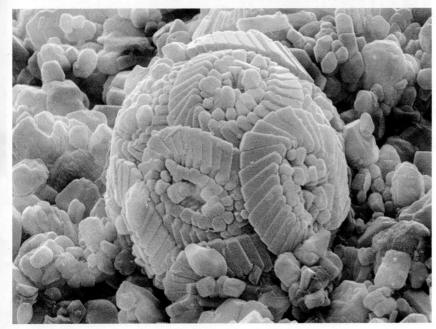

FIGURE 3: The shells of microscopic creatures like these are formed using carbon dioxide from the oceans.

Carbon that was once in the atmosphere as carbon dioxide became part of the bodies of the animals and plants living on the Earth. When these animals and plants died, their bodies sometimes became buried and eventually became the **fossil fuels** we use today.

Carbon dioxide dissolves in the oceans. Some of it reacts with salts in the seawater to form insoluble calcium carbonate. This forms **sediment** and eventually turns into **sedimentary** rocks. Sea creatures also use the carbon dioxide to build shells of calcium carbonate. When they die the shells sink to the bottom of the sea and become part of the sediment. Over billions of years thick layers of limestone and chalk have built up.

These different processes have removed almost all of the carbon dioxide that was in the early atmosphere, leaving the air with the composition we have today. Scientists have found evidence in rocks and fossils in different parts of the Earth to support these ideas. Most scientists agree that this was how the atmosphere changed.

## QUESTIONS

**3** How did the evolution of living organisms change the Earth's atmosphere?

**4** Describe one way that carbon dioxide was removed from the atmosphere that depends on living organisms and one way that does not involve living organisms.

## Changing ideas

Sixty years ago many scientists thought that the air in the early atmosphere was largely ammonia and methane. New discoveries of the composition of very old rocks convinced most scientists that this idea was not correct and that the atmosphere was largely carbon dioxide.

Today many scientists think that humans are causing changes in the atmosphere by burning fossil fuels and releasing carbon dioxide back into the air.

### Did you know?

Iron is produced from iron ore which was formed millions of years ago by the reaction of oxygen with iron in the oceans. The oxygen came from photosynthesis by living organisms.

## QUESTIONS

**5** Why are scientists fairly certain that they understand the changes that took place in the atmosphere?

**6** Why did ideas about the composition of the early atmosphere change?

early atmosphere changes

# Humans and the air

## We are learning to:
- explain how gases have been added to the air by humans
- explain how humans have changed the amount of some substances in the air
- explain how side effects of technology have affected air quality

## How can we know that the air is changing?

There are many substances in the air in tiny percentages which are very difficult to measure. In 1958 Charles Keeling set up a new instrument on Mauna Loa in Hawaii. It could measure the amount of carbon dioxide in the air precisely enough to see changes year after year.

FIGURE 1: Winds and weather systems mix the air, so the air over Mauna Loa is typical of the air elsewhere.

## How has human activity changed the air?

Farms, factories, power stations, shops and offices, homes, lorries, buses, cars, trains, ships, aircraft – all these places and types of transport produce gases that find their way into the atmosphere. Since humans started farming, burning fuels and using machines we have changed the amount of some gases such as **carbon monoxide**, **nitrogen oxides** and **sulfur dioxide** in the air and added some extra substances. As the population increases we are making more changes to the atmosphere.

### QUESTIONS

1. Name a type of transport that releases gases into the atmosphere.
2. State something you do that adds gases to the air.

## Gases and particles added to the air

Burning fuels releases carbon dioxide. In the last 50 years, since Charles Keeling began his measurements, the amount of carbon dioxide in the air has increased by about 25%.

Modern agriculture makes a lot of use of machines that give out gases, but ploughing land and rearing livestock also releases gases into the atmosphere. For the last 200 years the increasing amount of industry, electricity generation and transport has also produced gases.

Particulate matter is tiny bits of solid that float in the air. Dust from volcanoes, desert sand storms and soot from forest fires are natural sources of **particulates**. Soot is made up of tiny bits of carbon. In cities there can be four times the normal amount of particulates due to the soot and other dust produced by burning fuels. Burning down forests to make more farmland also produces particulates and carbon dioxide.

FIGURE 2: How does this change the composition of the air?

### Did you know?
Particulates in the atmosphere stop some of the Sun's light reaching the Earth's surface. The increasing amount of particulates is causing what is called 'global dimming'.

### QUESTIONS

3. Name four gases that human activity releases into air.
4. State one natural and one human source of particulate matter in the atmosphere.

Keeling    Mauna Loa    carbon dioxide

# C1 Air quality

## Measuring air quality

The amount of carbon dioxide in the air is measured in parts per million (ppm). A reading of 1 part per million means that there is 1 g of the substance in 1 million grams of air. Today the amount of carbon dioxide in the air is about 390 ppm, which is 0.039% of the air. The other pollutant gases – carbon monoxide, nitrogen oxides and sulfur dioxide – are measured in parts per billion (ppb), which is a thousand times smaller. Although these figures appear small, over the whole atmosphere there are many thousands of tonnes of these gases.

Air quality monitoring stations across the UK measure automatically the amount of particular substances in the air at regular intervals. The data is transmitted to a central computer for recording and analysis. This data can be accessed on the internet.

### Watch out!

Make sure you look at the units given in tables and graphs. The amounts of carbon monoxide, nitrogen oxides and sulfur dioxide in the air are very much less than carbon dioxide but are still important.

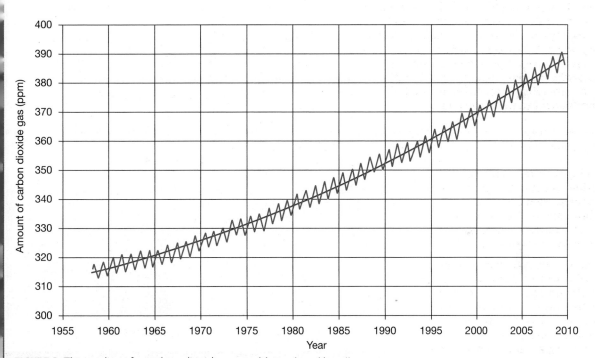

**FIGURE 3**: The readings for carbon dioxide gas on Mauna Loa, Hawaii, from 1958 to the present day in parts per million.

### QUESTIONS

**5** By how much in ppm has the amount of carbon dioxide in the air increased since 1958?

**6** Amounts of nitrogen oxides and sulfur dioxide are often higher in cities than in the countryside. Why is this?

| Pollutant | Amount in μg/m³ |
|---|---|
| carbon monoxide | 700 |
| sulfur dioxide | 13 |
| nitrogen oxides | 321 |
| particulates | 35 |

Typical air quality readings for a large city in the UK (μg = microgram).

🔍 UK National Air Quality Archive

# Air quality and health

**We are learning to:**
> explain how health is affected by pollutants in the air
> explain how the environment is affected by pollutants in the air
> explain why there is a correlation between air pollutants and health

## How did the Beijing Olympics improve air quality?

In the weeks before the Beijing Olympic Games in July 2008, the Chinese government prevented heavy trucks entering the city and people were only allowed to use their cars every other day. Coal-burning power stations were shut down. These actions were taken because air quality was often poor in Beijing.

**FIGURE 1**: Air quality in Beijing improved during the Olympic Games because less fossil fuel was burned.

## Pollution and air quality

A **pollutant** is a substance that is harmful to health or the environment and which is produced by human activity. Air quality is 'good' if there are very few pollutants in the air. If there are large amounts of pollutants in the air then the air quality is 'bad' or 'poor'. When air quality is poor, people with asthma and other lung diseases have difficulty breathing and may die. Carbon monoxide in the air is one cause of severe breathing difficulties. Inhaling it reduces the amount of oxygen that the blood can carry to the organs. If it is present in large quantities it can quickly lead to death.

### QUESTIONS

**1** Which of the following gases are pollutants in the air?
nitrogen, sulfur dioxide, oxygen, carbon monoxide

**2** What would you expect to happen to people with asthma if air quality changes from good to poor?

## Air quality, disease and the environment

Many pollutants are released into the atmosphere by human activities. Where the amount of pollutants in the air from burning fossil fuels is high there are more deaths from asthma, and heart and lung diseases. Scientists say there is a **correlation** between air quality and health.

Mexico City has grown from about 2 million people in the 1950s to over 20 million now. In the 1950s, when there was little fuel burned, the air was clear and air quality was good. Now there are many days when there is **smog** and air quality is poor. Many more young children and older people die from heart and lung diseases today than in the 1950s.

**FIGURE 2**: Smog forms over Mexico City when the amount of pollutants is high.

air quality health correlation

C1 Air quality

Sulfur dioxide and nitrogen oxides make rain acidic. **Acid rain** damages plants and kills fish in rivers and lakes.

The increasing amount of carbon dioxide in the air is causing climate change and this is a factor in the extinction of many plants and animals.

### QUESTIONS

**3** Why has the number of deaths in every million people from heart and lung disease increased in cities such as Mexico City since 1950?

**4** State two ways in which poor air quality affects the environment.

## Correlation and cause

Breathing air containing a little sulfur dioxide will not give you asthma or heart disease, but if you have these diseases already then the pollutant may make it more difficult for you to breathe. Many studies have shown that as the amount of pollutants breathed in increases so the number of people dying from various heart and lung diseases also increases. Poor air quality is therefore linked to a high death rate from these diseases. This kind of connection between one factor and another is called a *correlation* even though the pollutants are not a *cause* of the disease.

If acid rain or climate change damages crops then people will go hungry. Damage to the environment caused by air pollution will therefore affect health indirectly.

**Watch out!**
Some pollutants are waste materials but others are useful substances that have escaped into the atmosphere.

**FIGURE 3**: Industrialisation in China is affecting the environment.

### Did you know?
London is allowed by European Union rules to have 35 'bad air quality' days a year but usually has many more than this, possibly leading to over 4000 extra deaths a year.

### QUESTIONS

**5** What evidence is there for a correlation between air quality and health?

**6** Would you expect the death rate from heart and lung disease to be greater in cities or in the countryside? Explain your answer.

air quality health correlation

# Burning fuels

## Can you light a fire under water?

Fire fighters use water to put out flames, but it is possible to make a fuel burn under water. Underwater welders carry tanks of fuel and oxygen. Together they make a flame that is hot enough to cut through or weld metal.

FIGURE 1

**We are learning to:**
> describe the burning of fossil fuels and name the products
> explain the role of oxygen in burning
> explain how new ideas develop when observations are interpreted in new ways

## What happens when fuels burn

Coal is a fossil fuel. It is mainly carbon atoms. When coal burns the carbon combines with oxygen in the air to make carbon dioxide gas and energy (heat) is released.

Other fossil fuels such as petrol, diesel, fuel oil and natural gas are made of compounds called **hydrocarbons**. A hydrocarbon is made up of carbon and hydrogen atoms joined together.

When a hydrocarbon burns it joins with oxygen. The carbon in the hydrocarbon produces carbon dioxide, and the hydrogen produces hydrogen oxide, which is water.

### QUESTIONS

1. Which two substances are formed when fossil fuels burn?
2. What is in a hydrocarbon that joins with oxygen to make water?

### Did you know?

Every year, on average, every person in Europe uses the energy from burning 4 tonnes of fossil fuel.

## The oxygen theory of combustion

When you watch wood or charcoal burning, the fuel seems to disappear, and smoke and ash form. People in the past explained burning by simply describing what they saw. They did not think that the air was reacting with the fuel. Antoine Lavoisier, a French chemist, was the first person to suggest that an invisible gas in the air, that he had named *oxygen*, was combining with the fuel. We can write the reaction as a word equation.

**hydrocarbon fuel + oxygen → carbon dioxide + water (+ energy)**

With the help of his wife, Lavoisier published a book explaining his ideas. He said that oxygen can join up with many substances. We give the name **oxidation** to any reaction where substances combine with oxygen. Reactions where oxygen is lost are called **reduction**. Burning or **combustion** of a fuel is an oxidation reaction. Gradually more and more people began to agree with Lavoisier because more observations could be explained by his ideas than by the old descriptions of burning.

The English chemist Joseph Priestley never agreed with Lavoisier's theory of combustion. However he did show that fuels burn more rapidly and more brightly in pure oxygen than in air.

FIGURE 2: What can you see happening when something is burning?

Lavoisier oxygen theory   combustion

C1 Air quality

## QUESTIONS

**3a** Write a word equation for burning methane, the main hydrocarbon fuel in natural gas.

**b** Why is this called an oxidation reaction?

**4** Why didn't everyone accept Lavoisier's ideas straight away?

**Watch out!**
Combustion is another word for burning. It is used for the reaction of any substance with oxygen that gives out energy.

## Burning in pure oxygen

Today the gases in air can be separated and used for various purposes. Pure oxygen is used with fuels such as acetylene, so that they burn at a high enough temperature to melt steel, even under water (see Figure 1).

acetylene + oxygen → carbon dioxide + water (+ energy)

The combustion of hydrogen gas is much faster with pure oxygen than with air. The reaction is used in rockets such as the NASA Space Shuttle.

hydrogen + oxygen → water (+ energy)

$2H_2 + O_2 \rightarrow 2H_2O$

## QUESTIONS

**5** Why is pure oxygen and not air used in rockets such as the Space Shuttle?

**6** Explain how an oxy-fuel torch can burn under water.

**Hint:**
You won't be expected to write a symbol equation in your exam.

**FIGURE 3**: A really fast reaction between hydrogen and oxygen.

Lavoisier oxygen theory    combustion

115

# Rearranging atoms

## Can atoms ever be destroyed?

You are made of atoms formed billions of years ago in distant stars. In their time on Earth the atoms may have been in compounds in rocks, the sea or the air, as well as making up your body. The atoms will continue to exist for billions of years to come.

**We are learning to:**
> describe the arrangement of atoms in reactions
> explain why the masses of reactants and products of a reaction are the same
> explain that an experimental measurement is never exactly the true value

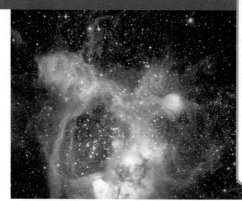

FIGURE 1: Atoms of elements are made by nuclear reactions in stars and are spread across the galaxy when stars die.

## What happens to atoms in chemical reactions?

When a piece of coal or a drop of petrol burns, the atoms do not change. Instead there is a change in the way the atoms are joined together. New substances are formed with the atoms joined up in different **molecules**. These new substances are often gases. This is how atoms that were in a fuel can become part of molecules in the air. The new gases in the air can affect air quality.

### QUESTIONS

**1 a** Carbon dioxide is formed when coal is burned in air. Do the carbon atoms in the carbon dioxide come from the coal or from the air?

**b** Where do the oxygen atoms in the carbon dioxide come from?

## Reactants and products

In a chemical reaction the atoms from the **reactants** always rearrange themselves in a particular way to make the **products**. For example, when methane burns completely the carbon and hydrogen atoms from the methane molecules separate and join with oxygen atoms to make carbon dioxide and water. The number of carbon, hydrogen and oxygen atoms in the products is the same as in the reactants.

The mass of the reactants and products is made up of all the atoms. We can weigh a sample of the reactants before the reaction and then weigh the products after the reaction. The mass stays the same because the number of atoms of each **element** has not changed. We say that the mass has been *conserved*.

If you do an experiment to weigh the reactants and the products of a reaction you will probably find that sometimes the mass increases a little bit and sometimes it falls a tiny amount. This is to be expected because there is always variation in measurements. There is always some error in the reading. Perhaps a breeze has blown the balance or a drop of a solution was spilled. We can never record the true value of a quantity.

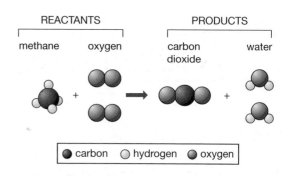

FIGURE 2: Methane burns in air to form carbon dioxide and water.

rearranging atoms

C1 Air quality

## QUESTIONS

**2** Look at Figure 2. Count the number of carbon, hydrogen and oxygen atoms in the reactants and in the products. What do you notice?

**3** Draw a similar diagram, showing what happens when hydrogen gas burns. Hydrogen molecules are made up of two hydrogen atoms.

**Watch out!**
Oxygen molecules in the air are made up of two atoms of oxygen joined together.

## Conservation of mass (Higher tier only)

When a fuel such as alcohol is burned it seems to disappear because all the products are gases and no solid or liquid is left behind. If we carry out the reaction inside a closed container so that no gases escape, we can show that the mass is conserved and no atoms are lost. That is why all the carbon atoms in all the fossil fuels that have ever been burned are still present in the environment, as carbon dioxide, carbon monoxide or particulates. The **law of conservation of mass** is the reason why air quality has changed when we burn fuels.

**Did you know?**
In the 1770s Antoine Lavoisier did a lot of precise experiments which showed that mass was conserved in chemical reactions. Since then it has been important for chemists to record masses accurately.

FIGURE 3: Mass is conserved in a chemical reaction.

## QUESTIONS

**4** 65 g of zinc reacts with 32 g of sulfur. What mass of products would you predict will be formed? Explain your answer.

**5** If you weighed the products, would you expect to get the exact answer you predicted? Explain your answer.

rearranging atoms conserving mass

# Reactants and products

**We are learning to:**
- describe how the properties of reactants and products are different
- explain why sulfur dioxide is formed when fuels containing sulfur are burned
- explain why some of the ways we use fuels have unexpected results

## What is sulfur?

Sulfur is an element. It is a yellow solid often found near volcanoes. Sulfur forms compounds such as sulfuric acid and calcium sulfate, which is used as plaster. Sulfur is needed by all living organisms. Bad eggs smell because they give off hydrogen sulfide.

**FIGURE 1**: Sulfur is easily recognisable by its colour.

## Burning sulfur

When you watch a chemical reaction you can often see things happening. This is because the substances are changing. The **products** which are formed have different properties from the **reactants** taking part in the reaction. If you burn some sulfur it changes from a yellow solid to a colourless gas. The product is sulfur dioxide. Sulfur does not dissolve in water. Sulfur dioxide does dissolve in water and makes a solution that is an acid.

### Did you know?

A sulfur-containing compound called mercaptan is added to natural gas to give the gas a strong smell. This is necessary so that gas leaks can be detected easily.

### QUESTIONS

1. Write down two differences between sulfur and sulfur dioxide.
2. How can you tell that a reaction has happened when you burn sulfur?

**FIGURE 2**: What can you see happening in this picture of sulfur burning?

sulfur dioxide

## Sulfur in fuels

New substances are formed by chemical reactions. Elements may join together to form **compounds**, or compounds may have split up. The products have a different arrangement of atoms to the reactants and so their properties are different.

Fossil fuels are formed from the remains of dead animals and plants. The sulfur which was in the organism is trapped in the fossil fuel. When the fuel is burned the sulfur burns too and forms sulfur dioxide. Each atom of sulfur joins with two atoms of oxygen to make sulfur dioxide.

sulfur + oxygen → sulfur dioxide
S          $O_2$              $SO_2$

 +  →

FIGURE 3: Formation of sulfur dioxide.

Fossil fuels include coal, oil, petrol and natural gas. Coal often contains the most sulfur, so burning coal can give off a lot of sulfur dioxide gas.

**Watch out!**
Sulfur is a natural impurity in fossil fuels. It is not used as a fuel itself.

### QUESTIONS

**3** Choose a fuel and make a list of the differences between it and the products formed when it burns.

**4** How can scientists tell that natural gas contains less sulfur than coal?

## A shocking discovery

Coal containing sulfur was burned in steam engines and in power stations. Fuels containing sulfur were burned in road vehicles, trains and ships. Few people worried about the smelly sulfur dioxide that was given off, but as we have seen, sulfur dioxide is a pollutant that harms health and the environment.

In the 1970s people noticed that forests were dying and lakes and rivers were losing fish. Scientists explained that the damage was caused by sulfur dioxide blown by the wind from industrial cities. The sulfur dioxide had reacted with the air to form sulfuric acid dissolved in rain.

FIGURE 4: Acid rain damage occurs a long way from where the fossil fuels are burned.

### QUESTIONS

**5** Why do you think the environmental damage caused by sulfur dioxide became more noticeable in the 1970s?

**6** Why has natural gas replaced coal as the main fuel used in factories?

fossil fuels

# Sources of pollutants

## Should we burn fossil fuels?

Smog, acid rain, global warming – in recent years we have become familiar with the problems caused by pollutants in the air. Now that we know the problems are caused by burning fossil fuels should we still be using them?

**We are learning to:**
- explain how different pollutants are produced by burning fossil fuels
- describe the arrangement of atoms in molecules
- describe what scientists have learned about the damaging use of fossil fuels

**FIGURE 1:** Vehicle exhaust is the source of a big fraction of all pollutants.

## How pollutants are formed

Power stations and transport produce most of the pollutants because they use most of the fossil fuel that is burned. Burning fossil fuels produces pollutants in various ways.

- Sulfur dioxide is produced if there is sulfur in the fuel.
- Carbon dioxide is always formed.
- Carbon monoxide is formed when there is not enough air to burn the fuel completely to carbon dioxide. It is poisonous.
- Particles of carbon, called **particulate** carbon, are produced when there is not enough air to burn all the fuel.
- Nitrogen oxides are formed when nitrogen and oxygen in the air react together. Nitrogen is not a fuel but the high temperature in furnaces and engines makes the gases react.

**Watch out!** Water is also formed when fossil fuels are burned, but it is not a pollutant.

### QUESTIONS

1. List all the pollutants formed by burning fossil fuels.
2. What pollutants are formed if there is not much air mixed with the fuel?

### Did you know?

Brown coal is one of the worst fuels for producing pollutants. It produces more carbon dioxide, particulates and sulfur dioxide than other fuels for the same amount of energy released. Yet it is still used in power stations across the world.

## Complex reactions

As the use of electricity and transport increased, people did not realise that damage was being done to health and the environment by burning fossil fuels. They did not know what all the chemicals being produced were, nor how they interacted to form damaging pollutants.

Figure 2 shows the atom arrangement of some of the pollutants (and water) produced by burning fossil fuels.

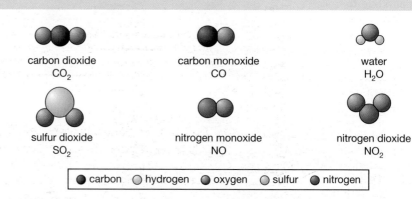

carbon dioxide $CO_2$ — carbon monoxide $CO$ — water $H_2O$

sulfur dioxide $SO_2$ — nitrogen monoxide $NO$ — nitrogen dioxide $NO_2$

● carbon ○ hydrogen ● oxygen ○ sulfur ● nitrogen

**FIGURE 2:** The formula of a compound shows the number of atoms of each element that are joined together in a molecule. The properties of a compound are related to its formula and atom arrangement.

air pollution sources

The effects of the combinations of pollutants are very complex. It is only in the last 50 years that scientists have discovered how the different pollutants are formed in furnaces and engines and also how they react with the air to produce smog, acid rain and climate change.

## QUESTIONS

**3** Why do carbon monoxide and carbon dioxide have different properties?

**4** Give one similarity and difference in the atom arrangement of carbon dioxide and sulfur dioxide.

## What's $NO_x$? (Higher tier only)

Scientists have discovered that nitrogen monoxide is formed in furnaces and engines when nitrogen and oxygen in the air are heated to about 1000 °C. When the nitrogen monoxide is released into the atmosphere it cools down and then reacts with more oxygen to form nitrogen dioxide. Nitrogen dioxide is a toxic and acidic brown gas.

$N_2 + O_2 \rightarrow 2NO$ nitrogen monoxide

$2NO + O_2 \rightarrow 2NO_2$ nitrogen dioxide

As there is more than one oxide of nitrogen, they are called nitrogen oxides or $NO_x$, where the x stands for the number of oxygen atoms (1 or 2, and so on).

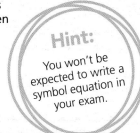

**Hint:** You won't be expected to write a symbol equation in your exam.

The reaction of nitrogen with oxygen can have a damaging effect on health and the environment.

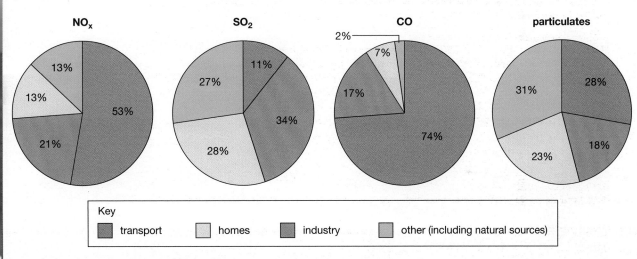

**FIGURE 3**: The fraction of each pollutant produced by various sources.

## QUESTIONS

**5** Look at the charts in Figure 3.

  **a** Which source produces most of the nitrogen oxides?

  **b** Which source makes the biggest contribution to sulfur dioxide?

  **c** For which pollutant does the combined production by transport and industry make up the largest fraction of the total?

**6** Describe the conditions required to form nitrogen monoxide and nitrogen dioxide.

# Removing pollutants

**We are learning to:**
- explain what happens to pollutants in the atmosphere
- explain why results in a set vary and the importance of the mean and range
- decide what to do about results that are at the edge of a range

## Why isn't there smog every day?

Air quality can be poor in all big cities and every day more pollutants are produced. Nevertheless, even in cities with very bad air pollution there are some days in the year when the smog disappears and the air is clear. Where does the smog go? What happens to all the pollutants?

**FIGURE 1**: The 1952 smog in London lasted for 5 days and may have killed over 10 000 people, but eventually it blew away.

 ## Air quality and the weather

Scientists measure air quality in cities every day. They repeat measurements many times because the results vary. A gust of wind may blow more or less pollutant over the sensor.

In a city on hot dry days when there is no wind, air quality gets worse. The amount of pollutants in the atmosphere increases. Mixtures of particulates and gases produce a cloud called smog. When the weather changes, a wind may blow the smog and the pollutants somewhere else. Rain washes the pollutants from the air on to the land or into the oceans.

###  QUESTIONS

**1** Why do people in polluted cities often hope for wind and rain?

**2** In the UK, air quality measurements are sometimes taken every 15 minutes. Why are they taken so frequently?

air pollution

# Where does pollutant carbon go?

Climate scientists take the **mean** of many measurements for each pollutant. The mean is a good estimate of the 'true' value. The difference between the highest and the lowest result is called the **range**.

Pollutants may be removed from the air but they have not disappeared completely.

Particulates settle on surfaces such as the leaves of plants, the ground or on buildings. Leaves that are coated with soot and dust cannot carry out photosynthesis properly so the plants may grow more slowly. Buildings covered with soot look dirty.

Carbon dioxide is used in photosynthesis but plants cannot use all the carbon dioxide added to the atmosphere. Some dissolves in rain water and in the oceans. Carbon dioxide solution is a weak acid.

**FIGURE 2**: There are boxes containing air quality sensors like these in many locations across the UK.

## QUESTIONS

**3** What happens to the smoke from a power station burning coal?

**4** In London the readings at one air quality sensor for particulates were 18, 15, 16, 17, 19 ($\mu g/m^3$). What is the mean and range of these readings?

**Watch out!** Conditions must stay the same for a set of readings if calculating the mean is to be valid.

# Acid rain

Sometimes a measurement may seem to be very different from the others, for instance, one sulfur dioxide reading may be much higher than others. If it can be shown that this reading is an error it can be left out as an **outlier**. Ignoring results which appear different can lead to mistakes. The reading could be due to a single vehicle burning fuel that contains sulfur and so be a correct result. The variation in the data could be due to a pattern.

### Did you know?

The most acidic lake in the world has a pH of around 0 and is in Costa Rica. It is in a volcano and the acid was formed naturally and not by pollutants.

Sulfur dioxide and nitrogen dioxide react with oxygen and water in the atmosphere to produce a mixture of sulfuric acid and nitric acid in rain water. This acid rain can have a pH as low as 3. Acid rain can damage plants that it falls on and it also makes rivers and lakes too acidic for fish and water plants to survive.

## QUESTIONS

**5** Why does acid rain damage sometimes occur in places a long way from where the pollutants were released into the atmosphere?

**6** The hourly readings of nitrogen dioxide at a roadside in London were 40, 58, 59, 69, 40, 30 ($\mu g/m^3$). Were any of these readings outliers? Explain your answers.

acid rain

# Improving power stations

**We are learning to:**
> explain how to reduce or remove the pollutants produced by burning fossil fuels in power stations
> explain how scientists can improve technology to reduce pollution and waste

## Why is there still acid rain?

The amount of sulfur dioxide over the Earth peaked in about 1980. It has since fallen in Europe and America but is still increasing in eastern Asia. While there is less damage from acid rain in some places, there is more in others.

**FIGURE 1**: The white clouds are harmless water vapour. Any sulfur dioxide being given off is invisible.

## Reducing our use of electricity

If we use less electricity, then power stations will burn less fossil fuel. Scientists have found ways of reducing the electricity used by new fridges, washing machines and other electrical goods.
Low energy lights have replaced the older, wasteful types of light bulb. We must learn to turn off appliances such as flat-screen TVs when they are not needed.

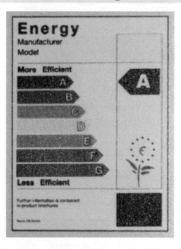

**FIGURE 2**: A washing machine with an A rating uses the least amount of electricity to get the job done.

### QUESTIONS

1 List three ways of cutting down the amount of electricity you use.

2 Name two ways that scientists have found to reduce the amount of electricity needed.

## Reducing the pollution from power stations

Even if we reduce electricity consumption we will still need fossil-fuel power stations. In the last 30 years scientists have developed ways of making power stations produce less pollutants.

Some power stations are burning low sulfur fuels. In Europe and America fuel oil and natural gas have replaced a lot of the coal used. Before oil and natural gas is burned, sulfur can be removed and sold to other industries.

It is difficult to remove sulfur from coal. Power stations burning coal produce the most sulfur dioxide and particulates but these can be removed before the waste gases are released.

**Watch out!**
The cleaning up of coal-fired power stations that has happened already has not reduced the amount of carbon dioxide they give out.

saving electricity

Solid particulates are removed by an electrostatic filter. This gives the particles an electric charge, which makes them collect on a charged metal grid. They can then be easily removed.

Sulfur dioxide can be removed from power station waste gases by **flue gas desulfurisation**. This is a chemical process that relies on sulfur dioxide being acidic.

### QUESTIONS

**3** Why has the amount of coal used in power stations been cut recently?

**4** What effect do you think making improvements to power stations has on the cost of electricity?

## Removing the sulfur dioxide (Higher tier only)

There are two methods of flue gas desulfurisation. The most common method uses calcium oxide (lime) which is obtained by heating limestone. Carbon dioxide gas is given off. The lime is mixed with water to make an alkaline slurry. Sulfur dioxide is washed or 'scrubbed' out of the flue gases with water and reacts with the lime. The compound formed is reacted with oxygen in the air to make calcium sulfate. This can be sold to make plaster for walls and this helps to pay for the process.

### Did you know?

Some scientists have plans for trapping the carbon dioxide given off by burning coal and then burying it. Other scientists are doubtful about whether this would be safe or too expensive.

**FIGURE 3**: Calcium sulfate stored at a power station. Coal-fired power stations in the UK produce over a million tonnes of calcium sulfate by flue gas desulfurisation each year.

The second method is used in coastal power stations where seawater is used as a coolant. Seawater is naturally slightly alkaline and will react with the sulfur dioxide. Air is then pumped through to oxidise the product to sulfate. Carbon dioxide is given off and the seawater is returned to the sea. A large volume of seawater has to be used for this process, otherwise it would become acidic.

### QUESTIONS

**5** Why might some people say that the lime method of removing sulfur dioxide means more damage to the environment?

**6** What part have scientists and engineers played in reducing the pollution from power stations?

power station desulfurisation

# Reducing carbon dioxide

**We are learning to:**
> explain how we can minimise the carbon dioxide released from fossil fuel use
> describe alternative fuels
> discuss how scientists can help us to use natural resources in a more sustainable way

## Is a zero carbon future possible?

The scientists at the Centre for Alternative Technology think so. They have produced a plan that examines new ways in which we could reduce our reliance on fossil fuels, such as using wood and grass grown on land unsuitable for food crops.

## Cutting carbon dioxide

When we burn fossil fuels, the carbon that was trapped in the fuels is released into the air as carbon dioxide. If we want to reduce the amount of carbon dioxide we put into the air we should burn less fossil fuels. We can do this in a number of ways:

> using alternative ways to produce electricity that do not involve burning fossil fuels

> improving insulation in buildings so that we use less fuel or electricity to keep warm

> walking, cycling or using public transport instead of using private cars.

FIGURE 1: 'Boris Bikes'. These bicycles in London can be hired, so commuters can use energy from food instead of fossil fuels to get about.

###  QUESTIONS

1 How does insulating your house reduce the amount of carbon dioxide released into the air?

2 What advantages are there in encouraging people to walk or cycle instead of using their cars?

## Replacing fossil fuels

Instead of burning fossil fuels we can use **biofuels**. Biofuels are made from plants. As plants grow, they remove carbon dioxide from the air through the process of photosynthesis. The carbon becomes part of the plant. When the biofuel is burned the same amount of carbon dioxide is released again. Biofuels include wood chips, palm oil and alcohol made from sugar.

A very large area of land would be needed to grow enough biofuel to replace the fossil fuels that we use today.

FIGURE 2: Plants grown for biofuel production take up large areas of land.

alternative fuels

C1 Air quality

## QUESTIONS

**3** Explain why the burning of biofuels does not affect the amount of carbon dioxide in the atmosphere.

### Watch out!
The energy needed to cultivate and transport biofuels must be considered before deciding if they are a good substitute for fossil fuels.

## Dash to gas

One part of the solution to reducing carbon dioxide emissions in developed countries like the UK has been to reduce the amount of coal burned in power stations. Other fossil fuels have a lower carbon content and so produce less carbon dioxide for the same amount of energy released. The fuel with the least carbon is natural gas (methane). Many gas-fired power stations have been built in recent years.

Scientists realise that stocks of natural gas will not last long and it is not renewable. Also, although it is cleaner than coal, it still produces carbon dioxide when burned. For these reasons it is not a **sustainable** source of energy.

Hydrogen gas is an alternative fuel that produces no carbon dioxide when it is burned, but there are technical difficulties with its production and supply.

**FIGURE 3**: It's not known how long this pipeline will be carrying natural gas.

Percentage of carbon by mass in some fuels.

| Fuel | % carbon |
|---|---|
| coal | approx. 100 |
| diesel | 86 |
| petrol | 84 |
| methane | 75 |
| alcohol | 52 |
| hydrogen | 0 |

## QUESTIONS

**4** Which of the fuels listed in the table are not fossil fuels?

**5** Which of the fuels in the table is a biofuel? Why is the percentage of carbon in these fuels not important in deciding whether it should be used?

**6** Suggest what contribution scientists are making to the decisions about future fuels.

### Did you know?
Natural gas has to be brought by ship or pipeline from countries such as Russia. There are concerns that the supply of gas could be affected by political disagreements, or the pipelines could be attacked by terrorists.

🔍 natural gas low carbon fuel

# Improving transport

**We are learning to:**
- explain how pollution produced by vehicles has been reduced
- explain the benefits and problems of alternatively powered vehicles
- explain what is meant by a correlation between two factors

## Is air pollution from cars getting worse?

The number of cars, vans and lorries on the roads keeps on increasing, particularly in the developing countries of Asia and South America. The use of fossil fuels is still increasing but in many countries air quality has actually improved in the last 30 years. This is down to advances in technology.

**FIGURE 1**: Heavy traffic in Bangkok.

## Reducing air pollution

Air pollution from vehicles can be reduced by:
1. encouraging people to drive cars which use less fuel
2. using cleaner fuels
3. removing pollutants from the exhausts of vehicles
4. encouraging people to use public transport such as buses and trains by increasing parking charges and charging for using some roads.

### QUESTIONS

1. What might persuade people to take fewer journeys in cars?

## Removing the pollutants

Modern vehicles have been designed with more efficient engines, which use less fuel to cover the same distance. The fuel is burned more completely so less carbon monoxide and particulates are formed.

In the UK, the Government encourages people to use cars that burn less fuel by increasing the tax on fuel, and by increasing taxes on vehicles that use more fuel. Vehicles are only allowed on the roads if their exhaust emissions are within the limit allowed by law. This legal limit is enforced by the MOT vehicle test.

Since 1980 new cars have been fitted with a **catalytic converter**. This contains a platinum catalyst that makes the pollutant gases react with each other.

carbon monoxide + nitrogen monoxide → nitrogen + carbon dioxide
$\quad\quad CO \quad\quad\quad\quad\quad NO \quad\quad\quad\quad\quad N_2 \quad\quad\quad CO_2$

In this reaction carbon monoxide gains an oxygen atom and is **oxidised**. The nitrogen monoxide loses an oxygen atom and is **reduced**.

Sulfur damages the platinum catalyst, so low sulfur fuels are used. This also reduces the amount of sulfur dioxide produced.

**Watch out!** Catalytic converters do not reduce the amount of carbon dioxide produced by motor vehicles. Only using less fossil fuels can reduce carbon dioxide emissions.

catalytic converter

C1 Air quality

Since 1980 the amount of carbon monoxide in vehicle exhausts has fallen in developed countries. It could be that the use of catalytic converters has *caused* the fall in carbon monoxide emissions but there may be other factors such as improved efficiency of engines. We can only say that there is a **correlation** between the use of catalytic converters and reduced carbon monoxide emissions.

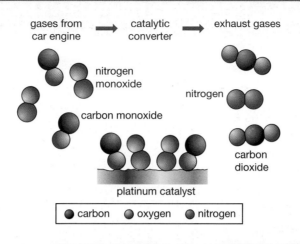

FIGURE 2: In the presence of the catalyst, oxygen atoms move from the nitrogen monoxide to form carbon dioxide.

### Did you know?

Platinum and other metals from catalytic converters have been found in ice on the Greenland icecap and all over the Earth's surface. Scientists are not sure yet whether this platinum pollution is harmful.

### QUESTIONS

**2** How does the annual MOT test on a vehicle's emissions help air quality?

**3** What is meant by the efficiency of a vehicle engine?

**4** In 1990 the amount of carbon monoxide given out by vehicles in the USA was 60% of its value in 1980. Explain why this shows a correlation between carbon monoxide emissions and catalytic converter use.

## Transport without fossil fuels (Higher tier only)

There are a growing number of electric cars which do not give out any pollutant gases at all. Unfortunately electric cars cannot travel far and it takes a long time to charge up the batteries. Electric vehicles are only zero carbon if the electricity used to charge them is produced using renewable energy sources or nuclear power.

Biofuels are an alternative to fossil fuels for powering vehicles. These include plant oils which replace diesel fuel, and alcohol which is fermented from sugar and replaces petrol. Burning biofuels does not contribute to global warming because the carbon released was taken in when the plants were grown. Vehicle engines do not have to be modified and vehicles can fill up with fuel at service stations in the normal way. Unfortunately, to replace all fossil fuels used in transport with biofuels, far more farmland would be needed to grow the fuel plants than we have available. Already vast areas of natural forest have been turned into plantations of crops such as palm oil, destroying habitats.

### QUESTIONS

**5** What is the benefit of using electric or biofuel-powered vehicles?

**6** What problems do
  **a** electric and
  **b** biofuel-powered vehicles have?

FIGURE 3: Electric cars can just plug in to recharge their batteries.

electric cars benefits disadvantages    biofuel GCSE

# Preparing for assessment: Evaluating and analysing evidence

*To achieve a good grade in science, you not only have to know and understand scientific ideas, but you need to be able to apply them to other situations and to analyse evidence. These tasks will support you in developing these skills.*

## Particulates in Beckenham

Beckenham is part of Bromley in South London. As in all London boroughs and large cities, the air quality is carefully monitored and recorded.

Careful measurements are made of various pollutants in the air. The data is then displayed on a graph so that levels and changes can be seen.

'PM10' particulates are those that have a diameter of 10 micrometres or less. The level of these is significant, because larger particulates are usually filtered out by nasal hair. This graph shows the way that PM10 particulates varied during a period of about 4 weeks.

The vertical axis is marked in hourly mean concentration. This means that every hour the readings of the concentration of the particular pollutant are averaged and displayed. The horizontal axis is marked in dates; the vertical line above a date is midnight at the start of that date.

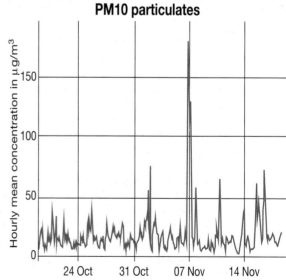

# C1 Evaluating and analysing evidence

## ✹ Task 1

> Describe any pattern you can see in the variation of particulate level during the course of one day. Suggest what might cause this.

> Is it the same pattern day by day? Explain your answer.

## ✹ Task 2

> You might think that there would be a change in the readings for 5th November. Why? Is there such a variation?

> In fact, there is quite a change in the readings but at another point on the graph. Can you suggest why that should be?

**Bonfire Night in Beckenham**
**06 November 2010**

Croydon Road Recreation Ground, Beckenham
Gates open and bonfire lit at 6pm, fireworks at 8pm
A great evening of fireworks and fun for all the family

## ✹ Task 3

> Explain why PM10 particulates are of particular concern.

> Suggest a good way of explaining to someone how big a PM10 particulate is.

## ✹ Task 4

> Why do you think the results are plotted from the hourly average instead of taking and displaying instantaneous readings?

> If you tried to gather evidence yourself to support the results shown, by placing double-sided sticky tape on poles and positioning them outside, what do you think you would see on examining the tape under a microscope?

> Should this evidence be used to ban Bonfire Night?

## ✹ Maximise your grade

These sentences show what you need to include in your work to achieve each grade. Use them to improve your work and be more successful.

**E**
For grade E, your answers should show that you can:
> understand that particulates may be released into the atmosphere by humans, and that these can affect air quality
> understand how burning fossil fuels pollutes the atmosphere with particulates
> identify trends in data displayed in a graph

**C**
For grades D, C, in addition show that you can:
> understand how human activity has added extra carbon dioxide and small particles of solids (e.g. carbon) to the atmosphere
> recall that particulate carbon is deposited on surfaces
> suggest explanations for trends in data displayed in a graph
> explain the scale of measurement

**A**
For grades B, A, in addition show that you can:
> understand that some of these substances are harmful to the environment and so cause harm to humans indirectly
> effectively convey a sense of scale
> suggest how data can be manipulated to indicate underlying trends without being distorted by instantaneous variations

# C1 Checklist

## To achieve your forecast grade in the exam you'll need to revise

Use this checklist to see what you can do now. Refer back to pages 106–129 if you're not sure.

Look across the rows to see how you could progress – **_bold italic_** means Higher tier only.

Remember you'll need to be able to use these ideas in various ways, such as:
> interpreting pictures, diagrams and graphs
> applying ideas to new situations
> explaining ethical implications
> suggesting some benefits and risks to society
> drawing conclusions from evidence you've been given.

Look at pages 312–318 for more information about exams and how you'll be assessed.

| To aim for a grade E | To aim for a grade C | To aim for a grade A |
|---|---|---|
| recall that air forms part of the atmosphere, and is made up of nitrogen, oxygen and argon, plus small amounts of water vapour, carbon dioxide and other gases; recall that this mixture of gases consists of small molecules with large spaces between them | recall that the relative proportions of the main gases in the atmosphere are about 78% nitrogen, 21% oxygen and 1% argon | |
| recall that the Earth's first atmosphere of mainly carbon dioxide and water vapour was probably formed by volcanic activity, and that when the Earth cooled water vapour condensed to form the oceans | understand that geological processes and living organisms have changed the oceans and the atmosphere over time | explain how photosynthesising organisms added oxygen to and removed carbon dioxide from the atmosphere, and how carbon dioxide was removed by dissolving in the oceans and forming sedimentary rocks and fossil fuels |
| understand that human activity has added small amounts of carbon monoxide, nitrogen oxides and sulfur dioxide to the atmosphere | explain how human activity has added extra carbon dioxide and small particles of solids (for example carbon) to the atmosphere | |
| describe how some of these substances, called pollutants, are directly harmful to humans | understand that some of these substances are harmful to the environment and so may cause harm to humans indirectly | |
| recall that coal is mainly carbon and that petrol, diesel and fuel oil are compounds of hydrogen and carbon, called hydrocarbons; recall that hydrocarbons burn to form carbon dioxide and water | understand that combustion reactions involve oxidation | explain why a substance chemically joining with oxygen is an example of oxidation, why loss of oxygen is an example of reduction, and how this relates to combustion |

| To aim for a grade E | To aim for a grade C | To aim for a grade A |
|---|---|---|
| understand understand that fuels burn more rapidly in pure oxygen than in air | explain how a supply of oxygen can be used to control combustion rates | |
| understand that atoms are rearranged during a chemical reaction, and that the numbers of atoms of each element must be the same in the products as in the reactants; understand that the properties of the reactants and products are different | interpret drawings showing the rearrangement and conservation of atoms during a chemical reaction | understand that mass is conserved in a chemical reaction, and that the conservation of atoms during combustion reactions means that some atoms in the fuel may react to give products that are pollutants |
| understand how burning fossil fuels pollutes the atmosphere with carbon dioxide, sulfur dioxide, carbon monoxide, particulate carbon (from incomplete burning) and nitrogen oxides | relate the formulas for carbon dioxide, carbon monoxide, sulfur dioxide, nitrogen monoxide, nitrogen dioxide and water to drawings of their molecules | recall that nitrogen monoxide is formed during the combustion of fuels in air, and is subsequently oxidised to nitrogen dioxide, and that NO and $NO_2$ are jointly referred to as '$NO_x$' |
| understand that atmospheric pollutants cannot just disappear, they have to go somewhere | recall that particulate carbon is deposited on surfaces; carbon dioxide is used by plants in photosynthesis and dissolves in rain and sea water | recall that sulfur dioxide and nitrogen dioxide react with water and oxygen to produce acid rain |
| understand how pollution caused by power stations can be reduced by using less electricity | understand how power station pollution can be reduced by removing sulfur from natural gas and fuel oil, and by removing sulfur dioxide and particulates from flue gases of coal-burning power stations | understand how sulfur dioxide is removed from flue gases by reaction with lime or by using seawater |
| understand that the only way of producing less carbon dioxide is to burn less fossil fuels understand how vehicle pollution can be decreased by burning less fuel, using cleaner fuels, and by encouraging the use of public transport | understand how vehicle pollution can be decreased by using more efficient engines, using low-sulfur fuels, using catalytic converters in which nitrogen monoxide is reduced and carbon monoxide is oxidised, and having legal limits to emissions | understand the benefits and problems of using alternatives to fossil fuels for vehicles, particularly biofuels and electricity |

# Exam-style questions

## Foundation level

**1** This question is about the composition of air.

**a** Match the gases in the air to their percentage in dry air. [3]

| Gas | % in dry air |
|---|---|
| oxygen | 78 |
| carbon dioxide | 21 |
| nitrogen | about 1 |
| argon | very much less than 1 |

**b** Which of the gases in the air was produced by plants photosynthesising? [1]

**c** How has the composition of the air been changed by human activity? Your answer should include:
- how the amount of carbon dioxide has changed
- the names of other gases that have entered the air
- what caused these changes.

The quality of written communication will be assessed in your answer to this question. [6]

[Total: 10]

## Foundation/Higher level

**2** Patio heaters are used to keep warm outdoors on cool evenings. They burn propane gas ($C_3H_8$), a hydrocarbon fuel which is stored in metal bottles.

**a** Name the products of burning propane gas. [2]

**b** Draw a diagram to show how the atoms are arranged in the products formed when propane burns. Diagrams of propane and oxygen molecules are provided to help you. [2]

$C_3H_8$
propane

$O_2$

**c** Burning propane is an example of a chemical reaction. Which *two* words describe the reaction when propane burns?

photosynthesis  oxidation
combustion  reduction [2]

**d** At the end of an evening after the patio heater has been burning the propane, the gas bottle weighs 20 g less than when the heater was lit. Explain what has happened to the missing 20 g. [2]

[Total: 8]

**3** Look at these graphs.

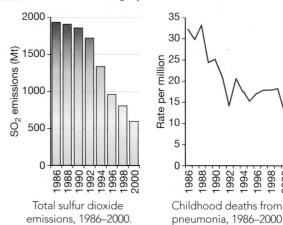

Total sulfur dioxide emissions, 1986–2000.
Childhood deaths from pneumonia, 1986–2000.

**a** The graphs show a correlation between two factors. Describe this correlation. [2]

**b** The graphs do not prove that sulfur dioxide causes deaths from pneumonia. What other information would you need to prove this? [2]

[Total: 4]

**4** Oxides of nitrogen are formed when fuels are burned in vehicle engines. Nitrogen oxides are removed from vehicle exhausts by a catalytic converter.

**a** Which of the following is the source of the nitrogen in the nitrogen oxides?

fuel  air  lubricating oil
water in radiator [1]

**b** The reaction in the catalytic converter is

carbon monoxide + nitrogen monoxide
$\rightarrow$ carbon dioxide + nitrogen

The nitrogen monoxide is said to have been *reduced*. What does this mean? [1]

**c** The catalyst only begins to work when it is hot. What are the consequences of this for air quality on housing estates where many people use their cars to commute to work? [3]

[Total: 5]

## Higher level

**5** Give one reason why each of the following may improve air quality, and one reason why it may not.

**a** Car manufacturers are producing an increasing number of electric cars to meet demand. [2]

**b** Fuel suppliers are adding biofuel to some petrol and diesel. [2]

[Total: 4]

# C1 Air quality

## Worked examples

**1** Sulfur is present in all fossil fuels when they are obtained from the ground. When the fossil fuel burns the sulfur reacts with the air to form sulfur dioxide gas. Sulfur dioxide is toxic and produces acid rain.

AO1 **a** Complete the diagram to show how the atoms of sulfur and oxygen rearrange.

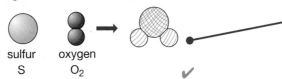
✓
[1]

### How to raise your grade
Take note of the comments from examiner. These will help you to improve your grade.

> Make sure you can draw molecule diagrams for the molecules mentioned in this topic.

AO1 **b** How does sulfur dioxide produce acid rain? [2]

*Sulfur dioxide reacts with water to form sulfuric acid.* ✓

> This answer is worth 1 mark. For 2 marks the student should have mentioned that oxygen is required in the reaction as well as water.

AO2 **c** Developing countries such as China have large stocks of coal. How does this affect the air quality in developing countries? [2]

*Burning large quantities of coal in power stations will release sulfur dioxide gas so the air quality will be poor.* ✓

> 1 mark for noting that burning coal in power stations will release a lot of sulfur dioxide and that this will have an effect on air quality. The second mark would be given for recognising that coal is the problem because the sulfur is difficult to remove from coal.

## Higher

AO1 **2** There are two methods for removing sulfur dioxide from flue
AO3    gases of coal-fired power stations.
1. The flue gases are reacted with wet lime and then oxygen. The lime is obtained from quarried limestone. The product is calcium sulfate, used as plaster. Carbon dioxide is a waste product.
2. The flue gases are dissolved in seawater and air is blown through the solution. Carbon dioxide is given off and the seawater is returned to the sea.

Identify the benefits and disadvantages, and compare the environmental impact of each method. The quality of written communication will be assessed in your answer to this question. [6]

*In the first method, there is a useful product: calcium sulfate which is used as plaster. Quarrying the limestone damages habitats, and transporting it to the power station releases pollutants. The process also releases carbon dioxide which causes climate change. In the second method, no quarrying and transportation of rock is needed. But there is no saleable product, and the method can only be used in power stations on the coast. Carbon dioxide is also given off in this process.*
*Overall method 2 produces less environmental damage than method 1.*

> The answer correctly states the benefits, disadvantages and environmental impact of both methods, and compares the environmental impacts.

> All the information is relevant, clear and presented in a structured way. Correct and appropriate scientific terms are used, such as limestone, calcium sulfate, habitat, pollutant, carbon dioxide and climate change. Grammar and spelling are accurate.

> This answer is worth the full 6 marks. (See the example banded mark scheme on page 317.)

135

# C2 Material choices

## What you should already know...

### Properties of materials vary

Solids, liquids and gases have particular properties that can be explained by the arrangement of the particles that they are made of.

Changes of state, for example from liquid to gas, are reversible and depend on temperature.

The state of some materials can be difficult to classify.

Some mixtures can be separated by distillation.

 What happens to the particles in a liquid when its temperature reaches its boiling point?

### Chemical reactions produce new substances

Many chemical reactions cannot be reversed.

Elements can be joined together to form many different compounds.

Models of molecules can be used to show similarities and differences between substances, and the changes that take place in chemical reactions.

 How are the atoms arranged in molecules of water, $H_2O$?

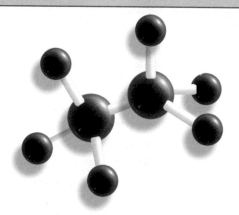

### There are patterns in the behaviour of substances

Word and symbol equations show the patterns of change in chemical reactions.

Groups of substances made up of similar elements react in a similar way.

 Name two substances that produce the same products when they are burned.

# In C2 you will find out about...

- the properties that materials have that make them useful
- the properties of rubbers, plastics and fibres

- the sources of natural and synthetic materials including metals, ceramics and polymers.
- the substances that are in crude oil
- separating useful substances from crude oil
- why modern polymer materials have replaced older materials

- making polymers from small molecules
- what gives polymers their properties
- how polymers can be changed to improve their properties

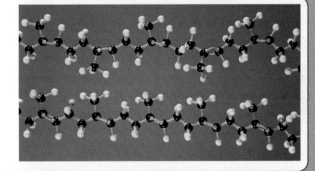

- how nanotechnology uses very small particles
- the sources and properties of nanoparticles
- uses of nanoparticles
- using nanoparticles safely

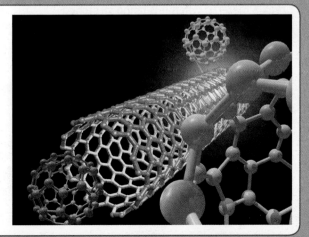

# Using materials

**We are learning to:**
> explain why materials are used for particular purposes
> use data to compare the properties of materials
> explain how the true value of a quantity is estimated

## Why are trainer soles made from rubber?

What does the sole of a shoe have to do? It has to protect your foot from rough paths and hard stones. So the sole has to be made of a material that is hardwearing but is a bit springy. Rubber is one material that does the job well.

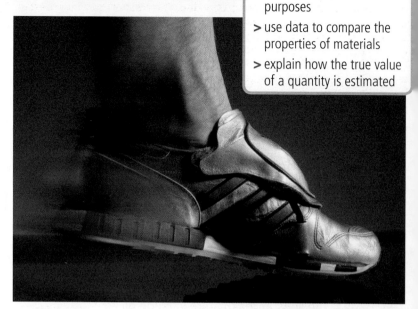

**FIGURE 1**: What properties does the material need to do its job?

### Comparing materials

Rubber is used for car tyres because it is springy, or **elastic**. **Plastics** keep their shape when moulded, which makes them useful for making washing up bowls. **Fibres** are used to weave cloth for clothes because they are strong when they are stretched. Each material has properties that make it suitable for the job it is doing.

### QUESTIONS

1. What property does rubber have that makes it useful as a rubber band?

2. Why do mountaineers use ropes made of nylon?

### Measuring properties

Properties describe how a material behaves.

> **Melting point** is the temperature at which a solid turns into a liquid.

> **Tensile strength** is the force needed to break a material when it is being stretched (in tension).

> **Compressive strength** is the force needed to crush a material when it is being squeezed.

> **Stiffness** is the force needed to bend a material.

> **Hardness** is a comparison of two materials where one can or cannot scratch the other.

> **Density** is the mass of a given volume of the material.

**FIGURE 2**: Why is plastic used for the casing of electronic gadgets?

tensile strength materials

Many measurements of a property need to be taken because readings vary from the **true value**. The **mean** of a set of measurements should give a result that is close to the true value. The smallest to the largest of the measurements is the **range**. Any measurements that appear unusual or very different from the rest of the results are called **outliers** and should be checked to find out why they are different.

### Did you know?

Spider silk is five times stronger than steel and much more elastic. It is also about one-tenth the diameter of silk. It would be very useful if we could get enough of it.

**FIGURE 3**: Why are sails made from woven fibres?

### QUESTIONS

**3** Material A scratches material B but not material C.

  **a** Will material B scratch material C?

  **b** Put the three materials in order of hardness with the softest first.

**4** Sophie measured the melting point of a plastic material five times. The readings in °C were 138, 141, 137, 143 and 141. Explain why the true value is close to 140 °C.

## Properties and units

Some properties depend on the size and shape of the material being tested. For instance the strength and stiffness of a material depends on its cross-sectional area and is measured in pascals (Pa).

Density is the mass contained in a specific volume of the material and is measured in $g/cm^3$ or $kg/m^3$. An object made of a material with a high density will feel heavier than an object of the same size made of a lower density material.

### Errors and variation (Higher tier only)

Errors in measurements produce **variation** in data. Unless there is a repeated error in the readings such as using a ruler with the end broken off, the true value will fall in the range of the measurements. **Outliers** can be discarded if it can be shown that there was a particular error in the measurement.

**Watch out!**
Note the units used for each property carefully. You will find different units on US websites.

### QUESTIONS

**5** A broken window needs to be replaced by glass which has density 2.5 $g/cm^3$, or Perspex which has density 1.2 $g/cm^3$. The window made of which material would have the smaller mass?

**6** Polypropene has a tensile strength of 35 MPa while that of polyethene is 15 MPa. Why is polypropene preferred for making buckets?

C2 Material choices

rubber properties uses

139

# Choosing materials

> **We are learning to:**
> - link the usefulness of a product with the properties of its materials
> - choose materials for a job based on information about their properties
> - use data to support our decisions

## What makes a bullet-proof vest so strong?

A bullet-proof vest has to stop a bullet which is moving very fast. The fibres in the vest have to be very strong but the vest must not be too heavy to wear for a long time. At the moment Kevlar fibres have the best set of properties to do this job well.

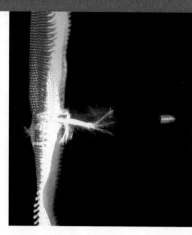

**FIGURE 1**: Kevlar will stop a bullet moving at 220 metres per second.

## Perfect plastics

A manufacturer is producing a new smart phone. The choice of material for the casing is very important.

> It must be possible to make it into the shape suggested by the designers.
> It has to be strong enough to protect the electronics inside.
> It must wear well so that it keeps its good looks for a long time.
> It must not make the phone too heavy and it must be relatively cheap.

Plastics can have all of these properties. There are many different types of plastic. The manufacturer chooses the one with the properties that best match the requirements.

### QUESTIONS

**1** Why would a phone be more hard-wearing if it was made of a hard plastic rather than a soft one?

**2** What is the effect of having to use more of a material to make the product last longer?

## Comparing plastics

Most things that we buy are made from a number of materials, each with properties that are needed for the job they do. Engineers have a choice of materials for each job and they must compare data on the properties of the material to make their decisions.

The table on page 141 compares the properties of a number of plastic materials. The **hardness** shows how well the material stands up to wear, with the hardest of the plastics given the value 5. The strength of the plastics is also given a value from 1 to 5, with 5 being the strongest. The useable temperature shows how hot the material can be before it starts to soften. This data comes from tests carried out many times on the materials.

*Watch out!*
Materials are often mixed together to combine their best properties.

**FIGURE 2**: When a plastic is heated it softens and can be moulded or blown into a sheet.

uses of plastics

# C2 Material choices

| Plastic | Hardness (1–5) | Tensile strength (1–5) | Density (g/cm$^3$) | Maximum useable temperature (°C) |
|---|---|---|---|---|
| HDPE (high-density polyethene) | 1 | 2 | 0.97 | 120 |
| LDPE (low-density polyethene) | 1 | 1 | 0.94 | 100 |
| nylon | 3 | 5 | 1.15 | 140 |
| PET (polyethentetraphthalate) | 5 | 4 | 1.4 | 140 |
| PP (polypropene) | 2 | 2 | 0.91 | 130 |
| polystyrene | 4 | 3 | 1.05 | 80 |
| PMMA/acrylic (polymethylmethacrylate) | 5 | 4 | 1.2 | 90 |

Properties of different plastics.

## QUESTIONS

**3** Look at the table. Explain your answers to the following questions.
  **a** Which plastics float on water? (The density of water is 1 g/cm$^3$.)
  **b** Which plastics would soften if put in boiling water? (The boiling point of water is 100 °C.)
  **c** Which type of polyethene is more suitable for making shopping bags?
  **d** Why is PET the plastic chosen for drinks bottles?
  **e** Why is polystyrene only suitable for making cups for cold drinks?
  **f** PMMA is a clear plastic. What other properties make it suitable for shatter-proof windows?

## Rubbers and fibres

Rubbers have greater elasticity than other materials – they bounce back when a force is removed. Different rubbers also have different amounts of compressive strength and hardness. For example, the tyres used for racing cars on wet track are hard, but those used when the track is dry are soft.

Some materials can be drawn into long, thin filaments which have a greater tensile strength than the original material. Many filaments can be spun together to make fibres and the fibres can then be woven into cloth. Ropes are made by winding fibres together. The strength of a rope depends on the material and the number of fibres wound together.

### Did you know?

Bungee jumpers rely on the combination of the strength and elasticity of strands of rubber braided together, which are not affected by changes in temperature or humidity.

## QUESTIONS

**4** Use the data in the table above to explain why nylon fibres are used to make ropes.

**5** Polypropene fibres are woven into the cloth used to make cheap 'fleece' jackets. Discuss why polypropene is suitable for this use.

**6** Explain whether you should rely on the data in the table above.

**FIGURE 3**: Tyres are tested to check that their properties are as expected.

types of rope

# Preparing for assessment: Applying your knowledge

To achieve a good grade in science, you not only have to know and understand scientific ideas, but you need to be able to apply them to other situations. These tasks will support you in developing these skills.

### ✳ Vorsprung durch Technik – progress through technology

In the 1990s Audi was the first big car manufacturer to start mass-producing cars with aluminium bodies. Aluminium was used not only in the panels but also in the frame, called the ASF, or Audi Space Frame. The first model to use the new technology was the luxury A8 saloon. Today Audi continue to research new materials at the Aluminium and Lightweight Design Centre, in Germany.

Two major factors that influence car design are fuel economy and passenger safety. With fuel becoming more expensive and pollution more of a problem, car designers have to find ways of making cars travel further for each litre of fuel used.

For fuel economy, cars need to be light, but still strong. Audi found that making both the frame structure and the body panels from aluminium was a good way of achieving this is.

Cars also have to be as safe as possible. Each new design is thoroughly crash-tested before it is put on sale.

Another advantage of aluminium is that it does not rust and so does not need the protection that steel does.

## C2 Applying your knowledge

### Task 1

A more common material used to make car bodies is steel (mainly made up of iron). Discuss and decide in your group whether steel or aluminium, or both, would satisfy the following design features.

> Car bodies should conduct electricity as this can then form part of the electrical circuits, such as to light bulbs.

> Car bodies should not corrode, as this weakens them and spoils their appearance.

> Car bodies have to be shaped from flat sheets, so the material has to be workable (this is called malleability).

### Task 2

A car design team considers possible materials and tests them. They come to the conclusion that the aluminium-bodied car would be lighter, corrosion free but more expensive. Discuss and explain why each of these factors is significant.

### Task 3

Another motor company decided to follow the example of the Audi A8 and build its car bodies from aluminium. It did this, and after several years noticed that its cars:

> lasted longer than other cars

> used less fuel than similar steel cars.

Discuss and explain these two observations. Then try to decide whether the motor company is likely to be pleased with each of its findings.

### Task 4

The stiffness of aluminium is around one-third of that of mild steel; so too is its density.

> Decide how important a feature stiffness is in car construction. Describe how you might compensate for a material not being as stiff.

> The difference in density means that a steel block would weigh three times as much as an identically sized block of aluminium. Explain how significant this would be.

> Discuss how these various pieces of information might influence the choice of a construction material for car bodies. Explain the advantages and disadvantages of both metals and justify an overall conclusion.

### Maximise your grade

These sentences show what you need to include in your work to achieve each grade. Use them to improve your work and be more successful.

**E**
For grade E, your answers should show that you can:
> describe similarities and differences between materials
> explain that materials are used because they have suitable properties
> describe advantages and disadvantages of building car bodies from a particular material
> describe how the performance of a product is related to the materials used to make it

**C**
For grades D, C, in addition show that you can:
> interpret information about how materials can differ in properties
> interpret information about the properties of materials to assess the suitability of these materials

**A**
For grades B, A, in addition show that you can:
> combine information of different types to produce a considered and justified overall view

# Natural and synthetic materials

**We are learning to:**
- explain that many of the materials we use come from living organisms
- explain that other materials are found in the Earth
- explain that new materials can be made from chemicals

## What materials would a castaway find on a tropical island?

The castaway would be surrounded by useful materials. Trees provide wood that can be turned into paper, and other plants, such as cotton, make fibres. The sand could be used to make glass and coral turned into concrete. A lucky castaway might find rocks containing ores from which metals could be extracted.

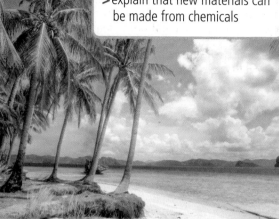

FIGURE 1

## Chemicals everywhere

Everything around us, every living thing, every rock, every pool of water, is made up of chemicals. Many of those chemicals are useful to us.

> **Metals** are shiny, **malleable** and conduct electricity.

> **Ceramics** include clay, glass and cement, and are hard and strong.

> **Polymers** are large molecules that can be made into rubbers, plastics and fibres.

Many of the materials we use are **mixtures** of chemicals. Bronze is a mixture of copper and tin, and concrete is a mixture of cement and sand.

### QUESTIONS

1. Name something you use made from each of these materials:

   a a metal   b a ceramic   c a polymer.

2. Give examples of the types of materials used to make a car and what the materials are used for in the car.

## Sources of materials

Cotton, paper, silk and wool all come from living organisms – cotton and paper are made from plants and silk and wool are obtained from animals. They are **natural materials** which need little processing to be useful. These materials are all polymers.

Other natural materials come from the rocks in the Earth's crust, for example limestone is turned into cement and haematite is a source of iron.

**FIGURE 2**: Wool is a natural fibre. It is made of a polymer called keratin which also makes up your hair and nails.

natural / synthetic materials

C2 Material choices

Many of the materials we use today are **synthetic materials**. They have been manufactured by chemical reactions which join simple chemicals together to make new materials. The raw materials that provide the simple chemicals often come from the Earth. The most important raw material is crude oil (or petroleum) found in rocks. Polyethene is an example of a synthetic material made from simple chemicals found in crude oil.

### QUESTIONS

**3** Name two materials which are polymers that are obtained from plants.

**4** Explain why polyethene is a synthetic material while wool is a natural material.

### Watch out!

All materials are made up of atoms that make up our Universe. Synthetic materials are made by manufacturing processes that do not occur naturally.

## Synthetic versus natural

Synthetic materials have taken the place of some natural materials in our modern world. This is because synthetic materials can be designed to provide the properties needed for a particular purpose. There may be a shortage of natural materials and the raw materials for making synthetic materials are often cheaper and available in greater amounts. Synthetic materials can be made in whatever quantity is needed. For many uses, neoprene and silicone rubbers have taken the place of natural rubber, which is obtained as latex sap from rubber trees. Nylon was invented as a synthetic substitute for silk produced by silk larvae. Bakelite was the first synthetic plastic and replaced wood for many uses.

**FIGURE 3**: Cotton fibres are formed in the seed head of the cotton plant. Cotton is made of the polymer, cellulose, which forms plant cell walls.

### Did you know?

We used to use many more natural fibres from plants, such as flax for making linen and hemp for making sacks and ropes, but these have been replaced by cotton and synthetic materials.

### QUESTIONS

**5** Explain why synthetic materials have replaced natural materials for many purposes.

**6** Why do some synthetic materials perform better than natural materials for some uses?

**FIGURE 4**: A nylon fibre is synthesised by reacting chemicals in two solutions together. The solutions do not mix and the nylon is formed at the interface between them.

natural / synthetic materials

# Crude oil

## Is crude oil too valuable to burn?

Chemicals in crude oil are used to synthesise important substances such as polymers, drugs and dyes. These substances would be more difficult and expensive to manufacture from other sources. Perhaps we should be saving crude oil for making these materials instead of burning most of it.

> **We are learning to:**
> - explain that crude oil is a mixture of different hydrocarbon molecules
> - explain how in reactions of hydrocarbons, atoms are rearranged but never destroyed
> - remember that the biggest use of crude oil is as fuels

### Did you know?
The world uses about 85 million barrels of crude oil every day. That is about 14 billion litres per day or about 2 litres for every person living on the Earth.

**FIGURE 1**: Crude oil is usually a thick black liquid, but it varies in colour and thickness depending on where it comes from.

## What is crude oil?

**Crude oil**, or 'petroleum', is a mixture of thousands of different compounds. Most of these compounds are **hydrocarbons**. Hydrocarbons are compounds of just carbon and hydrogen.

Most of the hydrocarbons in crude oil are used as fuels. When the fuel burns the carbon and hydrogen atoms join up with oxygen atoms and are rearranged into carbon dioxide and water molecules. The number of each type of atom in the products is the same as in the reactants.

Nearly 90% of crude oil is used as fuel. Less than 3% is used in the chemical synthesis (manufacture) of other chemicals.

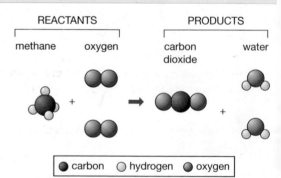

**FIGURE 2**: Methane burns in air to produce carbon dioxide and water.

### QUESTIONS

**1** Write a few sentences saying what you understand by the terms crude oil, hydrocarbon, fuel, synthesis.

**2** Figure 2 shows what happens when a hydrocarbon called methane burns. Describe what happens to the carbon and hydrogen atoms in methane.

# C2 Material choices

## Carbon chains

There are so many different hydrocarbon molecules in crude oil because carbon atoms can join together in chains with hydrogen atoms attached. Crude oil is made up of hydrocarbons with up to 100 carbon atoms. As it is a mixture, the actual composition of crude oil varies from place to place. Some sources of crude oil have more of the smaller molecules, some have more of the larger molecules.

A group of hydrocarbons known as the **alkanes** is a series of compounds with similar properties including methane, $CH_4$, ethane, $C_2H_6$, and propane, $C_3H_8$. All members of the alkane series have the formula $C_nH_{2n+2}$, where $n$ is any whole number. Alkanes are found in fuels. Octane, $C_8H_{18}$, is part of petrol.

*Watch out!* Hydrocarbon molecules contain only carbon and hydrogen atoms.

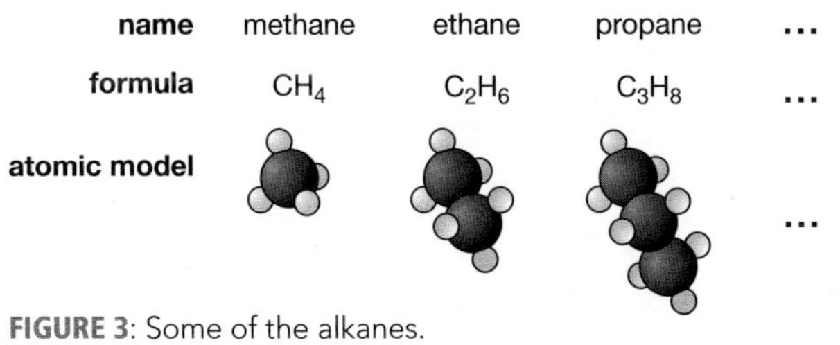

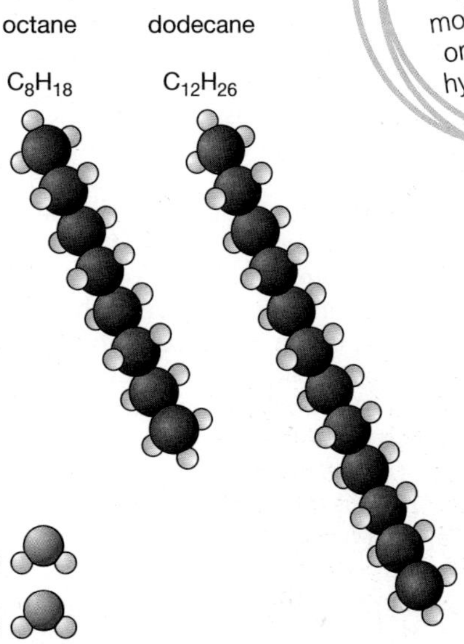

| name | methane | ethane | propane | ... | octane | dodecane |
|---|---|---|---|---|---|---|
| formula | $CH_4$ | $C_2H_6$ | $C_3H_8$ | ... | $C_8H_{18}$ | $C_{12}H_{26}$ |
| atomic model | | | | ... | | |

**FIGURE 3**: Some of the alkanes.

When hydrocarbons burn, the hydrogen atoms always join up with oxygen to form water. The carbon atoms join up with oxygen to form carbon dioxide or carbon monoxide, or sometimes the carbon atoms are left on their own as particulate soot.

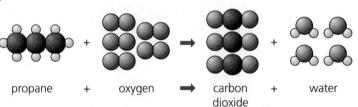

propane + oxygen → carbon dioxide + water

**FIGURE 4**: Propane, $C_3H_8$, is a hydrocarbon used in bottled gas.

## QUESTIONS

**3** Count the atoms of carbon, hydrogen and oxygen on each side of the reaction in Figure 4. What do you notice about the numbers? Explain your answer.

**4** Suggest the formula of the fourth member of the alkane series with four carbon atoms.

## Useful hydrocarbons

A small percentage of the substances in crude oil, mainly the smaller hydrocarbon molecules, is used to make a huge range of other chemicals in the petrochemical industry. For example, a hydrocarbon molecule called ethene, $C_2H_4$, can be combined with water to form ethanol. This is the first stage in making many more complex molecules.

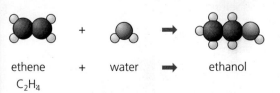

ethene + water → ethanol
$C_2H_4$

**FIGURE 5**: Formation of ethanol.

## QUESTIONS

**5** The reaction of ethene and water is called an 'addition' reaction. How does the rearrangement of the atoms (Figure 5) justify the name?

**6** Why is crude oil an important raw material in the petrochemical industry?

hydrocarbons GCSE    crude oil

# Separating hydrocarbons

**We are learning to:**
- explain how crude oil is separated into its different compounds
- explain that the different hydrocarbons can be separated because their boiling points differ
- explain that in investigations all but two factors must be kept constant

## How useful is crude oil?

Crude oil was first used as a substitute for whale oil in lamps, but its flame is very smoky and some crude oil is very difficult to light. It was only when the substances in crude oil were separated out that it became the useful resource that we rely on for so much today.

## Separating out the substances in crude oil

Crude oil is separated by a process called **fractional distillation**. Crude oil is heated to about 400 °C when all the hydrocarbons in it are turned into gases. The gas is passed into a tall tower, called a fractional distillation column, which is cooler towards the top. The gases rise up the column, cool, and **condense** to liquid at their boiling points. Some have high boiling points so condense low down in the column. Others with lower boiling points rise up the column before they condense. Some stay as gases even at the top of the column. At seven or eight points up the column the substances are collected and piped off. Each substance is a mixture of hydrocarbons with similar boiling points called a **fraction**.

### QUESTIONS

1. What is meant by a 'fraction' of crude oil?
2. Why do some gases condense higher up the fractional distillation column than others?

**FIGURE 1**: A tower at the Esso oil refinery at Fawley, near Southampton, where crude oil is distilled.

## Boiling points and molecules

The hydrocarbons in each fraction have boiling points within a particular range of temperatures. The boiling points are similar because the hydrocarbon molecules are a similar size – they have a similar number of carbon and hydrogen atoms. For example the top fraction, the petroleum gases, all have boiling points below room temperature and have from one to four carbon atoms. The naphtha fraction includes hydrocarbon molecules with five to ten carbon atoms, and these boil between about 40 °C and 170 °C. This fraction provides petrol for fuel but also the chemicals used for synthesis in the petrochemical industry.

In an oil refinery there are a lot of distillation columns and the different fractions are separated a number of times before the final products are ready for use.

### Did you know?

There is not enough of the naphtha fraction of crude oil to supply the demand for petrol. Molecules in other fractions are broken or cracked to produce more of the hydrocarbons used in petrol.

Exxon Fawley virtual tour    fractional distillation petroleum

# C2 Material choices

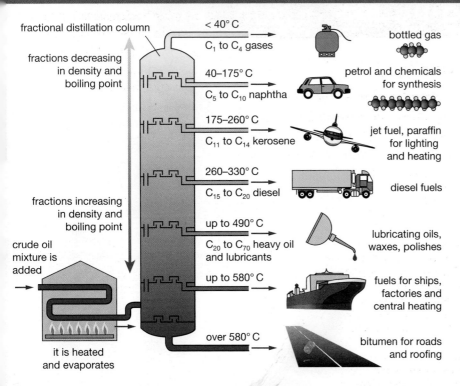

FIGURE 2: Crude oil at 400 °C enters the distillation column and is separated into fractions.

**QUESTIONS**

**3** Why is diesel fuel collected lower in the column than naphtha?

**4** What is the boiling point of the fraction used as lubricants?

**Watch out!**
A fraction is a mixture of hydrocarbons that boil and condense over a *range* of temperatures.

## Investigating boiling points

To investigate how boiling points of hydrocarbons vary with the size of the molecule, all other possible factors that could affect the measurement must be kept constant. Figure 3 shows boiling point data for members of the alkane series of hydrocarbons, where the carbons form a chain. It can be seen from the shape of the graph that there is a correlation between the boiling point and the number of carbon atoms in a molecule.

There are attractive forces between molecules which hold them together. Force is needed to separate the molecules. When heated, the molecules gain energy until at the boiling point they have enough energy to overcome the forces holding them together and the molecules become a gas. The graph shows that as the size of the molecule increases, the force between the molecules increases.

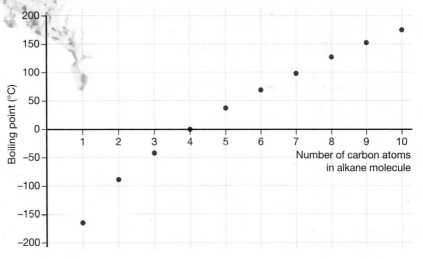

FIGURE 3: A graph of the boiling points of some alkanes against the number of carbon atoms in the molecules. The greater the number of carbon atoms, the larger the molecule.

**QUESTIONS**

**5** What is the boiling point of the alkane heptane, $C_7H_{16}$, found in petrol?

**6** Can we say that the boiling point of a hydrocarbon is *caused by* the number of carbon atoms in the molecule? Justify your answer.

Exxon Fawley virtual tour    fractional distillation petroleum

# Making polymers

## Is polyethene a dangerous pollutant or a valuable material?

Millions of polyethene carrier bags have been thrown away to litter the countryside and pollute the oceans, which is an environmental issue. Polyethene also has more important uses such as the sheeting used in 'polytunnels' for growing fruit and vegetables and for insulating electrical wires.

**FIGURE 1**: Polytunnels – a good use of polyethene.

**We are learning to:**
> explain how large 'polymer' molecules are formed from small 'monomer' molecules
> explain why there is a large variety of different polymers

## Very large molecules

A **polymer** is a very large molecule. It is made by joining together lots of similar small molecules, called **monomers**, like putting together the links of a chain. The process of making a polymer is called **polymerisation**. A polymer has a chain of thousands of carbon atoms, each joined to the next. Polyethene is the simplest polymer and is made up of carbon and hydrogen atoms.

### QUESTIONS

1. Why is a polymer like a chain?
2. What is polymerisation?

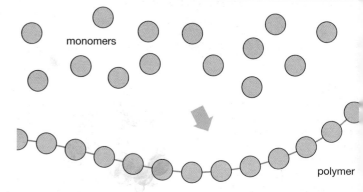

**FIGURE 2**: Monomer molecules join up to form longer and longer polymer molecules.

## Joining molecules

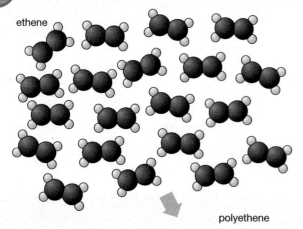

The monomer used to make polyethene is the hydrocarbon called ethene. Ethene has the formula $C_2H_4$. When ethene polymerises, the carbon atoms of one molecule join up to the carbon atoms of the next. Between 20 000 and 200 000 ethene molecules can join together to form each polymer molecule.

The reaction can be summarised by:

many ethene molecules → a polyethene molecule made up of many similar repeated units

$nC_2H_4$ → $[C_2H_4]_n$

Here $n$ stands for a large number in the thousands.

**FIGURE 3**: Polymerisation of ethene.

polymer polymerisation    polyethene development history

C2 Material choices

Different monomer molecules produce different polymers. Nylon is also a polymer but is made from joining two different monomers together in a chain.

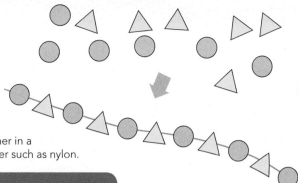

FIGURE 4: Two different monomer molecules join together in a regular pattern to form a polymer such as nylon.

**Watch out!**
Remember that every polymer molecule is made up of thousands of the same repeated units of the monomer.

## QUESTIONS

**3** Why do we write $[C_2H_4]_n$ to represent a molecule of polyethene?

**4** What is different about the way that nylon is formed compared to polyethene?

**Did you know?**
During the Second World War the process for making polyethene was top secret as polyethene was used to insulate the electronics in radar units used on British aircraft.

## Changing monomers

Today we have a huge variety of polymers made from different starting monomers. Propene is a hydrocarbon similar to ethene and forms the polymer polypropene. It is similar to polyethene but with short branches made up of a carbon and three hydrogen atoms at regular intervals along the chain.

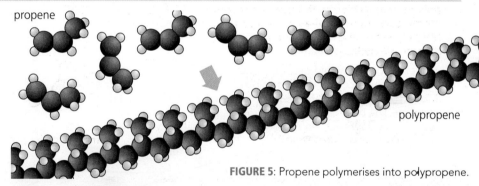

FIGURE 5: Propene polymerises into polypropene.

Other polymers are made by replacing one or more of the hydrogen atoms in the monomer with other atoms or groups of atoms. For example polytetrafluoroethene, commonly called Teflon, is made from a monomer where all the hydrogen atoms in ethene are replaced by fluorine atoms:

$$nC_2F_4 \rightarrow [C_2F_4]_n$$

Today there are hundreds of different polymers that are manufactured starting from different monomers. Each polymer has its own set of properties and uses.

| Monomer | | Polymer | |
|---|---|---|---|
| Name | Formula | Name | Formula |
| ethene | $C_2H_4$ | polyethene | $[C_2H_4]_n$ |
| propene | $C_3H_6$ | polypropene | $[C_2H_3(CH_3)]_n$ |
| tetrafluoroethene | $C_2F_4$ | polytetrafluoroethene (Teflon) | $[C_2F_4]_n$ |
| chloroethene | $C_2H_3Cl$ | polychloroethene (PVC or vinyl) | $[C_2H_3Cl]_n$ |
| phenylethene | $C_2H_3(C_6H_5)$ | polyphenylethene (polystyrene) | $[C_2H_3(C_6H_5)]_n$ |

Monomers and the polymers that they form

## QUESTIONS

**5** Draw a diagram to show a section of the molecule polychloroethene.

**6** If an ethene molecule were 1 cm long, what would be the length, in metres, of a polyethene molecule with 25 000 repeat units?

polymer polymerisation    polyethene development history

# Better materials

> **We are learning to:**
> > describe how modern materials have replaced old ones

## What more can polymers do?

We use polymers in almost every part of our lives, but scientists are still looking at new ways of using polymers. Soon you may be able to buy a polymer TV screen that you can roll up, or flexible polymer batteries and electronic components that can turn your clothes into a wearable computer.

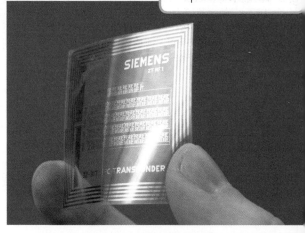

**FIGURE 1**: Polymer 'chips' like this could be combined with the material used to make clothes to tell the washing machine the correct wash temperature.

## Improved materials

Fifty years ago water buckets were made of iron. They were difficult and expensive to make, heavy to carry and soon rusty holes caused leaks. Today buckets are made out of polypropene, a synthetic polymer. Polypropene buckets are tough, strong, light and last for many years. Also polypropene can be easily moulded into complex shapes when it is being made.

No tennis player would get far today with a wooden tennis racquet, but it is only since the 1970s that other materials have been used. Modern racquets have frames made from carbon fibre mixed with a polymer. The new racquets are stronger, lighter for their size than wooden racquets and do not go out of shape if they get wet.

### QUESTIONS

1 Give two reasons why polypropene has replaced iron for buckets.

2 Why can modern tennis racquets be larger than the old wooden ones?

**FIGURE 2**: A tennis racquet has to withstand a large force when hitting the ball.

## Synthetic polymers

Many of the materials that have been in use for thousands of years such as wool, silk, cotton, leather, wood and rubber are natural polymers. These materials are still used for a lot of purposes. However, modern synthetic polymers have replaced them, and metals and ceramics, for many uses.

An important reason why synthetic polymers have been chosen for a purpose is because they are cheaper and available in larger quantities than the natural material that they replace. This is why nylon replaced silk and wool for stockings in the 1940s. Synthetic polymers have many other properties that make them useful and also their properties can be adjusted to make them suit the job they have to do.

A synthetic polymer called PET (polyethylenetetraphthalate) has replaced glass for bottled water and soft drinks. Like glass, PET is clear. PET is a strong polymer which does not shatter like glass. It has a much lower density than glass so lorries can carry many more full PET bottles than glass bottles.

### Did you know?

Animal furs have been replaced by fake furs made from a polymer called acrylic. Acrylic furs look and feel like the real thing but do not insulate as well. This is a case of a polymer not being quite as good in every way as the material it replaced.

### QUESTIONS

**3** A PET bottle is much thinner than a glass bottle of similar size. What does this tell you about PET?

**4** Think about other objects made of polymers that in the past may have been made from wood, metals or glass. What advantages does the polymer have?

## Choosing the right material

When sailing boats carried trade goods across the oceans their sails were made of a cotton cloth. Now racing yachts use sails made of a variety of synthetic polymer fibres including nylon and Kevlar. The table shows a comparison of the properties of cotton, nylon and Kevlar and explains why Kevlar is the material chosen by many sailors. Kevlar does have some disadvantages, however, as the ultraviolet (UV) light in sunlight causes it to break down more rapidly than some materials. Scientists are continuing to search for better materials for many different applications.

| Sailcloth material | Tensile strength (MPa) | Density (g/cm$^3$) |
|---|---|---|
| cotton | 225 | 1.54 |
| nylon | 616 | 1.14 |
| Kevlar | 3400 | 1.45 |

Properties of sailcloth materials.

**FIGURE 3**: Modern sailing yachts travel much faster because their sails can make better use of the wind.

### QUESTIONS

**5** Why has Kevlar replaced cotton in sailcloth? Explain your answer.

**6** What advantage does nylon have over Kevlar for sails that may have to be put up or pulled down by hand?

### Watch out!

Modern materials may appear to have all the advantages over natural materials, but this is not always true. Synthetic polymers can be difficult to dispose of.

Kevlar history

# Polymer properties

**We are learning to:**
- explain how the arrangement of polymer molecules explains their properties
- explain how the melting point of polymers depends on the forces between the molecules

## Are all polymers the same inside?

Some polymers are soft, waxy and melt at low temperatures, while others are hard and strong and do not melt even at very high temperatures. All polymers have very large molecules, but if we could look at polymers with a powerful microscope we would see that they are not all the same.

**FIGURE 1**: What a polymer might look like if you could see the polymer chains.

### Attractive molecules

If you hold two magnets close together and then let go, they will snap together. If you hold them further apart and do the same they may not move at all. There is a force between magnets that is stronger the closer they are together. The same happens with polymer molecules, although they are not magnets. There are small forces between molecules that attract them to each other. The closer the molecules are to each other, the stronger the force needed to pull them apart.

When we heat substances we give the molecules energy. The stronger the force holding the molecules together, the more energy is needed to separate them and the higher the melting point of the substance.

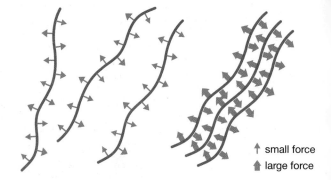

↑ small force
⬆ large force

**FIGURE 2**: Forces are stronger when molecules are closer together.

### QUESTION

**1** Copy out the following sentences and fill in the missing words.
There is a .................. between polymer .................. that pulls them together. The .................. the molecules are to each other the stronger the force. Polymers with a .................. force between their molecules melt at .................. temperatures.

**Watch out!**
Greater forces of attraction between molecules mean that more energy is needed to pull them apart or make the polymer melt.

### Polyethene differences

The first samples of polyethene were made in the 1930s. It was a waxy material that softened in hot water. In the 1950s a new way was found to make polyethene but the new material was harder, stronger and melted at a temperature above the boiling point of water. What was different about the two types of polyethene?

**FIGURE 3**: Low density polyethene has branches between the molecular chains which reduce the attractive forces between the chains.

polymer forces melting point

It was found that the early type, now called LDPE (low density polyethene), had long molecules with branches. The branches kept the molecules apart so that the forces between them were weak. The later version, called HDPE (high density polyethene), did not have branches so the polymer molecules could lie close to each other, making the forces between them stronger. Not only did this explain the differences between the two types of polyethene, but it also showed how the properties of polymers could be adjusted.

## QUESTIONS

**2** Why do branched chain polymers soften at a lower temperature than polymers without branches?

**3** Why do you think LDPE has a lower density than HDPE?

FIGURE 4: LDPE is used for plastic carrier bags.

HDPE is used for water pipes.

## More force (Higher tier only)

HDPE is highly **crystalline**. This means that there are a lot of areas in the material where there is a regular pattern of molecules lined up alongside each other (as in Figure 1). Highly crystalline polymers are strong and have higher melting points, although they can be brittle. Other polymers have been designed so that their molecules can line up even more regularly and close together.

Other ways of increasing the force of attraction between molecules are also used. Polymers with atoms such as nitrogen, fluorine, chlorine and oxygen on the chains also have greater forces between the molecules, which means that higher temperatures are needed to soften them. Nylon and PVC are examples of these kinds of polymer.

## QUESTIONS

**4** Later in the 1950s MDPE was developed which had a few branches along the polyethene molecule. How do you think the properties of MDPE compare with LDPE and HDPE?

**5** Look back at the table on page 141. What evidence is there that PET has strong forces between its molecules?

### Did you know?

Hula hoops (large rings for exercising with, not the snack!) were invented in 1958 to use up a batch of HDPE that was too brittle to be used for pipes and bottles. It did not matter if the hula hoops cracked and snapped after a few months.

# Improving polymers

> **We are learning to:**
> > explain how the properties of polymers can be changed

## What are designer polymers?

Some polymers, such as Teflon, were originally made by accident. Today scientists can produce polymers with a set of properties to suit their purpose. These are called 'designer polymers' and scientists have a variety of ways to modify polymers to get just the right product.

### Making longer molecules

Changing the properties of a polymer means changing the forces between the molecules. One way of doing this is to make the molecules longer. The longer the molecules, the greater the force needed to separate them. UHMWPE is a form of polyethene with very long molecules. It has chains with half a million carbon atoms joined together. UHMWPE is strong and wears so well it is used for artificial hip joints and chopping boards.

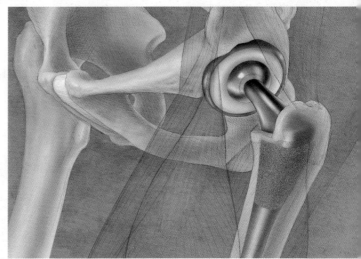

FIGURE 1: UHMWPE is used as the cup in which the ball of an artificial hip joint rotates.

#### QUESTIONS

1. Why are hip joints and chopping boards made out of UHMWPE?
2. Why is LDPE (see page 155) softer and weaker than UHMWPE?

### Harder or softer?

Sometimes we need a polymer to be softer and more flexible. PVC is strong and rigid and is used for window frames and gutters on buildings. When it is used in clothing it is modified. A **plasticiser** is added. Plasticisers are small molecules that fit between the polymer molecules, keeping them apart and weakening the forces between them. Plasticised PVC is still hard-wearing but is much more suitable for waterproof clothes.

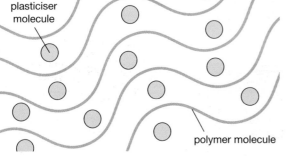

FIGURE 2: The plasticiser molecules make the material less rigid.

FIGURE 3: A raincoat coated with plasticised PVC is waterproof.

PVC

Most polymers soften when heated to a certain temperatures and can be moulded into shape. They are **thermoplastic**. If the polymer molecules are joined together by **cross-links** then the material becomes rigid, strong and does not soften at all. It is called a **thermosetting** material. Cross-linking locks molecules together so they cannot melt but may still leave empty space in the structure so the material is like a sponge. In natural rubber, sulfur atoms cross-link the molecules. This makes the rubber tough and elastic. Cross-linked polymers are used for electric plugs and sockets because they can withstand high temperatures.

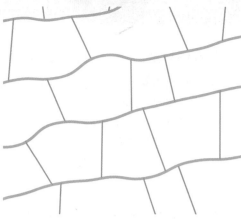

**FIGURE 4**: Cross-links (blue) connect all the polymer chains (green) into one giant molecule.

## QUESTIONS

**3** Why was plasticised PVC used for babies' squeezy toys?

**4** Why do cross-links make a polymer thermosetting?

## Making it even stronger (Higher tier only)

The more crystalline a polymer, the stronger it is. Crystallinity can be increased by reducing the number of branches on the main polymer chain. It can also be improved by making the polymer chains themselves as flat and rigid as possible so that they can line up neatly very close together. This is what makes Kevlar a strong polymer and suitable for making bullet-proof vests and high-performance sails.

When a polymer is turned into a fibre, the heated material is drawn through a tiny hole. This makes the polymer molecules line up and become more crystalline, giving the fibre a higher tensile strength than the original plastic.

**FIGURE 5**: Kevlar vests are very strong but also light to wear.

### Did you know?

The strongest fibre known is a polymer made of just carbon atoms in the form of nanotubes. When scientists are able to make the fibres long enough they will be woven into extremely strong, lightweight cloth.

### Watch out!

Actual materials make use of more than one of the ideas suggested here and often are a mixture of polymers.

## QUESTIONS

**5** Give two reasons why Kevlar fibres are amongst the strongest known.

**6** Polystyrene can be highly crystalline and is used for CD cases. Why is polystyrene suitable?

# Nanotechnology

> **We are learning to:**
> - describe nanotechnology as the use and control of objects about the size of a molecule
> - develop a sense of the scale of nanoparticles

## Is nano the new micro?

A few years ago 'micro' was used by advertisers to promote the newest goods – microphone, microprocessor, microchip. Now there are lots of products with 'nano' in their name or they contain 'nanoparticles'. Is nano better? What can nanotechnology do that microtechnology cannot do?

### Seeing small

What is the smallest thing you can see? You can certainly see a human hair. That is about one-tenth of a millimetre thick. If your eyesight is really good you might just see the largest human cells. Ten of the largest cells would fit across a hair. To go any smaller and see inside cells you need a powerful microscope.

Electron microscopes can see structures inside the nucleus of a cell, but atoms are a thousand times smaller. Objects the size of atoms and small molecules are what is called 'nanoscale'. The first machine for producing images of atoms was built by IBM in the 1980s. It was a type of electron microscope called a scanning tunnelling microscope (STM).

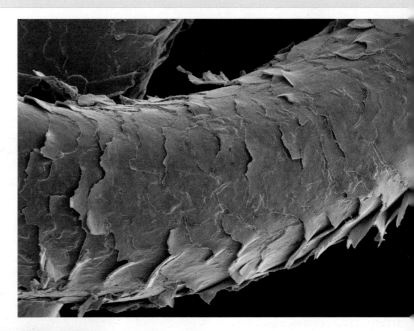

**FIGURE 1:** A human hair magnified 700 times.

**FIGURE 2:** An STM picture of a heap of gold atoms on a sheet of carbon atoms.

### QUESTIONS

**1** How many human hairs side by side would be 1 millimetre across?

**2** Put the following objects in order of size with the largest first: human cell, gold atom, human hair, DNA molecule, fingernail.

nanotechnology

C2 Material choices

# Building small

Victorian engineers built things on a massive scale – ships, bridges, steam engines. Beginning with large lumps of iron or steel which they cut to shape, they made sure things fitted together properly by measuring the pieces to an accuracy of 1 millimetre. In the 1950s and 1960s, when the microcomputer age began, microtechnology engineers had to learn to make electronic components on pieces of silicon that were 1 micrometre in size, that is one-thousandth of a millimetre.

In 1990 the scientists at IBM found that the scanning tunnelling microscope could be used not only to see atoms but to *move* atoms. They succeeded in moving xenon atoms to form the company's logo. They had invented **nanotechnology**.

Now scientists working on nanotechnology are building more complex structures on the molecular scale, from 1 to 100 **nanometres (nm)** in size. An atom is about one-tenth of a nanometre in diameter. It is a bit like building a model with unimaginably small building blocks.

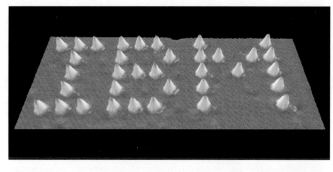

FIGURE 3: This was the first man-made structure at an atomic scale.

**Watch out!**
In nanotechnology individual atoms or small groups of atoms are used to build very small structures.

### QUESTIONS

**3** What is the actual width of the human hair in Figure 1?

**4** A molecule of polyethene is made up of about 10 000 carbon atoms joined together. Approximately how many nanometres long would the molecule be?

# Measuring smallness

A nanometre is one-thousandth of a micrometre, or one-millionth of a millimetre.

$$1\,000 \text{ millimetres} = 1 \text{ metre so, } 1 \text{ mm} = 1 \times 10^{-3} \text{ m}$$
$$1\,000\,000 \text{ micrometres} = 1 \text{ metre so, } 1\,\mu\text{m} = 1 \times 10^{-6} \text{ m}$$
$$1\,000\,000\,000 \text{ nanometres} = 1 \text{ metre so, } 1 \text{ nm} = 1 \times 10^{-9} \text{ m}$$

Ten billion atoms in a row would measure about 1 metre. Nanotechnology is about building structures up to a thousand atoms across, that is 100 nm, but many structures are smaller than this. The heap of gold atoms in Figure 2 is about 5 nm wide. Carbon 'nanotubes' may be as small as 3 nm in diameter.

### QUESTIONS

**5** Look back at your answer to question 4. What is the length of a polyethene molecule in metres?

**6** Gold atoms are 0.25 nm in diameter. How many fit across the bottom of the pile in Figure 2?

### Did you know?

In a 1960s film, *Fantastic Voyage*, a tiny machine was used to repair damage to cells in the human body. It was a fictional use of nanotechnology.

FIGURE 4: These nanostructures made from carbon atoms are called nanotubes.

nanotechnology

# Nanoparticles

## Why does the chemical industry need gold?

We know gold as a pretty metal that does not tarnish. But gold nanoparticles about 2 nanometres thick are a very effective catalyst for speeding up chemical reactions. Many branches of the chemical industry want to use gold nanoparticle catalysts in the future. Strangely, people have been using gold nanoparticles in stained glass for hundreds of years without realising it.

**We are learning to:**
- explain that nanoparticles are produced naturally as well as being manufactured
- explain that nanoparticles have special properties partly because of their large surface area

## Natural nanoparticles

What do we mean by **nanoparticles**? Nanoparticles are bits of material containing about a thousand atoms, or they are tubes or sheets that are just a few nanometres thick.

We did not realise it until recently, but nanoparticles are formed naturally. For example, the soot from a candle and the exhaust from vehicles contain nanoparticles of carbon. A visit to the seaside brings you into contact with nanoparticles of salt formed from seaspray. Now scientists are learning how to make nanoparticles. They have many uses because the very small size of the particles gives them different properties from larger pieces of the material.

**Watch out!** Nanoparticles are much smaller than most particles of dust and soot.

FIGURE 1: A model of a natural nanoparticle of carbon called a 'buckyball'.

### QUESTIONS

1. Where are nanoparticles formed naturally?
2. Why do nanoparticles have special properties?

## Balls, tubes and catalysts

Astronomers first found nanoparticles in space when tiny pieces of carbon containing just 60 atoms arranged in a football shape were discovered. Now these 'buckyballs' (Figure 1) are being made in laboratories, as are carbon **nanotubes** (see Figure 4 on page 27.

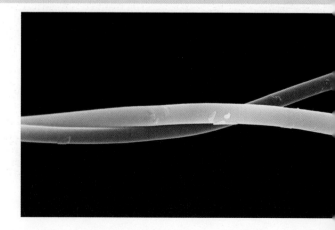

FIGURE 2: Carbon nanotubes can be grown to be millimetres long but are only a few nanometres in diameter. They are much stronger than other forms of carbon and have a tensile strength 100 times greater than steel.

nanoparticles    Buckminsterfullerene (buckyballs)

C2 Material choices

Nanoparticles of materials such as gold have been found to be very effective **catalysts**. Reactions take place on the surface of the nanoparticles. Nanoparticles have much more surface on which the reactions can take place compared to larger particles. A 1 cm³ cube of gold has a surface area of 6 cm². If the cube is cut in half it gains two new surfaces and has a total surface area of 8 cm². As it is cut into smaller and smaller pieces the surface area gets bigger and bigger. The surface area of a small amount of gold in the form of nanoparticles is huge and provides very many sites for reactions to take place.

## QUESTIONS

**3** Why are nanotubes a millimetre long called nanoparticles?

**4** Why do nanoparticles of gold or other materials act as good catalysts?

## Changing material properties with nanoparticles

Since Roman times gold has been added to molten glass to give it a red colour. The glass is found in the stained glass windows in old churches. Now it has been discovered that only gold nanoparticles produce the red colour, so the Romans were doing nanotechnology 2000 years ago. The small size of the particles means that they scatter light in a different way to larger pieces of gold.

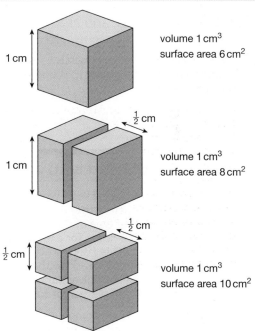

**FIGURE 3:** 1 cm³ of gold could make $1 \times 10^{21}$ nanoparticles each with a volume of 1 nm³. If each nanoparticle was a cube it would have a total surface area of $6 \times 10^{21}$ nm², which is $6 \times 10^{6}$ cm². This is one million times the surface area of the original piece of gold.

### Did you know?

Scientists in Canada have used light of different colours to control the size and shape of silver nanoparticles made in their laboratory.

## QUESTIONS

**5** How is the appearance of gold nanoparticles different from that of larger pieces of gold?

**6** Look at Figure 3. What would be the volume and surface area of nanoparticle cubes with sides of 10 nm?

**FIGURE 4:** This Roman cup looks red because of gold nanoparticles in the glass.

nanoparticles

# Making use of nanoparticles

**We are learning to:**
> explain how nanoparticles can be used to improve the properties of materials

## Do your socks get smelly?

Socks containing silver nanoparticles will never get smelly. It seems there is no limit to the possible uses of nanoparticles. Already they have found their way into medicines, cosmetics, paints and clothes, and are used to strengthen materials and make them more hard-wearing. It is the special properties of the nanoparticles that make them so useful.

**FIGURE 1:** Nanoparticles can keep your feet fresh.

### Silver socks and sunscreen

How do silver-impregnated socks work? The silver nanoparticles are added to the fibres before they are knitted into socks, or woven into the cloth for any other type of clothing. Silver and silver compounds have been used in medicines for a long time, because they are known to be toxic to bacteria. Silver nanoparticles are very effective at killing the bacteria that get into dirty socks without harming humans. The antibacterial properties of silver nanoparticles are also made use of in wound dressings and in plastics for food containers and packaging.

Titanium oxide is a white compound that has been used in sunscreens for a long time. It forms a white film on the skin and absorbs UV light. A new sunscreen containing nanoparticles of titanium oxide is transparent and so looks a lot cleaner on the skin. The nanoparticles also absorb UV light, but let visible light pass through.

### QUESTIONS

**1** Give reasons why socks with silver nanoparticles are a good idea.

**2** Why are silver nanoparticles useful in food packaging?

**3** How is a sunscreen containing nanoparticles better than older types?

### Mixing materials

In many of the uses of nanoparticles they are mixed with another material, such as a metal, ceramic or a plastic. This makes the combined material, called a **composite**, stronger and more hard-wearing. Carbon fibre reinforced composites, used for tennis racquets, racing car bodies and aircraft parts, combine the lightness of plastic with the strength of carbon. Using smaller carbon nanotubes, with the atoms bound together very tightly in tiny tubes, makes the composite even stronger, stiffer and very light for its size.

Nanoparticles added to rubber for use in tyres and tennis balls make the material more hard-wearing. The tennis balls stay bouncy for longer.

**Watch out!**
Nanotubes and nanofibres can be millimetres in length but have a diameter of only a few nanometres.

### QUESTIONS

**4** How will carbon nanotubes improve sports equipment such as tennis racquets?

antibacterial silver

# Super-strength materials

Graphite is a soft form of carbon used in pencils but it is made up of sheets of carbon atoms which are strong. Scientists take the individual sheets, one atom layer thick, called graphene and roll them up to make carbon nanotubes. They can form single tubes or tubes can be stacked inside each other.

The table compares the strength of single-walled and multi-walled carbon nanotubes with other materials.

| Material | Maximum tensile strength (GPa) |
|---|---|
| stainless steel | 1.5 |
| Kevlar | 3.8 |
| single-walled nanotubes | 53 |
| multi-walled nanotubes | 150 |

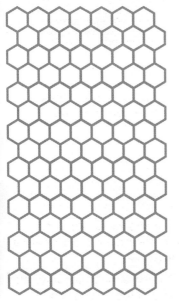
single layer of graphene

single-walled nanotube

multi-walled nanotube

**FIGURE 2:** Modern composite technology using graphene or nanotubes results in super-strength materials.

### Did you know?

A magnetic liquid, called ferrofluid, made of iron oxide nanoparticles coated with a detergent and mixed with a solvent, could find uses in frictionless motors, radar-reflective paint and adjustable curved mirrors for telescopes.

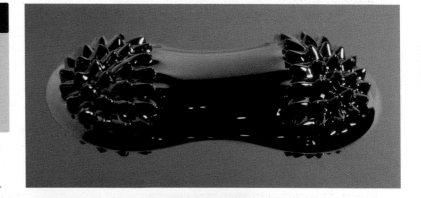

**FIGURE 3:** Ferrofluid placed near a magnet produces spikes that follow the lines of force.

## QUESTIONS

**5** How many times stronger are multi-walled nanotubes than stainless steel?

**6** What advantages are there in using multi-walled nanotubes in composites used for making the frames of racing cars or the masts of racing yachts?

nanoparticle uses

# Staying safe with nanoparticles

**We are learning to:**
- describe how nanoparticles could harm health and the environment
- understand that the risks introduced by a new technology need to be assessed
- understand why more tests are needed on the behaviour of nanoparticles

## Will nanoparticles turn the Earth into grey goo?

Some people fear that nanotechnology will produce tiny machines – nanorobots – that could replicate themselves and run wild, consume the Earth and turn it into a grey goo. The idea is fiction but nevertheless there are some real worries about the effects of nanoparticles that are in use today.

**FIGURE 1:** An artist's impression of how a nanorobot could inject a drug into an infected human cell.

## Are silver nanoparticles safe?

What happens to those antibacterial socks when they are washed? The socks contain silver nanoparticles that kill the bacteria that make feet smelly. When the socks are put in the washing machine, some of the silver gets washed out. The silver nanoparticles find their way to sewage treatment plants, where bacteria are used to break down the waste. The silver nanoparticles may kill these useful bacteria and could stop the sewage works from doing its job. If silver nanoparticles are used in other antibacterial cloths as well as socks, and if they escape into the environment, they could kill a lot of useful microorganisms.

**Watch out!**
In the presence of silver nanoparticles, microorganisms give out nitrous oxide, a gas that contributes to global warming and can damage the ozone layer.

### QUESTIONS

1. What special property do silver nanoparticles have?
2. How could silver nanoparticles get into the environment?

C2 Material choices

## Possible health hazards

Nanoparticles are used in a variety of cosmetics and sunscreens. They are made from materials that have been in use for years and tests have shown that they are not harmful to the skin. However, the nanoparticles of these materials are so small that they can slip through tiny pores in the skin and be absorbed into the blood where they could be carried to the organs in the body. It is not known whether some of these substances are harmful to tissues.

There is a lot of research into the uses of nanotechnology but little so far into the possible harmful effects of nanoparticles. One fear is that breathing carbon nanotubes into the lungs could cause diseases in the same way that tiny fibres of asbestos do. Asbestos was a widely used material until it was shown that it caused lung disease. We need to find out how hazardous nanoparticles are so we can assess whether they pose too great a risk to health to be used.

**Watch out!**
Hazard and risk are different things. A hazard is a possible danger while the risk is the chance of that danger occurring.

FIGURE 2: Sunscreen protects your skin from the Sun, but could it be a health hazard?

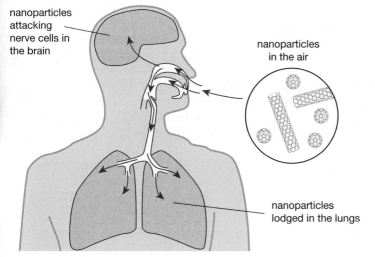

FIGURE 3: Nanoparticles may cause disease if they enter the lungs or the brain.

### QUESTIONS

**3** Why are nanoparticles in substances that are put on the skin a possible health hazard?

**4** Do you think we should stop using nanoparticles now? Explain your answer.

## Weighing the evidence

Some people think that as nanoparticles occur naturally, for example in soot and volcanic dust, they are no danger because we have evolved to live with them. Other people disagree because scientists are now making nanoparticles which do not occur naturally and can have very different properties from the same materials in bulk.

Can the nanoparticles added to coatings on windows and paintwork escape into the air to be breathed in? No one knows if carbon nanotubes might clump together in the brain or liver or what effect they may have on normal reactions in cells. Some environmental groups want more control of new nanotechnology and proof that nanoparticles are safe for health and the environment before they are released for general use.

### QUESTIONS

**5** Do you think naturally occurring nanoparticles are safe or do they affect health?

**6** What control could governments have on the use of nanoparticles?

nanoparticles health hazard

# C2 Checklist

## To achieve your forecast grade in the exam you'll need to revise

Use this checklist to see what you can do now. Refer back to pages 138–165 if you're not sure.

Look across the rows to see how you could progress – ***bold italic*** means Higher tier only.

Remember you'll need to be able to use these ideas in various ways, such as:
> interpreting pictures, diagrams and graphs
> applying ideas to new situations
> explaining ethical implications
> suggesting some benefits and risks to society
> drawing conclusions from evidence you've been given.

Look at pages 312–318 for more information about exams and how you'll be assessed.

| To aim for a grade E | To aim for a grade C | To aim for a grade A |
|---|---|---|
| recall that rubbers, plastics and fibres are used because they have particular properties | interpret information about how solid materials can differ in properties such as melting point, strength (in tension or compression), stiffness, hardness and density | |
| relate the performance of a product to the materials used to make it | interpret information about the properties of materials, such as plastics, rubbers and fibres, to assess their suitability for particular purposes | |
| recall that the materials we use are chemicals or mixtures of chemicals, and include metals, ceramics and polymers; recall that materials can be obtained or made from living things, and that there are synthetic materials which are alternatives to materials from living things | explain how raw materials from the Earth's crust can be used to make synthetic materials | |
| understand that in a chemical reaction the numbers of atoms of each element must be the same in the products as in the reactants | interpret representations of rearrangements of atoms during a chemical reaction | |
| recall that hydrocarbons are made from only carbon and hydrogen atoms; and that crude oil consists of hydrocarbon molecules of varying lengths | recall how the boiling point of a hydrocarbon depends on the length of the hydrocarbon chain | relate the strength of the forces between hydrocarbon molecules to the amount of energy needed for them to break out of a liquid and form a gas, and so to the boiling point |

| To aim for a grade E | To aim for a grade C | To aim for a grade A |
|---|---|---|
| understand that crude oil can be separated into different fractions, and that these fractions are used to produce fuels, lubricants and raw materials for chemical synthesis | understand that the fractions can be separated by distillation because of their different boiling point ranges; recall that most crude oil products are used for fuels and only a small percentage is used for chemical synthesis | |
| understand that small molecules called monomers join together to make very long molecules called polymers, in a process called polymerisation | understand that a wide range of materials may be produced by polymerisation; recall examples of materials that because of their superior properties have replaced materials used in the past | |
| understand that the properties of polymers depend on how their molecules are arranged and held together | relate the strength of the forces between the molecules in a polymer to the amount of energy needed to separate them from each other, and therefore to the temperature at which the solid melts | |
| understand how the properties of polymers can be modified by increasing the chain length and by cross-linking | understand how the use of plasticisers changes the properties of a polymer, and relate the structure of a polymer to its application | *understand how increased crystallinity affects a polymer's properties* |
| recall that nanotechnology involves structures that are about the same size as some molecules | understand that nanotechnology is the use and control of structures that are very small (1 to 100 nanometres in size) | |
| understand that nanoparticles can occur naturally, by accident and by design | understand that nanoparticles show different properties compared to larger particles of the same material | understand that one of the reasons for the different properties is the much larger surface area of the nanoparticles compared to their volume |
| give examples of the use of nanoparticles, including the use of silver nanoparticles to give fibres antibacterial properties | understand that nanoparticles can be used to modify the properties of materials, for example adding nanoparticles to plastics for sports equipment to make them stronger | |
| understand that some nanoparticles may have harmful effects on health | understand that there is concern that products with nanoparticles are being introduced before possible harmful effects have been fully investigated | |

# Exam-style questions

## Foundation level

**1** Crude oil is a very important raw material.

AO1 **a** Which of the following is the main use of substances obtained from crude oil? [1]

plastics    fuels    drugs    fertilisers

AO1 **b** The following sentences explain how crude oil is separated into fractions. The sentences are in the wrong order. Write the order in which the sentences should be written. [2]
1  Crude oil vapour enters the pipestill.
2  Crude oil is heated to over 400°C.
3  Liquids are piped off.
4  When substances in the vapour cool to their boiling point, they condense.
5  The vapours rise and cool.

[Total: 3]

## Foundation/Higher level

**2** Cotton was used for fishing lines but has been replaced by nylon. Nylon is more elastic and stronger than cotton.

AO1 **a** What does 'nylon is more elastic' mean? [1]

AO2 **b** The strength of a nylon fishing line was tested by putting weights of 1 and 10 newtons on a weight-hanger suspended from 50 cm of the line fastened to a clamp stand. The weight needed to break the line was recorded.

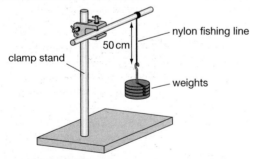

Here are the results:

| Test number | 1 | 2 | 3 | 4 | 5 |
|---|---|---|---|---|---|
| Weight needed to break fishing line (N) | 32 | 29 | 31 | 32 | 30 |

(i) Why was the measurement repeated five times? [3]
(ii) Give two reasons why the recorded breaking weights vary. [2]
(iii) What is the *mean* and the *range* of the results? [2]

[Total: 8]

**3** Scientists have applied for a patent on a lipstick containing gold nanoparticles. Gold nanoparticles have a red colour. It is claimed that these could replace pigments which may be toxic. It is also claimed that gold nanoparticles are beneficial to health because they kill some bacteria. But some people are worried about the effects of ingesting gold nanoparticles.

AO1 **a** What is different about the properties of gold in the form of nanoparticles compared to larger pieces of gold? [1]

AO3 **b** Suggest what action should be taken if gold nanoparticle-lipstick is to be sold to the public. [3]

[Total: 4]

## Higher level

**4** Mountaineers and polar explorers need to keep warm. They wear clothes that can be made from polypropene, PET or wool. The table below gives some information on these fibres.

| Material | Heat insulation | Water absorbance | Density |
|---|---|---|---|
| Polypropene | Good | Low | Low |
| PET | Good | Low | Low |
| Wool | Good | High | Medium |

AO1 **a** Which one of the materials in the table is not a synthetic material? [1]

AO2 **b** Why have fleeces made from polypropene or PET replaced wool in making clothing worn by explorers and mountaineers? [2]

AO2 **c** Polypropene is a polymer manufactured from hydrocarbons obtained from crude oil. This shows part of the polypropene molecule.

Draw the monomer from which polypropene is made. [1]

AO2 **d** PET is a polymer used for making drinks bottles. These can be recycled to make fleeces. The bottles are heated to make the PET soften and then it is drawn into fibres. PET fibres are stronger and have a higher melting point than in the sheet form used for bottles.
Describe the changes that take place when the PET is made into a fibre and explain the difference in properties.
The quality of written communication will be assessed in this question. [6]

[Total: 10]

# C2 Material choices

## ✳ Worked example

Rubber was originally used by the Mayans in South America. When it was brought to Europe one of its first uses was a 'rubber' to rub out pencil marks. Rubber was a soft material that became sticky when warmed, and hard and brittle when cold. It was after Charles Goodyear discovered hot 'vulcanised' rubber using sulfur, in 1839, that it became the tough, springy useful material we use today for the soles of shoes and car tyres. In the 20th century, synthetic rubber produced from crude oil took the place of the natural material obtained from trees.

**AO1 a** Rubber is a polymer. What is a polymer? [1]

*A large molecule*

**How to raise your grade**
Take note of the comments from examiners – these will help you to improve your grade.

Insufficient for the mark – a polymer is a large molecule made of many similar repeated units/monomers.

**AO1 b** Vulcanising rubber introduces cross-links. Sketch two diagrams showing rubber before and after vulcanisation. [2]

plain polymer    crosslinked polymer
                 crosslinks

Make sure that there are sufficient cross-links clearly marked.

**AO2 c** Explain how vulcanisation changes the properties of rubber. [2]

*Vulcanisation joins the polymer chains together with strong bonds. ✔*
*A lot of energy is needed to break these bonds so the rubber becomes stronger and needs a higher temperature to melt. ✔*
*When the rubber is stretched or squashed the polymer chains are moved, but when the force is released the cross-links pull the chains back to the position they were in before, so the rubber is elastic.*

The first sentence states what the cross-links do – 1 mark.

The other sentences give examples of how the cross-links modify the properties – 1 mark each.

But there is a maximum of 2 marks for the question, so only one change in property is required. Don't write more than is necessary in answers.

**AO1 d** Synthetic rubbers are polymers which have their properties modified. Describe two other methods of modifying polymer properties apart from vulcanisation, and state the change in properties. [2]

*Increasing chain length – increases the force between polymer molecules and gives them greater strength and a higher melting point. ✔*
*Adding a plasticiser – separates the molecules weakening the force between them and making the polymer softer, weaker and melt at a lower temperature. ✔*

Two correct methods are given with descriptions of how the properties are modified – 1 mark each.

**AO3 e** Suggest reasons why synthetic rubbers have largely replaced natural rubber for most uses. [3]

*Crude oil is (relatively) cheap so synthetic rubbers are cheaper than natural rubber. ✔*
*Rubber trees can only be grown in special environments so the supply of natural rubber is limited. ✔*
*There are many different types of synthetic rubber with properties that suit their uses. ✔*
*It is easier to modify the properties of synthetic rubbers for particular purposes when they are manufactured.*

Four points have been given – any three are sufficient for full marks.

# C3 Chemicals in our lives – risks and benefits

## What you should already know...

### The surface of the Earth changes

Igneous rocks are formed from molten rock beneath the Earth's crust

Rocks are changed and formed by the processes of weathering, erosion and sedimentation.

Heat and pressure can change the structure and properties of rocks.

Physical, chemical and biological processes are involved in the changes to rocks.

Rocks change over a very long time scale.

 How is a sedimentary rock, such as sandstone, formed?

### Substances can be separated by using their different properties

Some substances dissolve in solvents to form solutions.

Solutions can be separated from solids by filtration.

Substances can be separated from solutions by evaporating off the solvent.

 How can you show that salt dissolves in water?

### Elements combine to form compounds

Elements in different parts of the periodic table have different properties.

Compounds are often very different to the elements from which they are made.

Materials can be distinguished by their properties.

 How can you show that water is a compound?

### Chemical reactions follow patterns

Observations can be used to determine patterns in reactions.

Chemical reactions can be described in word and symbol equations.

Alkalis are a group of substances that react with acids.

 How can you tell that a substance is an alkali?

# In C3 you will find out about...

> the evidence that hundreds of millions of years ago the positions of the Earth's continents were very different

> how parts of Britain formed from different continents that have moved across the Earth's surface

> the processes that left deposits of useful minerals in parts of Britain

> how salt is obtained from seawater and from underground deposits

> the benefits and risks of adding salt to our food

> the properties and uses of alkalis

> the reactions of alkalis with acids

> the manufacture of alkalis in the 19th century and the problems it caused

> the benefits and risks of adding chlorine to water

> how electricity is used today to make alkalis and chlorine

> the health and safety issues of industrial chemicals

> risks of using the polymer PVC

> how to carry out a Life Cycle Assessment

# Moving continents

**We are learning to:**
- explain that rocks formed by processes that are still happening today
- explain the evidence that land making up Britain has moved across the Earth
- understand that scientific theories account for observations, make links and predictions

## What do rocks tell us?

A piece of rock from where you live can tell a lot. Chemists can analyse the substances that make up the rock; geologists can identify the type of rock and the fossils that are trapped in it; geophysicists can measure how long it has been since the rock was formed.

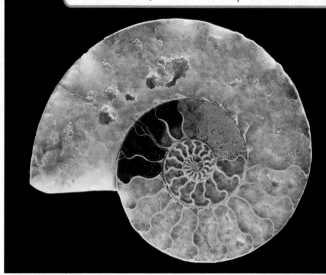

**FIGURE 1:** Fossils tell geologists when and where a rock was probably formed.

###  Looking at rocks

Geologists looking at the Earth's surface see changes that are occurring in rocks caused by the movement of **tectonic plates**. Plates of the Earth's crust are continually sliding past each other, moving apart or pushing together causing mountains to be lifted up. Geologists see evidence of the same processes in rocks that formed over hundreds of millions of years ago, including some of those in Britain.

Across Britain, there are lots of different types of rocks, from the white chalk along the south coast to the dark volcanic basalts in western Scotland. By studying the rocks, geologists can see evidence of how, when and where rocks were formed. They can tell that the land which makes up Britain today has, over many millions of years, moved across the surface of the Earth.

*Watch out!* The continents move extremely slowly – about as fast as your nails grow.

**FIGURE 2:** Geologists can tell that basalt rocks off the west coast of Scotland were formed at the same time as those in North America, about 60 million years ago.

###  QUESTION

1. Why do geologists think that rocks in Britain have been changed by the movement of tectonic plates?

# Making Britain

600 million years ago Britain wasn't even in one piece. England and Wales were separated from Scotland by an ocean and were parts of different continents. Both of these continents were close to the South Pole. The evidence comes from the different sea creatures trapped in the oldest rocks of each continent. Gradually the continents drifted towards each other and the ocean between them shrank. The continents crashed together to form a **supercontinent** joining Scotland to England and building a range of mountains.

Over the next 300 million years, the supercontinent moved northwards. The climate changed – there were deserts that formed red sandstone, tropical swamps with rainforest that produced coal, and chalk formed in the warm seas.

About 60 million years ago, the North Atlantic Ocean began to open up taking North America away from Europe and causing volcanoes to erupt along the western edge of Scotland. Ice ages during the last 3 million years brought glaciers down from the north eroding the highlands of Scotland, northern England and Wales.

**FIGURE 3:** Evidence in rocks show that parts of Britain were once separated and in different positions from where they are now.

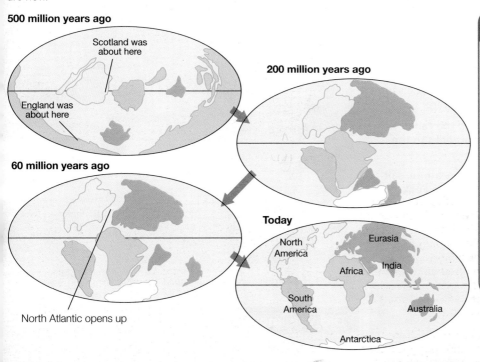

## QUESTIONS

**2 a** What evidence is there that different parts of Britain have experienced different climate conditions?

**b** What evidence is there that Britain was once closer to the Equator?

**3** How do scientists know that Scotland and England were once separated by an ocean?

# Stories in magnetism

The variety of different rocks found in Scotland gave a confusing picture. This has now been explained by the theory of **plate tectonics** and confirmed by readings of magnetism left in the rocks. When molten rock such as volcanic lava solidifies, **igneous rocks** are formed. As they solidify, the magnetic material in them records the direction towards magnetic north. When the rocks move, this direction changes. Also, the position of magnetic north changes (even reverses) over millions of years. Rocks in Scotland that formed at different times and in different places have different magnetic directions.

Geologists have been able to detect and compare the magnetic patterns of the rocks. They came to the surprising conclusion that Scotland was at one time in five separate chunks. The success of the theory has meant that predictions can be made about where similar rocks and similar fossils can be found.

## QUESTIONS

**4** Why is the magnetism in rocks of different ages lined up in different directions?

**5** What evidence might there be that Newfoundland in Canada was once joined to Scotland and England?

### Did you know?

The oldest rocks in Britain are 3 billion years old and are now exposed in the islands of Lewis and Harris off the west coast of Scotland.

geology Scotland

# Useful rocks

**We are learning to:**
> explain how resources in the Earth were formed
> explain how geologists discover how rocks were formed
> explain how the development of the chemical industry depended on resources being available nearby

## What can we do with rocks?

As well as food and water, there are lots of things that we need for a comfortable life. Look at all the containers in your kitchen and bathroom. Many of the products are made from substances found in rocks. The chemical industry grew up using those rocks to supply the products we buy.

### The beginnings of the chemical industry

The north-west of England is still one of the most important areas in Britain for the chemical industry. The industry began nearly 200 years ago near the River Mersey, between Liverpool and Manchester. There are a few reasons for this:

> There was a market for the products in the textile factories of Lancashire.

> The port of Liverpool and the canal system gave good transport links to the rest of the country.

> There was coal in south Lancashire, salt in Cheshire and limestone in the Peak District – the three resources needed most for chemical factories.

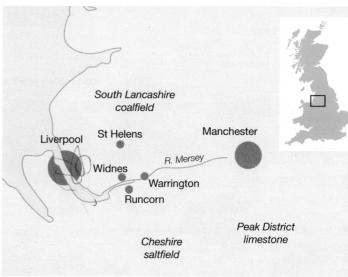

**FIGURE 1:** The chemical industry began in the north-west of England.

**FIGURE 2:** Limestone is a very useful resource for the chemical industry.

**Did you know?**
The chemical industry began to develop only after a tax on salt was removed in 1823.

**QUESTIONS**

1 a Name three materials needed by the chemical industry in the north-west of England.

b Why was it important that these materials were close to the factories?

### How limestone, coal and salt formed

350 million years ago, the north-west of England was covered by a sea. Calcium compounds in the water reacted with carbon dioxide in the air to form calcium carbonate. This, and dead shellfish, formed **sediment** on the seafloor, which hardened into the **sedimentary rock** limestone. Other rocks were laid down on top and then, 300 million years ago, tectonic plate movements pushed the limestone up to form mountains. Gradually the rocks above were **eroded** away until the limestone was exposed.

🔍 chemical industry history NW England

# C3 Chemicals in our lives – risks and benefits

About 290 million years ago, Lancashire was a warm, wet swamp on the edge of a sea where trees and other plants grew in abundance. When they died, they became buried under sediment and their carbon content gradually turned into coal. More sediment covered the coal forming more layers of rocks.

200 million years ago, Cheshire was covered by a shallow sea. Rivers flowing into the sea carried salts **dissolved** from rocks. The climate grew hot and dry so the sea **evaporated** leaving the salt mixed with sand blown in from the surrounding deserts. Later, other sediments were deposited over the rock salt, burying it.

**FIGURE 3:** Lancashire once looked like this before the plants died and became coal.

## QUESTIONS

**2** Which rocks used by the chemical industry in the north-west of England were formed or affected by the following processes?

sedimentation  evaporation  mountain-building  dissolving  erosion

**3** How did climate affect the way rocks were formed? Give examples.

## What geologists observe

We know how the limestone, coal and salt in the north-west of England were formed because geologists have found evidence.

> Coal contains fossils of the plants that formed it.

> Limestone contains bits of shell and fossils of other sea creatures that lived in the sea when the sediment was forming.

> Rock salt contains small, rounded sand grains which tell geologists that they were carried by the wind, with lots of collisions between the particles. Grains of rock carried by water tend to be larger and rougher.

They may observe other clues. For example, ripples in the sediment made by rivers or by waves on a seashore sometimes remain when the sediment dries or hardens. Geologists recognise these patterns when they look at layers of rock and can tell that they were formed under water.

**FIGURE 4:** Wind-blown sand grains are small and rounded.

**FIGURE 5:** What do ripples in a piece of rock tell geologists?

## QUESTION

**4** How do geologists know that:
  **a** rock salt formed during dry, desert conditions?
  **b** limestone was formed under the sea?

rock salt formation

# Salt

## How important is salt?

Salt has always been necessary for life. Roman soldiers were given salt as part of their pay, which became known as a 'salary'. Cities such as Salzburg in Austria were founded near sources of salt. It is still one of the most useful raw materials for the chemical industry today.

**We are learning to:**
> remember that salt has many uses
> understand that salt can be obtained from seawater or from underground rocks
> explain how salt is extracted from underground deposits and the effect this has on the environment

### Salt for our food

We need some salt in our food to stay healthy, but it is mainly used as flavouring. Many foods contain salt to boost the flavour, but many people add more table salt – particularly to vegetables. Salt is also a preservative and has been used for centuries to make food last longer – especially meat and fish. Much of the salt we buy is called 'sea salt'. It is made by collecting sea water and evaporating the water off either using the heat of the Sun or by heating the salt in tanks.

**Did you know?**

Salt helped to win wars. It was needed to preserve food for long campaigns when fresh food for soldiers was unavailable. The American Civil War was won by the Union partly because it had more secure supplies of salt.

 **QUESTIONS**

1. Give two reasons why salt is added to food.
2. How is sea salt obtained?

**FIGURE 1:** Using sunlight to evaporate seawater.

### Getting at salt

Every winter, tonnes of rock salt are spread on roads. Salt water has a lower freezing point than pure water so even if the temperature falls below 0°C the salty water on the roads will not freeze. The sand in the rock salt also helps vehicles grip the road. Rock salt has to be mined from underground deposits. There is just one salt mine in Britain – it is in Cheshire and a million tonnes of rock salt is dug out from its caverns each year.

**FIGURE 2:** The main use of rock salt is for de-icing roads.

sea salt evaporation

## C3 Chemicals in our lives – risks and benefits

Salt also has industrial uses. It is called sodium chloride, and chlorine and other chemicals can be obtained from it for use in many different chemical industries. Salt for the chemical industry is usually obtained from underground deposits of rock salt. Water is pumped down into the salt deposits. The salt dissolves to form a concentrated solution called brine. The brine is then pumped to the surface and piped to where it is needed. The process is almost automatic.

Solution mining is more convenient and cheaper than digging out rock salt, and the factory using the salt can draw as much salt as it needs from the underground source.

### QUESTIONS

**3** Why is rock salt put on roads in winter?

**4** Describe two ways that salt is obtained from underground deposits.

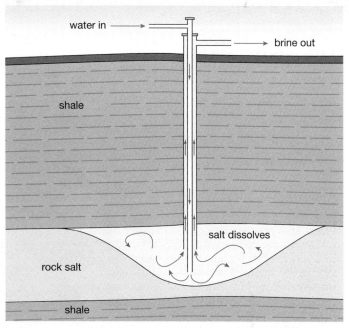

**FIGURE 3:** Solution mining – water is pumped at high pressure into the rock salt and the salt solution is pushed to the surface.

## Problems with salt extraction

Salt mining and extraction can have serious effects on the environment at ground level. In Cheshire, the rock salt is 200 metres below the surface. Since the early 19th century a huge amount of salt has been removed. This left large holes underground, which caused the ground to sink. Sometimes whole streets of houses sank into the earth. Now about half of the salt is left behind to support the roof of the mine. Subsidence could allow water into the salt deposits, which could then make wells become salty and undrinkable. Plants die if soil becomes contaminated with salt

In tropical areas, vast areas of sea shore are used for solar evaporation of sea water. The ponds in which the sea water lies become so salty that no animals or plants can survive.

**FIGURE 4:** Extracting too much salt from below the surface can have disastrous results.

**Watch out!** Mined rock salt contains sand – the salt obtained by solution mining only contains soluble substances.

### QUESTIONS

**5** Why do salt miners usually leave about half of the salt in a mine untouched?

**6** A chemical company wants to set up a new solar-evaporation site on the coast in a tropical country. Discuss the arguments for and against this proposal.

Q salt mine Cheshire

# Salty food – risks and benefits

## Do we all take risks?

Risk is part of life – we take a risk every time we step onto a road. There are also risks in the food we eat – eating too much or too little brings the risk of ill-health, but even the amount of salt in our food could have an effect on how fit we are.

FIGURE 1: Do you know how much salt you add to your food?

**We are learning to:**
- describe the purpose of adding salt to food
- understand the health problems linked to a high-salt diet
- use data on the health effects of salt in food
- understand how decisions are made about risks

## Salt for better or worse

Most people think that some foods taste better with a little salt added, such as potatoes and bread. But there are some foods that have a surprising amount of salt added to them, such as baked beans and tinned soups. Many foods have salt added to prevent bacteria and fungi from growing. Bacon and fish were preserved by coating them in salt. In the past, having salted food available during the winter stopped people from starving. Today we are more worried about health conditions that may be related to eating too much salt.

| Food | Amount of salt (g) |
|---|---|
| Can of vegetable soup | 3.5 |
| 2 large pork sausages | 2 |
| Portion of baked beans | 1.25 |
| Slice of pizza | 1 |
| Bowl of cornflakes | 0.5 |
| Slice of white bread | 0.4 |
| Glass of skimmed milk | 0.3 |

Amount of salt in foods.

### QUESTIONS

1. Name one food with a high salt content and one with a low salt content.
2. Why was fish stored in salt?

### Did you know?

Adults need about 4g of salt a day to remain healthy. Any extra is eventually removed by the kidneys in urine.

salt diet risk

# C3 Chemicals in our lives – risks and benefits

## Hazards and risks

Scientists know that eating a lot of salt raises blood pressure, which can lead to serious health conditions such as strokes. Salt is therefore a **hazard**. The chance of becoming ill because of a salty diet combined with the severity of the outcome is known as the **risk**. We can estimate the risk by looking at data from a large group of people over a long time. In Britain, men eat on average about 11 g of salt per day in their food, for women it is about 8 g/day. About one person in 1000 will die of a stroke each year. People who reduce their intake of salt to less than 6 g per day reduce their chance of having a stroke to about one in 2000, so low salt means low risk. Knowing the risk allows you to make decisions. One cup of a salty vegetable soup won't kill you but a lifetime of eating a diet of high-salt food gives you a higher risk of dying from a stroke.

**FIGURE 2:** Is this a high risk diet?

| Level of risk | What it means |
|---|---|
| High risk | Small chance of serious outcome, or large chance of a less serious outcome |
| Low risk | Small chance of outcome occurring |

Meaning of risk.

**Watch out!**
Make sure you know the difference between a hazard and a risk. A moving car is a hazard, but there is higher risk trying to cross a busy highway than a quiet side street.

### QUESTIONS

**3** If the population of Britain is 62 million, estimate the number of people who die from strokes in a year.

**4** Why is a high-salt diet considered to be a high risk?

## Helping to make decisions

The government Department for Health and the Department for Environment, Food and Rural Affairs help people to make decisions about their health and reduce their risk of death caused by salty diets. They examine foods for their salt content and calculate the risk that each food carries. This data may be published in leaflets and websites, or food manufacturers may be encouraged to put the information on the packaging. Government departments also run publicity campaigns to alert people to the risk of salt in diets and other dangers. Some people think that governments should make food manufacturers reduce the amount of salt 'hidden' in processed foods, but others think that will limit their freedom to eat what they want.

**FIGURE 3:** Food labels give the amount of salt in the food and the percentage of the recommended daily allowance.

### QUESTIONS

**5** What information can government departments give the public about risks to health?

**6** Governments have banned smoking in many places for health reasons. Do you think they should ban salty foods? Give your reasons.

**7** Why do people choose a high-salt diet?

recommended daily allowance

# Alkalis

## Were people dirty before chemically manufactured soap?

The modern chemical industry gives us a range of soaps and cleaners, and brightly coloured dyes for clothes and furnishings. Before there was a chemical industry, people had ways of making soap and of dyeing cloth using chemicals they found around them. So, what did they use?

**We are learning to:**
> understand that alkalis have been used for many purposes for centuries
> remember that early sources of alkalis were wood and urine
> explain that alkalis neutralise acids to make salts

**FIGURE 1**

## Recognising alkalis

**Alkalis** are chemicals that make **indicators** change colour – for instance, litmus turns blue in an alkali but red in **acids**.

Alkalis react with acids to form new substances called **salts**. The acid loses its acidity and becomes neutral – neither acidic nor alkaline. This is called **neutralisation**.

In the past, most people didn't know what an alkali was but they knew they could be useful and where to find them. 'Alkali' is an Arabic word for the ashes of burnt plants, one of the most useful alkalis. Another useful alkali is formed in urine, which used to be collected from people's homes.

**Watch out!**
To be an alkali, a substance must dissolve in water and neutralise acids.

**FIGURE 2:** Litmus paper turns blue in an alkali.

### QUESTIONS

1. Give two early sources of alkalis.
2. What happens when an alkali is added to an acid?

## Using alkalis

### Making soap

While soap may be useful for washing your skin, its main use in the past was for cleaning wool before it was spun and woven into cloth. Soap was made by first mixing the ashes of burnt wood or other plants with animal fat and water, boiling up the mixture and then adding salt. The soap formed a layer on the surface that could be scooped off. Vegetable oils could be used instead of animal fats.

**FIGURE 3:** A soapmaker collects the soap from the surface of a boiling vat of fat and alkali.

alkali soap middle ages

## Neutralising soil

Farmers have long known that mixing an alkali with acidic soils would improve their crops. Seaweed or seaweed ash is a source of alkali that could be used in coastal areas. A manufactured alkali that has been commonly used for a long time is lime, calcium oxide, obtained by heating limestone, calcium carbonate, in a lime burner. The lime can also be mixed with sand and water to make mortar, and also used as the whitewash painted onto houses.

## Making glass

Lime and seaweed ashes were also mixed with sand to make glass. Until the 18th century, glass was a valuable substance and few houses had glass windows.

## Dyeing cloth

Dyes obtained from plants have long been used to colour cloth. Dyers found that an alkali mixed with a mineral called alum made the dye stick to the cloth better. The alkali they needed was found in urine.

### QUESTIONS

**3** Give four uses of alkalis.

**4** Why was wood ash or seaweed ash a valuable material before the chemical industry began?

### Did you know?

The ashes of wood and land plants, known as potash, contain potassium carbonate; the ashes of seaweed contain sodium carbonate, which is called 'soda'.

**FIGURE 4:** A glassmaker heats a blob of glass on a steel pole and then spins it to flatten it for use in windows.

## Dyeing with alkalis

The alkali that forms when urine is left to stand for some time is called ammonia and gives it its familiar disgusting smell. Collecting urine became an important job for the dyers.

Ammonia solution reacts with alum to form aluminium hydroxide. When cloth is dipped in the alum followed by the ammonia and then in a dye, the aluminium hydroxide binds the dye molecules to the fibres in the cloth. The dye becomes 'fast' and is not washed out of the cloth.

### QUESTIONS

**5** How could you show that stale urine is an alkali?

**6** How could you prove that ammonia and alum are both needed to stick dyes to cloth?

**FIGURE 5:** Dyeing Chinese silk in the early 19th century.

mordant glass

# Reacting alkalis

**We are learning to:**
- explain why the industrial manufacture of alkalis increased in the 19th century
- remember which substances are classed as alkalis
- remember the patterns of reactions of alkalis with acids

## Are alkalis as reactive as acids?

We know that acids are corrosive and reactive substances, but the reason that alkalis are so useful is because they react with lots of chemicals too. A very reactive alkali is said to be 'caustic', which means it can burn or corrode skin and other materials.

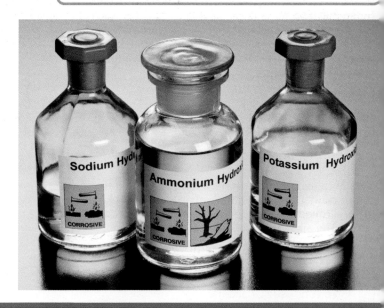

FIGURE 1: Strong alkalis are hazardous.

## Alkalis in demand

In the 1700s, the demand for manufactured goods increased in Britain and elsewhere as the population grew and became better off. Larger and larger factories were built to produce iron, textiles, glass and pottery in a greater quantity than had previously been possible. There was more demand for limestone for the ironworks and for alkalis for the textile, glass and pottery industries.

In Britain, trees were needed for ships and buildings so manufacturers looked for other sources of plants to be turned into ash. Huge amounts of seaweed from the coasts of Scotland were burned and shiploads of alkali from the ash of Canadian trees were carried to British ports. By the early 19th century, even these supplies of ash could not keep up with the demand for alkalis, particularly from the soapmakers – another source of alkali was needed.

FIGURE 2: The glass-making industry grew in the Industrial Revolution and increased the demand for alkali.

### QUESTIONS

1. Name two industries that grew during the 18th century.
2. Why was there a shortage of alkali in the early 19th century?

### Did you know?

The burning of wood to make alkali cleared land for agriculture in North America, and the wood-burning industry moved steadily westwards as the forests were cleared.

# Patterns of reactions (Higher tier only)

An alkaline solution is one with a **pH** greater than 7. It turns pH indicator blue or violet. Two groups of substances form alkalis when dissolved in water. One of these groups is the **soluble** hydroxides, such as sodium hydroxide. The other is the soluble carbonates, such as potassium carbonate. Hydroxides and carbonates that are not soluble, such as calcium carbonate, are not alkalis but are called **bases**. Bases react with acids in a similar way to alkalis, but do not affect indicators.

| Soluble hydroxides | | Soluble carbonates | |
|---|---|---|---|
| sodium hydroxide | NaOH | sodium carbonate | $Na_2CO_3$ |
| potassium hydroxide | KOH | potassium carbonate | $K_2CO_3$ |
| calcium hydroxide | $Ca(OH)_2$ | | |
| ammonia solution | $NH_4OH$ | | |

Examples of alkalis.

**Watch out!** There are other hydroxides and carbonates in addition to those listed here, but most are not soluble in water and are therefore not classed as alkalis.

When an alkali reacts with an acid, the hydroxide or carbonate part of the alkali reacts with the hydrogen of the acid – the acid is neutralised. For example:

sodium hydroxide + hydrochloric acid → sodium chloride + water
NaOH + HCl → NaCl + $H_2O$

**Hint:** You won't be expected to write a symbol equation in your exam.

The hydroxide, OH, and the hydrogen, H, of the acid join up to form water leaving the salt called sodium chloride, NaCl.

potassium carbonate + sulfuric acid → potassium sulfate + water + carbon dioxide
$K_2CO_3$ + $H_2SO_4$ → $K_2SO_4$ + $H_2O$ + $CO_2$

The carbonate, $CO_3$, and the hydrogen in the acid join up to make water and carbon dioxide leaving the salt called potassium sulfate.

The general patterns of the reaction of alkalis with acids are:

hydroxide + acid → salt + water
carbonate + acid → salt + water + carbon dioxide gas

The salts produced by different acids are shown in the table.

| Acid | | Salt | |
|---|---|---|---|
| hydrochloric | HCl | chloride | Cl |
| sulfuric | $H_2SO_4$ | sulfate | $SO_4$ |
| nitric | $HNO_3$ | nitrate | $NO_3$ |

Acids and their salts.

**FIGURE 3:** Sodium carbonate reacts with an acid giving off carbon dioxide gas.

## QUESTIONS

**3** Which of the following compounds are alkalis and which are salts?

| sodium chloride | potassium hydroxide | potassium nitrate | calcium chloride | sodium carbonate | ammonium sulfate |

**4** What are the products formed when:
  **a** potassium hydroxide reacts with hydrochloric acid
  **b** sodium carbonate reacts with nitric acid?

**5** Write word equations for the following reactions:
  **a** calcium hydroxide reacts with sulfuric acid to form calcium sulfate
  **b** potassium carbonate reacts with hydrochloric acid to form potassium chloride.

🔍 acid-alkali reaction

# Making alkalis

> **We are learning to:**
> - describe the first method of manufacturing alkali and the pollution problems it caused
> - explain how one of the pollutants became a source of a useful substance

## Why did Britain need a chemical industry?

By 1830, 35 000 tons of alkali per year were being exported from Canada to Britain and 10 000 tons of alkali from Scottish seaweed were being produced – but still the soapmakers and the glassworkers needed more. A method was needed that made use of the resources that were nearby – salt, coal and limestone.

### Nicolas Leblanc's bright idea

Nicolas Leblanc was a Frenchman who worked out a method of manufacturing alkali in 1787. An Irishman called James Muspratt took up Leblanc's ideas and settled in Liverpool where most of the raw materials needed, including salt, were available. A problem was that salt was too expensive, because of a 'salt tax'. Luckily for Muspratt the salt tax ended in 1823. Soon Muspratt's factories in north-west England and others in Glasgow were producing all the alkali that was needed. Their method used the 'Leblanc process' – reacting salt with sulfuric acid and then heating it with coal and limestone.

**FIGURE 1:** Leblanc alkali works – the tall chimneys 'disposed' of the waste gases.

#### QUESTIONS

1. Why was Muspratt's factory not built until after 1823?
2. Imagine you are Muspratt. What do you need to start your factory?

### A scene of desolation

The Leblanc process produced sodium carbonate, or 'soda', but the process also gave off a large amount of hydrogen chloride gas and produced huge heaps of solid waste called 'galligu'. To get rid of the hydrogen chloride, Muspratt built very tall chimneys to try to spread the acidic gas in the air – but most of it sank to the ground and killed plants and wildlife for miles around causing scenes of desolation. The galligu was also a serious pollutant because it gave off stinking and poisonous 'bad egg' fumes of hydrogen sulfide.

**FIGURE 2:** The Leblanc process produced huge amounts of waste called 'galligu', which piled up around the factories.

Part of this pollution problem was solved by William Gossage. He invented a tower in which the hydrogen chloride dissolved in water to make hydrochloric acid. In 1862, the Alkali Acts forced alkali manufacturers to use Gossage's method to stop the pollutant being released. By then it had been discovered that the hydrochloric acid was itself a useful product. It could be used to make chlorine, which was widely used as a bleach.

## QUESTIONS

**3** Why were workers in the alkali industry and their families often in poor health before 1862?

**4** Why were the Alkali Acts needed?

### Did you know?

In the 1880, the Solvay alkali process began to be used in north-west England and elsewhere. It produced much less waste than the Leblanc process, but some of the old Leblanc alkali factories carried on working into the 20th century.

## Making chlorine

Soon after chlorine was discovered in the 1770s it was found to be a bleach. Textile manufacturers had to bleach their cloth before dyeing it. The alkali industry would get a good price for chlorine if they could find a way of producing it from hydrochloric acid, the industry's by-product. The best method of getting chlorine was to react hydrochloric acid with manganese dioxide:

## QUESTIONS

**5** In what ways are hydrogen and chlorine different to hydrochloric acid?

**6** What were the consequences of chlorine being a substance in demand?

hydrochloric acid + manganese dioxide → chlorine + manganese chloride + water

In this reaction the hydrogen is taken away from the hydrogen chloride to form water. We say the hydrogen chloride has been oxidised because the hydrogen has joined with oxygen. Taking hydrogen from a compound is an oxidation reaction.

Chlorine is an element and is quite different from hydrogen chloride. It is a green gas, while hydrogen chloride is colourless. Chlorine is less soluble in water than hydrogen chloride. Chlorine is a powerful bleach; hydrogen chloride is not. Compounds have different properties from those of the elements they contain.

### Watch out!
Oxidation can be defined as the gain of oxygen or as the loss of hydrogen. Reduction is the opposite.

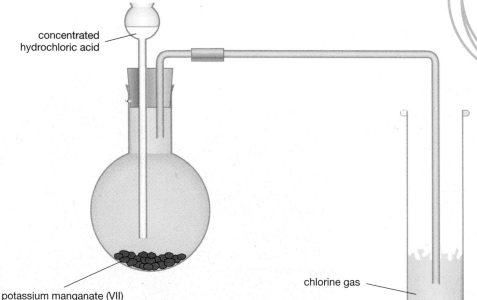

**FIGURE 3:** In the laboratory chlorine can be made using potassium manganate(VII) to oxidise hydrogen chloride.

# Chlorine in water – benefits and risks

## When was chlorine first added to a public water supply?

In August 1897, a typhoid epidemic hit Maidstone in Kent. By October over a thousand people had fallen ill and more than 40 had died. On 16 October the town council added chlorine to the reservoir supplying the drinking water, the first time it had been done. The number of typhoid cases fell quickly.

**FIGURE 1:** Safe water.

**We are learning to:**
> interpret data showing that the chlorination of drinking water improved public health
> describe the possible health risks of chlorine in water
> understand how people perceive risk and how governments assess risk

## Improving public health

More and more people lived in crowded cities in the 19th century. Sewage drained into rivers, which were also used to supply drinking water. Many died from diseases such as cholera and typhoid. Building sewers and sewage treatment works improved health but water supplies still often contained dangerous microorganisms. In 1908, Jersey City in the USA became the first city to add chlorine to its water supply all the time. Chlorine is a disinfectant because it kills microorganisms. Many cities in the US, Britain and Europe soon did the same. There was a great improvement in public health in the developed countries but water treatment remains a problem in less developed countries to this day.

### QUESTIONS

1. Why did many people become ill in the 19th century?
2. Why was chlorine used to treat drinking water?

## Proof in the data

The effect on public health of disinfecting water with chlorine can be seen in Figure 2, which shows the number of deaths due to typhoid in the USA between 1900 and 1960. Most cities in the USA began chlorinating water between 1908 and 1930. There appears to be a correlation between the chlorination of water and reducing deaths from typhoid. There were other improvements to living conditions during this period and some groups of people were vaccinated against typhoid. These factors also contributed to the falling death rate.

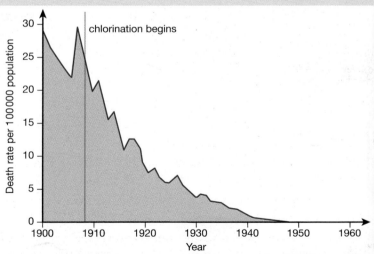

**FIGURE 2:** The start of water chlorination in the USA and the effect on the death rate from typhoid.

## QUESTIONS

**3 a** What was the number of deaths per 100 000 members of the population in the year that drinking water was first treated in the USA?

**b** How long did it take for the number of deaths to fall to a half of the maximum after chlorination began?

**4** Describe how the likelihood of dying from typhoid in the USA changed between the start of chlorination and 1930.

## Benefits and risks

Some people have always disapproved of the addition of chlorine to water. Chlorine is a toxic gas – as well as killing microorganisms it can affect human health if too much is present in water. Chlorine can react with organic material (chemicals from plants or animals) in the water supply. A number of compounds, called disinfectant by-products (DBPs), are formed that are toxic or carcinogenic. When this information was publicised, many people became worried about their health, and in some places local leaders stopped chlorine being adding to water supplies.

People have no choice in their supply of water. Because the stories about chlorine were new and unfamiliar, people perceived the risk as being higher than more familiar causes of death such as car accidents. Governments took advice on the problem and decided that the risk from DBPs could be kept low if the amount of chlorine and other materials in water was controlled carefully. They thought that the benefits of the disinfecting properties of chlorine outweigh the dangers.

### Watch out!

Remember that the level of risk of a chemical depends on how much harm it can cause and how much of it there is present.

**FIGURE 3:** The risk of being injured in a car accident is much higher than the risk of ill health from chlorine in drinking water.

### Did you know?

Alternatives to using chlorine to disinfect drinking water include ozone and ultraviolet light – but chlorine remains the cheapest and most effective and is used by most water companies.

## QUESTIONS

**5** A sample of drinking water is found to contain 0.05 mg/litre of chloroform, a DBP. Guidelines say that there should be less than 0.2 mg/litre. What percentage of the allowed amount of chloroform is in the water?

**6** Why might some people think that the risk from drinking chlorinated water is greater than, for example, drinking alcohol?

**7** 'Chlorine is a dangerous substance and should not be added to water.' What arguments could you make for and against this statement?

chlorine DBP risk

# Preparing for assessment: Evaluating and analysing evidence

*To achieve a good grade in science, you not only have to know and understand scientific ideas, but you need to be able to apply them to other situations and to analyse evidence. These tasks will support you in developing these skills.*

## What's in the water?

### The need for clean water

Everybody needs water and it has to be clean. Without clean drinking water we are likely to suffer some very unpleasant diseases and possibly won't live very long. Water that is unclean is a breeding ground for micro-organisms. Drinking the water introduces these straight inside the body, and some of them can cause illness and death.

The WHO (World Health Organisation) states that:

> In the European region, an average of 330 000 cases of water-related disease are reported every year.

> Access to an improved water supply and sanitation has increased across Europe, resulting in an 80% decrease in diarrhoeal disease in young children from 1995 to 2005. Nevertheless, more than 50% of the rural population in Eastern European countries still lives in homes not connected to a safe drinking-water supply.

> Sanitary equipment is insufficient in some areas of Europe: about 85 million people still lack toilets in their homes.

> Worldwide, infectious diarrhoea accounts for 1.7 million deaths per year. Water, sanitation and health intervention can reduce diarrhoeal diseases by 15–30%, and significantly reduce other diseases.
(Taken from *Water and sanitation: Facts and figures*, World Health Organisation European Regional Office)

### The case for chlorine

One of the most common methods of treating water is to add chlorine. This kills most water-borne pathogens, including those that cause cholera and typhoid fever. Adding extra chlorine will also kill organisms that arrive in the water after it has left the processing plant.

Chlorine is applied to water either as gas or as a chlorinating chemical, such as bleach.

### The case against chlorine

However, there is concern that chlorine also reacts with organic matter, naturally present in water, to form chemicals called THMs (trihalomethanes such as chloroform). Experiments and studies have shown that exposure to high levels of THMs leads to an increased risk of cancer. A recent study showed an increased risk of bladder and possibly colon cancer in people who drank chlorinated water for 35 years or more.

High levels of THMs may also have an effect on pregnancy. One study found that pregnant women who drank large amounts of tap water with high levels of THMs had an increased risk of miscarriage.

These studies do not *prove* that there is a link between THMs and cancer or miscarriage. However, they do show the need for further research in this area to confirm potential health effects.

Not all harmful bacteria are effectively disinfected by chlorine. The bacterium *Cryptosporidium*, found in the faeces of infected humans, cattle and other mammals, is highly resistant to chlorine at the levels normally found in drinking water.

### What are the alternatives?

There are other ways of treating water:

> One of these is to use ozone, which is very effective and does not form THMs. However, the effect of the ozone is much shorter lived than that of chlorine, so it won't kill pathogens that arrive at a later point in the supply system. Small amounts of chlorine or other disinfectants still must be added. Adapting water treatment plants so they can use ozone can be expensive.

> Ultraviolet (UV) radiation, generated by special lamps, is a non-chemical method of disinfecting water.

C3 Evaluating and analysing evidence

## Task 1

> Why is it important to clean drinking water?
> How is this usually done?
> What has the impact of this been?

## Task 2

> What does the data show about the effect of polluted water on health?
> How does the data support the case for the chlorination of drinking water?

## Task 3

> Why is there concern about the chlorination of drinking water?
> What are the alternatives?

## Task 4

> Considering the evidence provided what do you think about the case for chlorination?
> What other evidence would help you make an informed decision?

## Maximise your grade

These sentences show what you need to include in your work to achieve each grade. Use them to improve your work and be more successful.

**E** For grade E, your answers should show that you can:
> understand how the introduction of chlorination to treat drinking water made a major contribution to public health

**C** For grades D, C, in addition show that you can:
> interpret data about the effects of polluted water on health and the impact of water treatment with chlorine to control disease

**A** For grades B, A, in addition show that you can:
> recall the possible health problems from traces of chemicals formed by reaction of chorine with organic materials in the water

# Electrolysis of brine

**We are learning to:**
> remember that useful products are produced the electrolysis of brine
> explain how electrolysis produces new mater
> examine the environmental consequences of the electrolysis of brine

## Do we still need an alkali industry?

It is nearly two hundred years since factories using the Leblanc process began to appear in the north-west of England. The chemical industry is still there, although the processes have changed and attitudes to pollution have altered. The products of the chemical industry are as much in demand now as they were then.

**FIGURE 1:** The chemical works at Runcorn in north-west England has its own power station to generate electricity for the processes.

## How is electricity used to make new chemicals?

**Electrolysis** is a process in which electricity is passed through a molten salt or a solution of a salt. It causes chemical changes to take place. Brine is a concentrated solution of sodium chloride (salt). When electricity is passed through brine, three products are formed:

> chlorine
> hydrogen
> sodium hydroxide.

All three products have important uses so there is no waste.

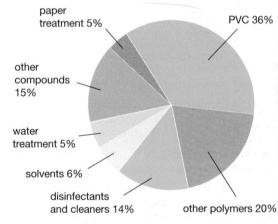

**FIGURE 2:** Uses of chlorine.

| Uses of sodium hydroxide | Uses of hydrogen |
|---|---|
| Soaps and cleaners | Fuel |
| Paper making | Margarine |
| Extracting aluminium | |
| Manufacturing other chemicals | |

Uses of sodium hydroxide and hydrogen.

## QUESTIONS

**1** Name the three products formed in the electrolysis of sodium chloride and give one important use of each.

**2** What percentage of chlorine production is used in making PVC?

C3 Chemicals in our lives – risks and benefits

## What happens in the electrolysis of brine?

The **membrane cell** shown in Figure 3 is one of the methods used to electrolyse brine in industry. Salt and water are the only raw materials.

A great deal of electrical energy is needed to make the chemical changes happen on a large scale and is the main running cost of the process.

Electricity enters the solution through two **electrodes**. Chlorine gas is given off and collected at the **anode**, the positive (+) electrode. Hydrogen gas is given off and collected at the **cathode**, the negative (−) electrode. Sodium hydroxide is left in the solution. Water is added to the membrane cell and the sodium hydroxide solution is piped off.

**Watch out!** In the electrolysis of solutions, the water is involved in the chemical changes that take place.

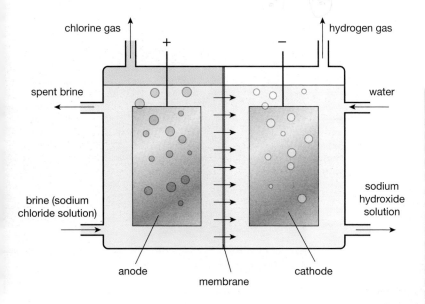

**FIGURE 3:** Hydrogen, chlorine and sodium hydroxide are the products of electrolysing sodium chloride solution in the membrane cell.

### QUESTIONS

**3** Write a word equation for the process that takes place in the membrane cell.

**4** Look at Figure 2. Europe produces 10 million tonnes of chlorine a year. How much is used to treat water?

**5** Electrolysis splits up sodium chloride solution, but where does the hydrogen that is formed come from?

## The environmental impact of brine electrolysis

The electrolysis of brine is one of the most widely used chemical processes and produces millions of tonnes of chlorine, sodium hydroxide and hydrogen each year. The industry has come a long way since three-quarters of the raw materials used in the Leblanc process ended up as highly polluting waste.

In the membrane cell, all the products have important uses – nevertheless, any loss of the products can harm the environment. Chlorine is toxic and corrosive, sodium hydroxide is a powerful alkali and hydrogen is flammable. Even a loss of salt could damage freshwater sources.

Until the membrane cell became the most commonly used process, the electrolysis of sodium chloride solution was often done using a liquid mercury cathode. In the 1960s, up to 200 g of mercury was lost for every tonne of chlorine produced. Europe was producing about 6 million tonnes of chlorine a year at that time. It was recognised that this loss of mercury to the environment caused a severe health risk to people living near the factories. As a result, steps were taken to limit the loss of mercury. By 2008 the loss of mercury was reduced to about 1 g per tonne of chlorine. By 2020 all mercury cathode processes will have been replaced by membrane cells.

### QUESTIONS

**6** Fossil fuels are used to generate most of our electricity. What is the environmental impact of the use of electricity in a brine electrolysis plant?

**7** What mass of mercury was lost to the environment in Europe in 1970 assuming that half the chlorine production was by mercury cathode electrolysis cells?

Pollution chlor-alkali industry

# Industrial chemicals

## Who assesses risks from chemicals?

A European Union law called the **R**egulation, **E**valuation, **A**uthorisation and restriction of **Ch**emical substances (REACH) makes businesses responsible for assessing the risks to health and to the environment of all the chemicals that they make and use.

**We are learning to:**
- discuss possible risks from products of the chemical industry
- explain that chemicals may remain in the environment for a long time
- consider how we make decisions about the risks from chemicals

### How do we know if chemicals are safe?

Chemicals are made of elements. Elements cannot be destroyed and so remain in the environment for ever. Most elements react to form compounds. Some compounds are more dangerous than others. We must carry out risk assessments to decide how dangerous substances are. Dangerous compounds must be controlled. There is a risk in using every chemical but we need to know if the risk is high or low. Even eating salt can harm health (see page 178) but a gram of sodium cyanide is far more dangerous than a gram of sodium chloride.

**FIGURE 1:** Mercury has been used for centuries but is poisonous. If it escapes into the environment it will remain as a hazard for ever.

> **QUESTIONS**
>
> 1 Why is chemical waste that was released from factories 200 years ago still a problem today?
>
> 2 Explain which would be the higher risk in your kitchen – a jar of sodium cyanide or a salt shaker.

### Chemicals and risk

Chemicals spread if they are in the environment – they are carried by wind and water whether they are gases, liquids or solids. Sulfur dioxide from British power stations was carried by winds to Norway and fell there as acid rain.

Chemicals may be absorbed by plants and animals and passed up the food chain. Animals may not be able to excrete some chemicals, so they accumulate until they reach dangerous concentrations. Mercury released by alkali factories accumulated in fish and has been found in the bodies of Inuit people in the Arctic.

**FIGURE 2:** Hazard warnings on a bottle of drain cleaner – how do we decide about the risk of using it?

CORROSIVE | OXIDISING | HARMFUL | HIGHLY FLAMMABLE

**BUSTER KITCHEN DRAIN CLEAR CONTAINS SODIUM NITRATE AND SODIUM HYDROXIDE.** Harmful if swallowed. Flammable. Contact with combustible material may cause fire. Contact with water liberates extremely flammable gases. Causes severe burns. Keep locked up and out of the reach of children. Avoid contact with skin and eyes. Wear suitable protective clothing gloves, eye/face protection. In case of contact with skin or eyes, rinse immediately with plenty of water and seek medical advice. In case of accident, or if you feel unwell, seek medical advice immediately (show this label where possible).

REACH chemical safety

# C3 Chemicals in our lives – risks and benefits

To decide the level of risk of a particular chemical we need to know:
> how much of it is needed to cause harm
> how much will be used
> the chance of it escaping into the environment
> who or what it may affect.

Having assessed the risk, we can decide if using the substance will have benefits that outweigh the risks, or if it can be replaced by another substance with a lower risk.

### Did you know?

Chlorofluorocarbons were used for years in aerosols before it was discovered that they destroy the ozone layer. CFCs were banned in the 1980s but will remain in the air for another hundred years. Propane is often used in aerosols today – it is flammable.

## QUESTIONS

**3** Explain how mercury got into the bodies of the Inuit people.

**4** Many batteries used today contain lithium. What do you need to know to decide whether or not lithium is a high risk to health and the environment?

## Testing, testing

Thirty years ago European laws made the risk assessment of every new chemical compulsory, but there are thousands of substances that have been used, sometimes for centuries, for which we have very little data about their hazards. REACH will eventually include all chemicals that are used today, but testing, involving the use of millions of laboratory animals, will take many years. Many familiar substances are hazardous but people may be more suspicious of substances whose names are new. The risks with new chemicals may be thought to be higher than traditional substances. There will be more test data available for the new chemicals so it will be easier to decide on their hazard levels.

### Watch out!
Remember that *everything* is made from chemicals and that all chemicals can cause harm if used incorrectly or in excessive amounts. Safety data enables a sensible estimate of risk.

## QUESTIONS

**5** Why do chemicals that have been in use for a long time need to be tested for safety?

**6** 'Traditional soaps are safer than modern formulations.' Do you agree or disagree with this statement? Explain your answer

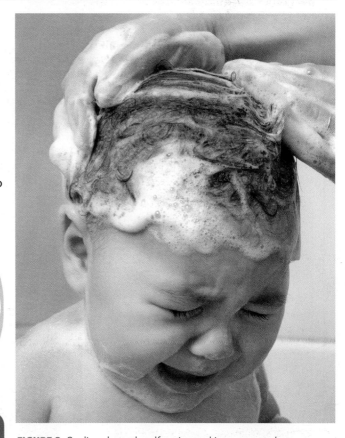

**FIGURE 3:** Sodium laureth sulfate is used in many modern shampoos and can cause irritation to the skin – but then so can traditional soap made by reacting sodium hydroxide with fats.

Chemical risk assessment

# PVC

## Should we worry about PVC?

PVC or 'vinyl' is one of the most used materials in modern life. We probably get close to something made of PVC every day. Although it has properties that make it useful, there is also evidence that it is a hazard to health and the environment – but making a decision is not easy.

> **We are learning to:**
> - explain that PVC is a polymer of carbon, hydrogen and chlorine
> - explain why plasticisers in some types of PVC may be harmful

FIGURE 1: Plasticised PVC is no longer used for toys that babies may suck.

## Useful PVC

**PVC** is a polymer containing the elements carbon, hydrogen and chlorine. It is made from chlorine gas and a hydrocarbon obtained from crude oil. PVC is a plastic – many plastics soften when heated and can be moulded into shape. PVC is a stiff, tough material that has many uses in buildings. It is hardwearing and will last for many years.

Small molecules called **plasticisers** are added to PVC to make it softer and more flexible. Plasticised PVC is used to insulate wires and cables, and as sheeting. It is also used instead of leather in clothes and seat coverings.

> **QUESTIONS**
>
> **1** Give one use for plasticised PVC and one use for PVC without plasticiser.
>
> **2** Why is plasticised PVC used in some chair coverings?

FIGURE 2: PVC is used for window frames, pipes and gutters, roof tiles and fencing. It is also used for water pipes.

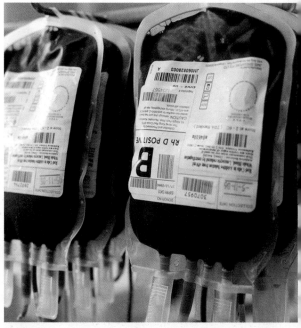

FIGURE 3: Plasticised PVC sheeting is used for bags for blood transfusions, shower curtains and floor coverings.

PVC uses properties

C3 Chemicals in our lives – risks and benefits

## Safety of plasticisers

Plasticised PVC can behave a bit like a sponge with soap in it. When the sponge is wetted the soap washes out of the sponge – this is called **leaching**. The small plasticiser molecules trapped between the large PVC molecules are not held very tightly. When the PVC is squeezed or stretched or wetted the plasticiser molecules may leach out.

The chemicals used as plasticisers have been tested for safety. Some have been shown to harm the health of test animals, such as rats, when fed to the animals in large amounts. It is also thought that some of the plasticisers can harm fish when they are leached into rivers and the sea. For these reasons, the use of plasticised PVC in children's toys has been banned in Europe and the USA.

**Watch out!**
Safety tests show how much of a chemical can kill a test animal. It is more difficult to prove or disprove other effects on human health.

### QUESTIONS

3 Explain why a baby's toy made of plasticised PVC could be harmful.
4 What evidence is there that plasticisers are harmful?

## PVC – high or low risk?

PVC has been used for over fifty years for many different purposes. At least two-thirds of the PVC produced is used by the building industry. There are environmental concerns about the manufacture of PVC. The process uses a lot of chorine and produces toxic by-products. If allowed to escape, these chemicals would be harmful pollutants. PVC is flammable and gives off highly toxic gases when it burns. It is a known cause of death in fires in buildings.

There is a lot of dispute about the risks from the plasticisers that are mixed with about one-third of all the PVC in use. Some people say that the amount that leaches out is small and cannot harm humans. Others disagree and say that plasticised PVC is too dangerous and should be banned completely. Other materials are available that could replace PVC for most, if not all, its uses.

**FIGURE 4**: PVC table coverings. Should PVC be banned for all or some uses?

### QUESTIONS

5 Suggest three ways that the manufacture and use of PVC presents a hazard to health and the environment.
6 Outline the risks and benefits of PVC in the modern world.

### Did you know?

Plasticised PVC clingfilm is 'clingier' than other types and uses plasticisers that are said to be low risk. But this type of clingfilm should not be used in a microwave oven because there are still fears that the plasticisers can be harmful if they leach into the food as it is cooked.

plasticisers PVC health concerns

# Life Cycle Assessment

> **We are learning to:**
> - explain what is meant by a Life Cycle Assessment and how it is done
> - use Life Cycle Assessments to compare the impact of products made from different materials

## What is a Life Cycle Assessment?

When you buy a soft drink think of what the bottle is made out of, where it has come from and how much energy was used to make it. Then when you have finished and dropped the bottle in the waste bin, consider what happens to it next. That's a simple Life Cycle Assessment.

FIGURE 1: Thinking about where it has come from and where it is going is part of a Life Cycle Assessment.

## Assessing the impact

We use many different materials for many purposes. All materials have to be made, turned into a product, which is used and then disposed of. Each of these stages uses the Earth's resources and causes pollution that may harm us and other organisms. A **Life Cycle Assessment (LCA)** helps to add up all the effects that a product has on the environment. It helps us to decide whether the product benefits us and the planet, or if there is a high risk of harm to health and damage to plants and animals.

FIGURE 2: The raw material for most polymers is crude oil. Drilling for oil can harm the environment.

### QUESTIONS

1. What is the source of the material used to make the polymer in a plastic cup?
2. Make a list of things you have used and thrown away in the last week.

## Preparing a Life Cycle Assessment

To produce an LCA for any product, its life must be broken down into stages – as shown in Figure 3:

We need to consider the following questions at each stage:

> How much natural resources are required?

> How much energy is needed or produced?

> How much water and air is used?

> How is the environment affected?

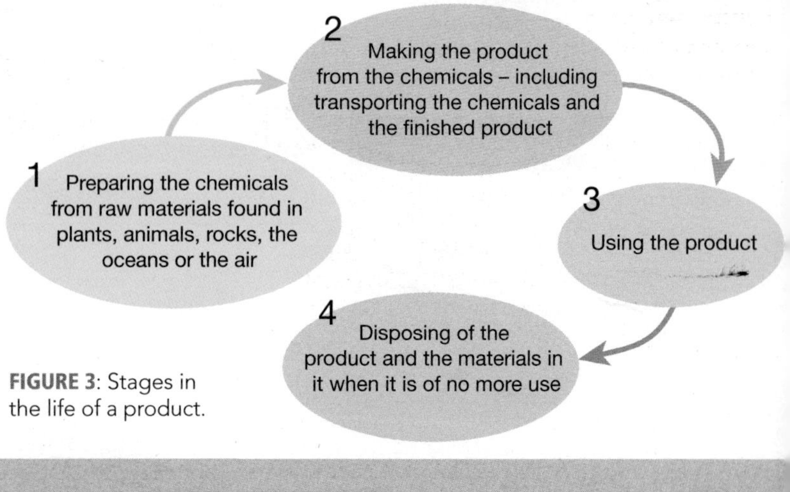

1. Preparing the chemicals from raw materials found in plants, animals, rocks, the oceans or the air
2. Making the product from the chemicals – including transporting the chemicals and the finished product
3. Using the product
4. Disposing of the product and the materials in it when it is of no more use

FIGURE 3: Stages in the life of a product.

Life Cycle Assessment

When all the answers to each stage have been combined, we have an overall picture of the costs and impact of the product on the Earth. Then alternative materials can be compared to see if they would have a lesser impact.

## QUESTIONS

**3** Look at the following list of materials. State if the raw material for each is obtained from rocks, plants or animals:

**a** cotton

**b** leather

**c** nylon

**d** paper

**e** steel.

**4** What do you think energy is needed for when a plastic bottle is manufactured from a polymer that softens when it is heated?

**FIGURE 4**: Waste plastic often ends up in the sea where it can harm fish.

### Did you know?

LCAs sometimes produce surprising conclusions – a paper cup may not be better for the environment than a plastic one; using a dishwasher may be better than washing dishes by hand.

## Totting up the figures

To produce a fair and accurate LCA requires data about quantities of materials used and pollutants produced, energy inputs and outputs and information about how substances affect plants and animals in the environment. If actual figures cannot be provided then scores can be awarded for each stage, giving an overall LCA rating for a product. When the LCAs of different materials are compared it is possible to make choices that reduce the impact of a product on the environment, reduce the quantity of natural resources and energy used, and hence make the product more sustainable.

**Watch out!**

Don't forget the energy needed to carry materials from one factory to another and to shops and the waste disposal site; and the water and air needed to dilute pollutants to concentrations allowed by law.

## QUESTIONS

**5** How do you think the energy requirements compare in the tasks given below:

**a** (i) Heating a polymer to 150 °C to soften it, (ii) melting aluminium at 600 °C.

**b** (i) Lifting and carrying a crate of 24 plastic bottles, (ii) lifting and carrying a crate of 24 glass bottles.

**6** Compare the environmental impact of:

**a** manufacturing drinks containers from a polymer or aluminium

**b** transporting drinks in plastic bottles or glass bottles.

energy costs plastic cup

# C3 Checklist

## To achieve your forecast grade in the exam you'll need to revise

Use this checklist to see what you can do *now*. Refer back to pages 172–197 if you're not sure.

Look across the rows to see how you could progress – **bold italic** means Higher tier only.

Remember you'll need to be able to *use* these ideas in various ways, such as:
> interpreting pictures, diagrams and graphs
> applying ideas to new situations
> explaining ethical implications
> suggesting some benefits and risks to society
> drawing conclusions from evidence you've been given.

Look at pages 312–318 for more information about exams and how you'll be assessed.

| To aim for a grade E | To aim for a grade C | To aim for a grade A |
|---|---|---|
| understand that movements of tectonic plates mean that the parts of continents that now make up Britain have moved over the surface of the Earth; and that different rocks formed in different climates | understand that geologists can explain past history of the surface of the Earth in terms of processes that can be observed today | understand that geologists use magnetic clues in rocks to track the slow movement of the continents over the surface of the Earth |
| understand that a chemical industry grew up in north-west England because resources such as salt, limestone and coal were available locally | understand how rock cycle processes have led to the formation of valuable resources in England; understand how geologists study sedimentary rocks to find evidence of the conditions under which they were formed | |
| understand the importance of salt (sodium chloride), for the food industry, as a source of chemicals and to treat icy roads; recall that salt can be obtained from the sea and also from underground salt deposits by mining or by solution | understand that the method of extraction may depend on how the salt is to be used; understand that all methods of obtaining salt can have an impact on the environment | |
| understand the advantages of adding salt to food as flavouring and as a preservative; recall the health implications of eating too much salt | evaluate data related to the content of salt in food and health; recall that Government departments have a role in carrying out risk assessments and advising the public about chemicals in food | |

| To aim for a grade E | To aim for a grade C | To aim for a grade A |
|---|---|---|
| understand that alkalis neutralise acids to make salts; recall that alkali used to be obtained from burnt wood or urine | recall that alkalis are used to neutralise acid soils, make chemicals that bind dyes to cloth, convert fats into soap, and make glass | *recall that soluble hydroxides and carbonates are alkalis, and recall the patterns in the reactions of these with acids* |
| understand that increased industrialisation led to a shortage of alkali in the 19th century; recall that the first process for manufacturing alkali used salt, coal and limestone | understand that the early alkali manufacturing process released an acid gas (hydrogen chloride) and created waste that released a toxic gas; understand that waste can sometimes be turned into useful chemicals; understand that oxidation can convert hydrogen chloride to chlorine, and that the properties of a compound are completely different from the elements from which it is made | |
| recall the uses of chlorine; understand how the chlorination of drinking water made a major contribution to public health | interpret data about the effects of polluted water on health and the impact of water chlorination in controlling disease | understand the possible health problems from traces of chemicals formed by the reaction of chlorine with organic materials in water |
| understand that an electric current can bring about chemical change (electrolysis); recall uses of sodium hydroxide, chlorine and hydrogen produced by electrolysis of brine | recall that chlorine is now obtained by the electrolysis of salt solution (brine); interpret data about the environmental impact of the large-scale electrolysis of brine | |
| understand that some toxic chemicals persist in the environment, can be carried over large distances and may accumulate in food and human tissues | understand that there is sometimes inadequate data to assess whether chemicals present a risk to the environment and/or human health | |
| understand that PVC is a polymer that contains chlorine as well as carbon and hydrogen | understand that the plasticisers used to modify the properties of PVC can leach out from the plastic into the surroundings where they may have harmful effects | |
| understand that a Life Cycle Assessment (LCA) is a consideration of the use of resources, energy input/output and environmental impact of each stage in the production, use and disposal of a product; compare and evaluate the use of different materials for the same purpose, given information from an LCA | | |

# Exam-style questions

## Foundation level

**1** The Haiti earthquake in 2010 destroyed most of the country's water treatment works. It was feared that there would be an outbreak of typhoid, but relief agencies were able to supply water treated with chlorine.

AO1 **a** Why is chlorine added to water supplies? [1]

AO2 **b** The relief agencies had to carry chlorine to Haiti to treat the water. Which of the following statements is/are true? [2]
  A Chlorine gas is toxic and corrosive
  B There is no risk if chlorine is carried in special containers
  C The benefit to the people of Haiti was more than the risk of harm caused by chlorine
  D Chlorine is a hazard, so the risk of a leak is high

AO3 **c** About 200 people are normally infected with typhoid in Haiti each year. What might have happened to the number of cases if the relief agencies had not supplied chlorinated water? [1]

AO1 **d** Look at the diagram showing the manufacture of chlorine gas from brine.

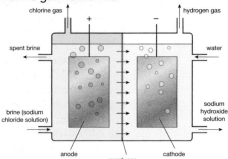

(i) The name of this process is:

distillation    electrolysis
polymerisation    condensation    [1]

(i) What are the other *two* products of the process? [2]
[Total 7]

## Foundation/Higher level

**2** The salt deposits in Cheshire, north-west England, have been used by the chemical industry for over 200 years. The salt is found mixed with sand grains within layers of rock formed more than 200 million years ago. The salt formed when shallow seas dried out and mixed with sand blown from deserts. Above the salt are layers of sand and gravel laid down when Cheshire was covered by glaciers during ice ages.

AO1 **a** What evidence is there that Cheshire has moved from a different part of the Earth? Choose the best answer from the following. [1]
  A Salt deposits have been covered by other rocks
  B The salt is mixed with sand
  C The salt was formed where the climate was hot and dry
  D The salt was formed 200 million years ago

AO2 **b** Describe how the layers of sand and gravel were formed that cover the salt deposits. [3]

AO1 **c** Describe the raw materials used to make alkali in the 19th century, the environmental problems that the process caused and how they were solved. The quality of written communication will be assessed in the answer to this question. [6]
[Total 10]

**3** Shower curtains are usually made from sheets of plasticised PVC ('vinyl'). Woven polyester is also used but this needs a waterproof PVC backing sheet. Polyethene is an alternative to PVC that does not need plasticisers but is less hardwearing.

AO2 **a** Which of the following properties make plasticised PVC suitable for shower curtains? [1]

   flexible    good insulator    waterproof    strong

AO1 **b** What effect does a plasticiser have on the properties of the PVC? [1]

AO3 **c** Many environmental groups are against the use of PVC because chlorine is needed in its manufacture. Chlorine is not needed to manufacture polyester or polyethene. Is polyester a good substitute for PVC for shower curtains? Explain your answer. [1]

**d** PVC and polyethene are polymers derived from crude oil. Compare the environmental impact of PVC and polyethene at these stages of their life cycle:
AO1 (i) Manufacture of the polymers. [2]
AO2 (ii) Manufacture of shower curtains. [2]
AO2 (iii) Use of shower curtains. [2]
AO1 (iv) Disposal of shower curtains. [2]
[Total 11]

## Higher level

**4** Calcium hydroxide is spread on acid soils.

AO1 **a** Why is calcium hydroxide used? [1]

AO2 **b** How can farmers test if they have added sufficient calcium hydroxide? [2]

AO2 **c** Nitric acid is one acid that may be present in soils. Write a word equation for the reaction of calcium hydroxide with nitric acid. [2]
[Total 5]

AO1 recall the science    AO2 apply your knowledge    AO3 evaluate and analyse the evidence

# C3 Chemicals in our lives – risks and benefits

## Worked examples

**AO1 1** An accident at a bauxite mine in Hungary in 2010 released a flood of alkaline waste containing poisonous materials that threatened to flow into the River Danube. Environmental scientists monitored the river to check that its pH remained at about 8.

**a** What did the scientists expect to happen to the pH of the river water if the waste had reached it? Choose the correct answer.
   A The pH would increase
   B The pH would stay the same
   C The pH would fall

   A ✔

> Correct – 1 mark awarded.

**b** Aluminium is extracted from bauxite after it has been treated with sodium hydroxide.

   (i) What raw materials are used to manufacture sodium hydroxide? [2]

   Salt ✔ electricity ✘

> Salt is correct – 1 mark; electricity is not a raw material so cannot be awarded a mark, the correct answer is water.

   (ii) All the sodium hydroxide used in treating the bauxite ends up in the waste reservoir. How could the treatment of bauxite be made more sustainable? [2]

   Recycle the sodium hydroxide in the waste. ✔

> Correct, re-using any resource makes it more sustainable – 1 mark.

**AO1 2** Sodium hydroxide is used in the manufacture of soap from vegetable oils. Describe the environmental impact of a bar of soap during its lifetime. The quality of written communication will be assessed in your answer. [6]

**AO2**

*Life cycle assessment*
- Preparation of raw materials:
  If salt is obtained by mining underground deposits it can cause subsidence. Electrolysis of sodium chloride also produces chlorine and hydrogen. A lot of energy is needed which, if derived from burning fossil fuels, releases pollutants. Chlorine leaks can damage the environment.
- Manufacture of soap:
  Vegetable oil and sodium hydroxide are heated. Soap is separated. Alkaline waste can damage the environment.
- Disposal.
  After use the soapy water passes into the sewage system. Soap made from vegetable oils is biodegradable, and is broken down in the sewage treatment process.

### How to raise your grade
Take note of the comments from examiners – these will help you to improve your grade.

> The answer covers three of the four stages of a Life Cycle Assessment and makes valid points about the environmental impact at each stage, but has left out stage 3 – transport and use of the soap. Packaging and transport uses resources and can release pollutants. Using soap requires water which has usually been treated with chlorine which can react with impurities to form toxic pollutants.
>
> Appropriate scientific language is used, and grammar and spelling are accurate. Information is relevant and is suitably split up into headed paragraphs.
>
> This answer meets the criteria for 5–6 marks (see the example banded mark scheme on page 317). The omission of stage 3 of the life cycle means it will get 5 marks.

# P1 The Earth in the Universe

## What you should already know...

### The Earth is one small planet in a vast Universe

The Earth is one of eight planets and many other objects orbiting the Sun.

These make up our solar system.

Objects remain in orbit because of the force of gravity.

The Sun is one of many billions of stars in the Universe.

  Which object in the solar system orbits the Earth?

### Artificial satellites are in orbit around the Earth

Satellites are used to observe the Earth and to transmit communications.

Orbiting space telescopes obtain information from the furthest parts of the Universe.

Unmanned spacecraft are used to explore our solar system.

  What sort of information can satellites give us?

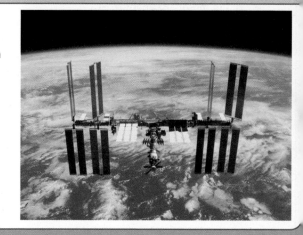

### The Earth is still changing

The Earth's rocks formed over a very long time. Different types of rocks form in different ways.

Igneous rocks are formed when magma (hot molten rock) cools.

Sedimentary rocks are formed from deposited fragments of eroded rock. Fossils can be found in sedimentary rock.

Rocks continually change as a result of weathering, erosion, transportation and sedimentation.

  How are metamorphic rocks formed?

# In P1 you will find out about...

> the different objects in our solar system

> the grouping of stars in vast galaxies

> how we observe stars and determine their distance from us

> the processes in stars that generate heat, light and new elements

> how the Universe formed and how it is still changing

> the processes of change in different rocks on Earth

> evidence that the continents have moved

> the causes of earthquakes, volcanoes and mountain formation

> the detection of earthquakes by the waves they produce in the Earth

> what these waves tell us about the structure of the Earth and its crust

> the transfer of energy by waves

> the behaviour of waves

> what we mean by wave speed and wave frequency

# Our solar system

**We are learning to:**
> describe the structure of the solar system
> develop a sense of scale

## Could we explore planets around another star?

The distances in space are vast. It would take a light signal 10.5 years to travel to the nearest star that has planets. Light travels 30 000 times faster than our fastest rockets, so humans could not reach another planetary system in their lifetime.

### The solar system

The **solar system** has the star called the Sun at its centre. It includes all the objects that travel around the Sun in paths called **orbits**. These objects are different sizes and have different paths:

> **Planets** – eight spherical planets travel around the Sun in near-circular orbits. The Earth is one of them.

> **Dwarf planets** – the five known dwarf planets are much smaller than planets. They also travel around the Sun in near-circular orbits.

> **Asteroids** – these are lumpy rocks in near-circular orbits.

> **Comets** – these objects have very elongated orbits that stretch far away from the Sun and may also approach very near to it.

> **Moons** – these are balls of rock that orbit a planet in near-circular orbits.

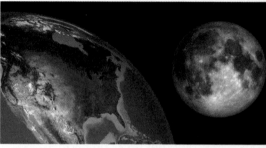

**FIGURE 1**: Earth has one moon, which we call the Moon.

## QUESTIONS

**1** What is the difference between a dwarf planet and a planet?

**2** What is the difference between a moon and a planet?

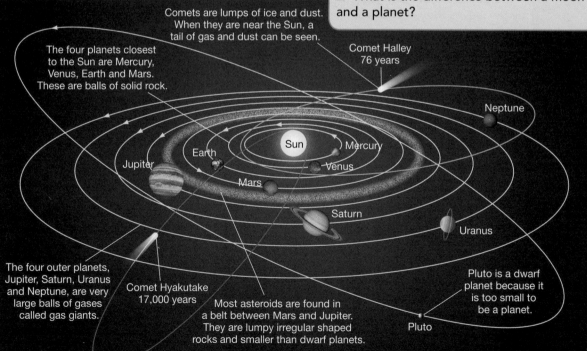

**FIGURE 2**: The solar system.

solar system

# P1 The Earth in the Universe

## Comparing objects in the solar system

The Sun contains over 99% of the solar system's mass. In order of size, the next largest objects are the planets (gas giants are bigger than the rocky planets), then dwarf planets, comets, asteroids and moons.

Distances in the solar system are so large it is hard to compare distances in kilometres. The table shows how the distances of the other planets from the Sun compare with the distance from Earth to the Sun. The Earth is 150 million kilometres from the Sun. The table also shows how their masses compare with the Earth's mass.

| Planet | Distance from Sun compared to Earth's distance | Mass compared to Earth's mass |
|---|---|---|
| Mercury | 0.4 | 0.06 |
| Venus | 0.7 | 0.8 |
| Earth | 1.0 | 1.0 |
| Mars | 1.5 | 0.1 |
| Jupiter | 5.2 | 317.8 |
| Saturn | 9.5 | 95.2 |
| Uranus | 19.2 | 14.5 |
| Neptune | 30.1 | 17.1 |

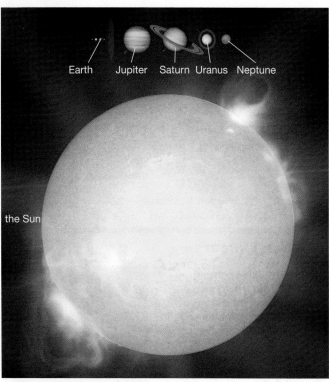

FIGURE 3: This shows how much larger the Sun is than anything else in the solar system.

### Did you know?

The mass of Earth is 6 million billion billion kilograms.

**Watch out!** A billion is a thousand million. In numbers this is 1 000 000 000.

### QUESTIONS

**3** Draw a bar chart that shows the distance of each planet from the Sun compared with the Earth's distance.

**4** Write down one disadvantage of using kilometres to measure distances in the solar system.

## Distances in space

Space stretches far, far beyond the planets. Distances are so large, they are measured in **light-years**. One light-year is 9.5 million million kilometres, the distance light travels in one year.

Our Sun is one of thousands of millions of stars in the Milky Way galaxy, and there are thousands of millions of galaxies in the Universe. We don't know how big the Universe is, but the furthest objects we can see are nearly 14 thousand million light-years away.

### QUESTIONS

**5** It took about 6 months for an unmanned space probe to reach Mars. Suggest some reasons why astronauts have not yet explored Mars or other planets.

**6** Explain why we have been able to find out more about Mars than Neptune.

solar system

# Observing stars

**We are learning to:**
> understand what makes up the Universe
> understand how information reaches us through radiation
> understand the differences between data, hypotheses or explanations, and theories

## How do we know the Sun is still shining?

If the Sun suddenly stopped shining one day, you wouldn't know for about 8 minutes. It takes that long for light from the Sun to reach Earth.

FIGURE 1

## The Universe

The **Universe** contains everything that exists. It contains thousands of millions of **galaxies** separated by enormous distances. Each galaxy contains thousands of millions of stars. The Sun, which is our nearest star, is just one of the thousands of millions of stars in our galaxy, the **Milky Way**.

 **QUESTIONS**

1 What does the Universe contain?
2 What does a galaxy contain?

**FIGURE 2**: Our galaxy is shaped like a spiral. The Sun is in one of its arms.

## Finding out about stars

Proxima Centauri is the star nearest to our Sun, nearly 40 million million kilometres away. In space, which is a vacuum, light travels at 300 000 kilometres every second. It still takes 4 years for light from Proxima Centauri to reach us. Since distances are so large in the Universe, astronomers measure distance using light-years. A **light-year** is the distance light travels in one year – 9.5 million million kilometres.

P1 The Earth in the Universe

**Watch out!** A light-year measures distance, not time.

It is too far for people to travel to the stars. The only way we can find out about them is from the light and other forms of **radiation** they send out. Distant stars are very faint, or invisible, to our eyes. Scientists use telescopes to detect different forms of radiation from the stars, including light, radio waves and X-rays. This gives scientists **data** about stars that they can use to write a **hypothesis**, which is a sensible guess or explanation for something they observe.

For example, a hypothesis may explain why a star's brightness varies with time. It cannot be proved right, but it can be proved wrong. The same data could have more than one possible explanation. A hypothesis becomes a **theory** if scientists test it over a period of time and their data keeps matching the hypothesis. Now we have several accepted theories explaining why a star's brightness varies.

**FIGURE 3**: Telescopes detect the radiation emitted as new stars form in the Milky Way.

### QUESTIONS

**3** Our galaxy is 100 000 light-years across. How long does it take light to travel from one side of the galaxy to the other? (You don't need to work anything out.)

**4** What is the difference between a hypothesis and a theory?

## Looking back in time

Proxima Centauri and Sirius are stars. You can see Proxima Centauri in the sky tonight as it was 4 years ago because it is 4 light-years away. You see Sirius as it was 8.5 years ago.

Distant stars are hard to detect. Telescopes that orbit Earth can take very detailed images of very faint, distant objects. This light has taken many millions of years to reach us. Telescopes looking into distant parts of the Universe see what it was like many millions of years ago. The Hubble space telescope has viewed galaxies formed billions of years ago – very soon after we think the Universe began.

### QUESTIONS

**5** Explain why we cannot see stars as they exist now.

**6** We can tell what the Universe was like in the past by looking at distant galaxies. Explain why.

### Did you know?

Some galaxies are so massive that they absorb nearby galaxies. This is called galactic cannibalism. Our nearest galaxy, Andromeda, is 'eating' one of its neighbours and may absorb the Milky Way in about 3 billion years.

NASA    Hubble space telescope

# Preparing for assessment: Evaluating and analysing evidence

*To achieve a good grade in science, you not only have to know and understand scientific ideas, but you need to be able to apply them to other situations and to analyse evidence. These tasks will support you in developing these skills.*

## Finding Neptune

The year was 1841. The solar system was known to contain seven planets. Cambridge undergraduate John Couch Adams, brilliant mathematician, noticed that the seventh planet, Uranus, was behaving rather oddly. It seemed to speed up for a few years and then slow down. He wondered what could be causing this movement and considered that it might be another planet, further out and as yet unseen.

In 1845, now a member of staff at the university, Adams calculated where he thought the new planet might be and gave his figures to the university observatory. It wasn't followed up, however, and the following year a French astronomer, Urbain Le Verrier, published his own calculations which were used by the Berlin Observatory to find the new planet, Neptune.

The diagram shows the position of Uranus between the years 1800 and 1850. The red arrows show the direction that the planet seemed to be pulled in by the unseen new planet.

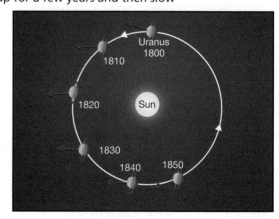

P1 Evaluating and analysing evidence

## Task 1

> The planet Jupiter has been known about for thousands of years. Why was it less than 200 years ago that Neptune was discovered?

> What would you expect conditions to be like on Neptune, in terms of temperature and length of year?

## Task 2

> Sketch a diagram showing the orbit of Uranus from 1800 to 1850. Add in the orbit and the likely positions of the new planet, bearing in mind:

> it didn't noticeably affect Uranus until 1800

> it accelerated Uranus until 1822

> it decelerated Uranus after 1822

> it was thought to be in orbit around the Sun, further out than Uranus.

> Explain how you've predicted the orbit of Neptune.

## Task 3

> Why did the observatories need the calculations of people like Adams and Le Verrier for the discovery of new planets?

> At the time of the discovery, the French newspapers claimed it as a victory for French science. Was this reasonable?

## Task 4

> Why did the observers not try to locate the position of the new planet by studying the orbit of Jupiter, being bigger and easier to see?

> Why is it only very recently that planets in other solar systems have been discovered?

## Maximise your grade

These sentences show what you need to include in your work to achieve each grade. Use them to improve your work and be more successful.

**E**
For grade E, your answers should show that you can:
> describe the main differences between different bodies in the solar system
> understand that light pollution and atmospheric conditions interfere with observations
> identify patterns in data

**C**
For grades D, C, in addition show that you can:
> explain the significance of the differences between bodies in the solar system
> explain why light pollution is a problem when observing the night sky
> explain how predictions can be made from patterns in data

**A**
For grades B, A, in addition show that you can:
> combine ideas about the scale of the solar system and the sizes of various bodies in it to construct explanations
> explain how the availability of evidence may be limited by technological considerations
> understand the limitations of predictions that can be made from patterns in data

# Distances to stars

## What links cancer diagnosis and distant stars?

Astronomers developed computer software to pick out very faint, distant stars from images sent in by telescopes. The same software is being used to help doctors screen tissue samples when a patient has cancer. This helps doctors decide on the best treatment.

**We are learning to:**
- arrange objects in the Universe in order of size
- understand how to measure distances to stars
- understand the causes of uncertainties in these measurements

## Sizes in the Universe

Sizes and distances in the Universe are so enormous, we measure them in light-years.

**Watch out!** The measurements are huge, but you don't need to remember exact numbers. You do need to be able to list objects in order of size.

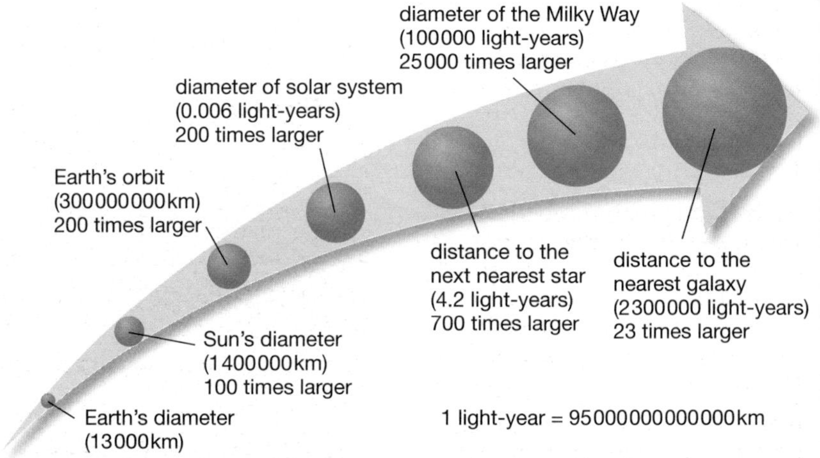

- Earth's diameter (13 000 km)
- Sun's diameter (1 400 000 km) 100 times larger
- Earth's orbit (300 000 000 km) 200 times larger
- diameter of solar system (0.006 light-years) 200 times larger
- distance to the next nearest star (4.2 light-years) 700 times larger
- diameter of the Milky Way (100 000 light-years) 25 000 times larger
- distance to the nearest galaxy (2 300 000 light-years) 23 times larger

1 light-year = 95 000 000 000 000 km

**FIGURE 1**: Comparing sizes and distances in the Universe.

### QUESTIONS

1. Why do scientists use light-years to measure sizes in the Universe?
2. If you drew a scale picture that showed the Earth 1 cm from the Sun, what size would you draw these?
   a. the diameter of the solar system
   b. the distance to the next nearest star

## Using star brightness to measure distance

When we look at stars, some seem brighter than others. This may be because these stars really are brighter, or because they are closer than the other stars. Closer stars seem brighter than stars that are further away.

To measure the distance to a star, scientists compare its brightness with another similar star a known distance away. They need to know:

- how bright each star really is – this is the **real brightness**
- how bright they appear to be – this is called **relative brightness**
- how far away one of the stars is.

These measurements are used to compare the distance of both stars. The unknown distance can then be calculated.

**FIGURE 2**: How can you tell which stars are closer to us?

parallax GCSE    measuring star distance

This method can only give an estimate of the distance because it relies on assuming that similar types of star have the same real brightness. It is also based on the distance to one of the stars, which we do not know precisely.

It has always been difficult to make accurate observations of stars. The night sky can be obscured by clouds, mist, fog or rain. Light from cities and towns, called **light pollution**, makes it hard to see the stars. By putting telescopes in space, scientists can now make much more accurate measurements than is possible from Earth.

## QUESTIONS

**3** The Earth's atmosphere makes it harder to study stars. Why do space-based telescopes help us make more accurate measurements?

**4** Why do scientists use a star's brightness to measure how far away it is?

## Using parallax to measure distance

Hold your finger out at arm's length. Move your head from side to side, focusing your eyes on your finger. As you look at your finger from a different angle, objects in the background seem to move. This effect is called **parallax**.

As the Earth moves around the Sun, we look at nearby stars from a different angle. The nearby star seems to stay in the same place. The stars behind it seem to change their position. This parallax is used to measure distances to nearby stars.

If an object is nearby, the parallax is greater. If it is further away, the parallax is smaller.

There are several reasons why it is hard to be sure that parallax measurements are correct.

> It is hard to get enough light from very dim stars and galaxies. Space telescopes like Hubble have helped with measurements like these.

> At long distances, parallax measurements are too tiny to measure accurately.

> We are not making direct measurements. Some information is obtained by comparing readings with readings taken for other stars.

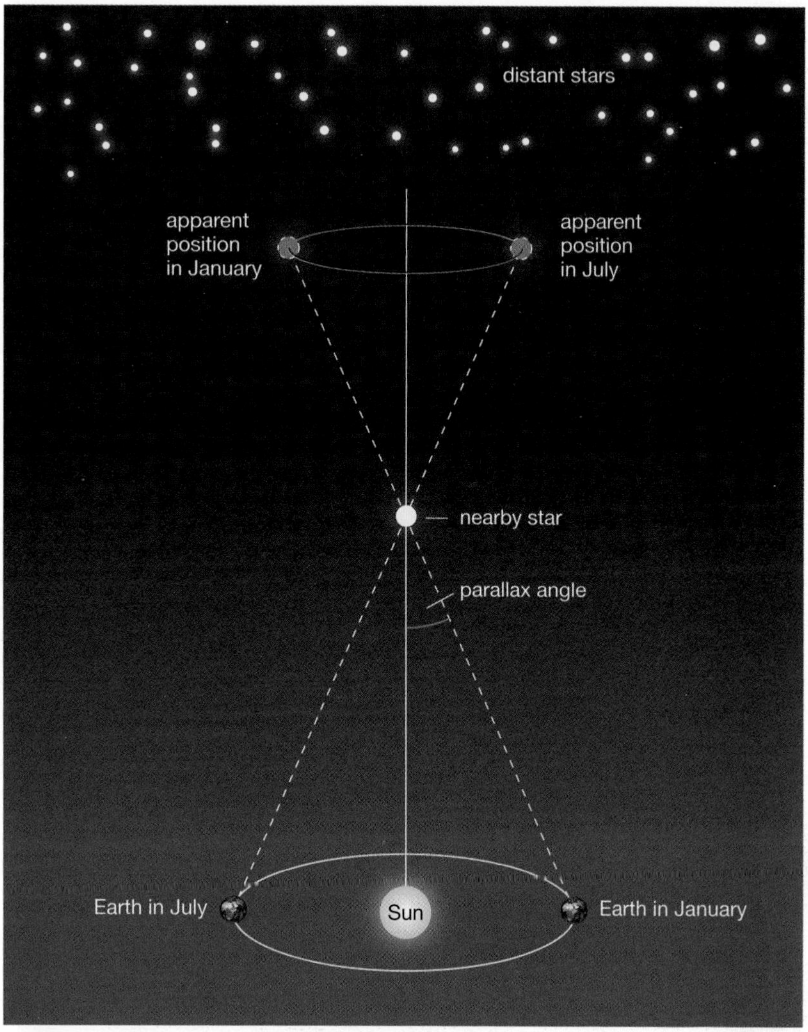

**FIGURE 3**: Parallax measurement on a nearby star.

### Did you know?

Stars are classified by colour. Certain colours of star are brighter than others and this helps scientists measure their real brightness.

## QUESTIONS

**5** Explain what is meant by parallax.

**6** Why is parallax no use for measuring very long distances?

# Fusion in stars

## Does our Sun make gold?

Centuries ago, alchemists tried to make gold and silver from metals like lead. They carried out many chemical experiments but never succeeded in making precious metals. We now know gold and other elements are made when massive stars explode, scattering dust and particles into surrounding space.

**We are learning to:**
> explain how nuclear fusion accounts for all elements
> understand that creative thinking is needed to develop explanations

**FIGURE 1**: Heavy elements form and scatter when a star explodes.

## Making new elements in stars

The Sun gives out huge amounts of energy in the form of heat and light. Its energy comes from hydrogen. Hydrogen nuclei are jammed together so hard that they combine in pairs to form a different element, helium. This process releases loads of energy and is called **nuclear fusion**.

**Watch out!**
Chemical reactions like burning do not involve an atom's nucleus. The same elements are present before and after. Nuclear fusion changes the nucleus and a different element is made.

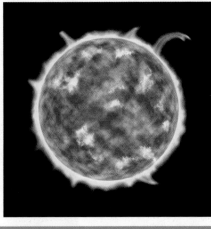

**FIGURE 2**: The source of the Sun's energy is fusion of hydrogen nuclei.

Very large stars end their life in a massive explosion. The energy involved is huge and spreads particles throughout space. These particles include all the elements found on Earth. Some of these elements formed during the star's lifetime, while others formed during the explosion.

### QUESTIONS

1. How does the Sun produce heat and light energy?
2. What new element is made when hydrogen fusion happens?

## Fusion in stars

In a star the temperature is so high and the density so great that nuclei collide at enormous speeds. When they hit each other, two or more nuclei are 'crushed' together in a fusion reaction to make a heavier nucleus. This means a different element forms. All nuclear fusion reactions give out heat and light energy.

Of all the atoms in the Sun, 94% are hydrogen. In the Sun, hydrogen nuclei fuse to make a helium nucleus plus heat and light energy. Hydrogen fusion is the source of the light and heat that we receive from the Sun.

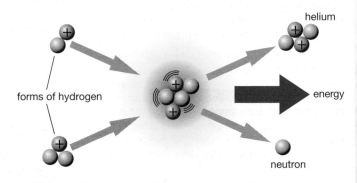

**FIGURE 3**: The fusion reaction that takes place in the Sun.

nuclear fusion GCSE

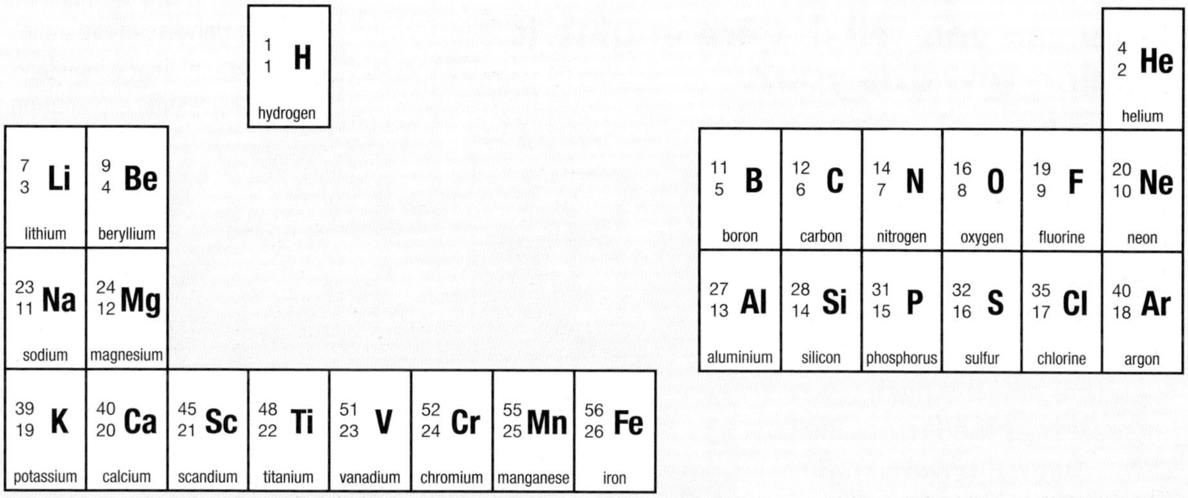

FIGURE 4: Fusion reactions in stars can form all the elements up to iron.

## Forming heavier elements

Some stars are much larger than others but all new stars contain hydrogen. As stars get older, fusion reactions in them change. Hydrogen nuclei fuse to make helium, then helium nuclei fuse, making beryllium and carbon. Other nuclear fusion reactions in different stars explain why we see other elements, as heavy as iron.

Scientists had to find an explanation for how elements heavier than iron formed. Stars more than eight times heavier than our Sun end their life as a massive explosion called a **supernova**. The explosion is so massive, lots of different elements are formed. Supernovas create all elements heavier than iron and spread them as dust and particles throughout the galaxy.

### QUESTIONS

**3** What conditions are needed for nuclear fusion to happen?

**4a** How are different elements formed in a star like the Sun?

**b** How are elements heavier than iron created?

## Elements in the Sun

Analysing light from the Sun shows it contains hydrogen and helium. It also contains about 2% heavier elements like silicon and magnesium. This is unexpected as the Sun still uses hydrogen as its main energy source. The fusion reactions needed to make silicon and magnesium are not happening in the Sun.

Scientists had to think creatively to explain this data from the Sun. Their explanation was that the Sun formed from dust and particles already present in the galaxy. The massive clouds of dust that form new stars contain particles from older stars that no longer exist. This means that the stars we see now contain elements that were formed in older stars that existed billions of years ago.

### Did you know?

The Sun's central core is at a temperature of more than 15 million °C. This is surrounded by cooler zones through which energy is transported. The outer photosphere is where sun spots and solar flares can be seen. Beyond that is the hot chromosphere, which is not usually visible.

### QUESTIONS

**5** What evidence is there that the Sun formed from the remains of older stars?

**6** Explain whether you feel that scientists have enough evidence to prove this theory.

nuclear fusion GCSE

# The expanding Universe

**We are learning to:**
> understand that other galaxies are moving away from us
> understand why we think the whole Universe is expanding
> understand what is needed from a scientific explanation

## How can you tell if a fire engine is coming towards you?

When a fire engine drives towards you, the pitch of its siren rises. As it drives away, the pitch falls. This happens because the sound waves are squashed as the fire engine approaches. The wavelength is shorter and the pitch increases. As the fire engine drives away, the wavelength is stretched. The faster the fire engine is moving, the greater the effect.

**FIGURE 1**: The pitch of the siren depends on whether the fire engine is coming towards you or moving away from you.

## Moving galaxies

Starlight coming from galaxies to Earth changes during its journey. The wavelength of the light waves increases, in just the same way as the wavelength of the sound waves from the siren of a fire engine zooming away from you. This stretching of the light waves is evidence that distant galaxies are moving away from us, very fast.

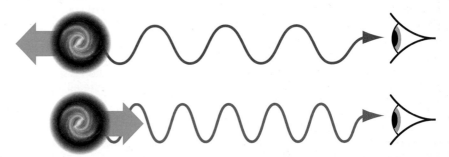

**FIGURE 2**: The wavelength of light waves from a star moving away from or towards you changes.

### QUESTION

1 How do we know that galaxies are moving away from us?

**FIGURE 3**: All of these galaxies are speeding away from us.

galaxies moving away

# Redshift (Higher tier only)

When light from stars in galaxies beyond our own is analysed, it appears redder than the light from similar stars in our own galaxy. Red light has a longer wavelength than blue light. This change in colour is caused by the stretching of light waves, and is called the **redshift**.

Light from nearly all galaxies is redshifted. Light from galaxies further away from us has a greater redshift. The amount of redshift depends on how fast the galaxy is moving away. It increases with the distance of the galaxy from us which tells us that the further away a galaxy is, the faster it moves away from us.

You cannot see the redshift using a telescope to look at distant galaxies. Instead, astronomers use an instrument called a spectrometer which analyses the light coming from galaxies. This gives enough information to tell them if the light has been redshifted. It can also measure the amount of redshift.

## Explaining the motion of galaxies

When redshift measurements were first made, scientists couldn't explain *why* galaxies were moving away from Earth. Assuming there is nothing special about the Earth, the results must mean that nearly all galaxies are moving apart from one another. One explanation is that all of space is expanding, like the surface of the balloon as it is inflated.

This explanation accounts for the data from starlight, but when it was suggested it presented a completely new way of looking at the Universe. Scientists have extended the theory to predict what the Universe will be like in future, and to consider what it was like in the past. They have also looked for more evidence to back up the theory.

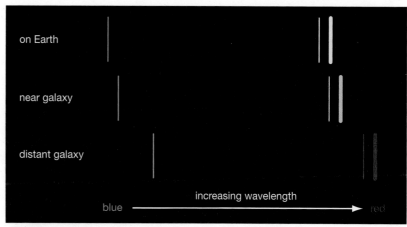

**FIGURE 4**: Images like this, called line spectra, are used to measure the amount of redshift in light from a galaxy.

## QUESTIONS

**2** What does a spectrum like those in Figure 4 tell us about a galaxy?

**3** Why do we think galaxies further away are moving faster than galaxies near us?

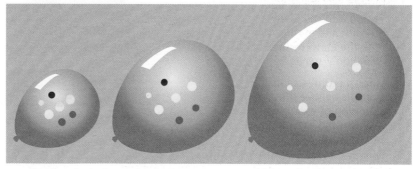

**FIGURE 5**: As the balloon expands, all the dots move further apart.

### Did you know?

Light from Andromeda, the nearest large galaxy to us, has a small blueshift because Andromeda is moving closer to the Milky Way.

### Watch out!

Don't confuse the redshift with the colour of stars. The colour of a star depends on its temperature. Bluer stars are hotter, redder stars are cooler. Redshift is a subtle effect that we cannot see with our eyes.

## QUESTIONS

**4** Describe the evidence that the whole Universe is expanding.

**5** Explain why scientists should look for more data to back up any new theory.

redshift GCSE

# The Big Bang

**We are learning to:**
> understand how the Universe formed and how old it is
> understand what may happen to the Universe in future
> understand why scientists may not agree on a new theory

## Can we see what happened billions of years ago?

Electromagnetic radiation created at the beginning of the universe is still out there. It has been detected by the 'COBE' satellite and analysed. The nature of the radiation and variations in it in different directions confirm the theory that the universe expanded very rapidly from a point, billions of years ago.

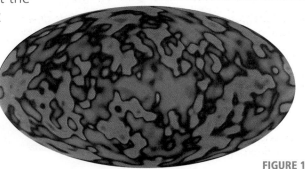

FIGURE 1

## How it all began

About 5 billion (5 thousand million) years ago, our Sun formed at the centre of a huge swirling cloud of gases and dust which collapsed on itself. A few million years later, the remaining matter swirling round the Sun formed into the planets, asteroids and moons. We think the Earth formed about 4.5 billion years ago.

As for the Universe itself, scientists believe the beginning of everything was when a rapid expansion started about 14 billion years ago. Matter and energy were flung outwards in what is called the **Big Bang**.

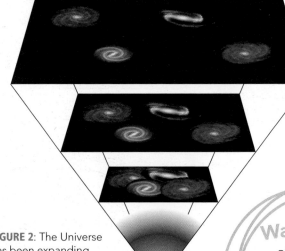

**FIGURE 2:** The Universe has been expanding since the Big Bang.

**Watch out!**
The Big Bang wasn't an explosion. When the Big Bang happened, space in the Universe started expanding. The Big Bang is closer to a balloon inflating than an explosion.

### QUESTION

1 How old are the Universe, the Sun and the Earth?

## Evidence for these ideas

Scientists have found evidence for how the solar system formed:

> Telescopes in space can see how other new planetary systems form.

> Scientists have been able to date meteorites that have fallen to Earth. These rocks formed when the solar system formed. They are about 5 billion years old.

> Since objects in the solar system orbit in the same direction we think the whole solar system formed at the same time.

Big Bang theory GCSE    age of Universe / solar system

## Evidence for how the Universe was formed

The Big Bang theory predicted that an 'echo' of the initial rapid expansion could be detected now, as radiation coming from all directions. The detection of this 'cosmic background radiation' (Figure 1) convinced many scientists that the Big Bang took place. It was first detected over 40 years ago and reported in conferences and journals. Since then, other scientists have also detected the radiation, so the theory became widely accepted. But not all scientists agreed.

The analysis of light from galaxies suggested that the Universe had been expanding for 14 billion years, but the data could be explained in different ways. Other theories about the Universe include the Steady State theory. This theory says that the Universe has always existed, and extra atoms are created as it expands. The Steady State theory explained the observed expansion but couldn't explain the discovery of cosmic background radiation.

Since most evidence fits in with the Big Bang theory, most scientists now believe this is correct.

### QUESTIONS

**2** Why are scientists confident that the ages of the Earth and Sun are accurate?

**3** Why was the discovery of cosmic background radiation important?

**4** Explain why most scientists believe the Big Bang theory is correct.

## The future of the Universe

We do not know what will happen billions of years in the future. The Universe may keep expanding, it could reach a fixed size, or it could collapse as a 'Big Crunch'.

It is hard to predict what will happen to the Universe, because of the difficulty of making reliable measurements. For example, we cannot measure the enormous distances accurately, nor the speed of the furthest galaxies.

### Different possibilities (Higher tier only)

Another thing we cannot measure accurately is the total mass of matter in the Universe. The total mass is crucial, because of the effect of gravity.

| Mass of Universe | What may happen |
|---|---|
| above a critical mass | Gravity starts to pull everything together again. The Universe reaches a maximum size then starts to shrink. |
| equal to the critical mass | The Universe reaches a fixed size. |
| below the critical mass | Gravity is not strong enough to stop galaxies moving apart. The Universe expands forever. |

Scientists think the mass of the Universe is near the critical amount. More accurate measurements are needed for a definite answer.

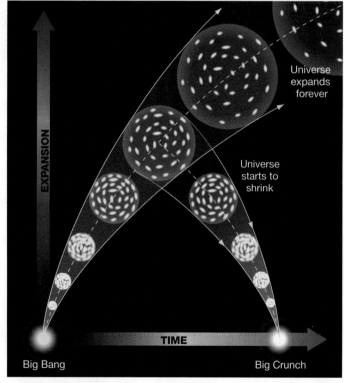

FIGURE 3: We do not know what will happen to our Universe in the future.

### QUESTIONS

**5** What may happen to the Universe in future?

**6** Explain why scientists cannot be confident about the future of the Universe.

# Rocks on Earth

**We are learning to:**
> understand how rocks tell us about changes in the Earth
> understand how rocks provide evidence of the Earth's age

## How smooth is the Earth's surface?

A snooker ball has a smooth shiny surface. If the Earth could be shrunk to the same size, it would be smoother than a snooker ball. It would be smoother still, if new rocks and mountains were not forming all the time.

### How we know the Earth changes

The Earth is constantly changing. Mountains are worn away by **erosion**. Eroded rock fragments are carried away and deposited as **sediments**, eventually forming new rock. The remains of animals and plants can form **fossils** if they are buried by sediments.

Some changes happen quickly, such as when a **volcano** erupts and the lava produces a new mountain or a crater. Movements of the Earth's crust over millions of years can cause rocks to fold, and can push rocks upwards to make new mountains.

**Geologists** study:

> the materials from which rocks are made

> the different types of rock and where they are found

> the fossils contained in rocks

> the shapes of mountains and landscapes.

Their findings provide evidence of how the Earth has changed.

**FIGURE 1**: These layers of rock have been folded by strong forces.

**Watch out!**
The Earth is constantly changing, but most of the changes in rocks and landscapes happen too slowly to be seen in a person's lifetime.

 **QUESTIONS**

1  Write down four causes of changes to the Earth.

2  Why do geologists study fossils?

### Erosion and sedimentation

Mountains, cliffs and other rock outcrops are constantly being broken down by weathering and eroded. This erosion is caused by:

> moving water (sea or rivers)

> glaciers (moving ice sheets)

> the wind (blowing particles away)

> gravity (landslides and rock falls).

Weathering and erosion break down and remove the surface rock, changing the size and shape of mountains and valleys over millions of years.

**FIGURE 2**: The layers of sedimentary rock are exposed on this rock face.

erosion

P1 The Earth in the Universe

Weathered mountains become smaller and smoother, and valleys can become deep as rivers cut into the rock of the river bed. Eroded rock fragments are transported by the wind, water and ice, broken up further, and deposited on riverbeds and in the sea. This is called **sedimentation**. Over millions of years, the sediments are crushed together to form layers of new sedimentary rock. Where this rock forms, the sea will become shallower, as the rock beneath becomes thicker.

> QUESTION

**3** Make a sketch to show what happens when a mountain is eroded and sediments are deposited.

## How old is the Earth?

It is likely that erosion and sedimentation have occurred all the time the Earth has existed. If no new rocks had been created, the large land masses would have been worn down to sea level. What causes the constant mountain building?

We now know that the Earth's outer layer – the **crust** – is constantly shifting.

> Volcanoes create new mountains when molten rock escapes from under the Earth's crust and solidifies on the land surface or under the oceans.

> Different parts of the Earth's crust move closer together over millions of years, pushing the rocks together and upwards to make new mountain ranges where the rocks collide.

FIGURE 3: This crater was formed by a volcano.

Geologists can find and study mountains of all ages in different places on Earth. This means they can find examples of rock processes taking place today that can account for past changes.

There are several methods for measuring the age of rocks. Geologists have found that the oldest rocks are about 4 thousand million years old. This means that the Earth is at least 4 thousand million years old. Some rocks are much younger, evidence that the Earth continues to change.

> QUESTIONS

**4** Why do we think the Earth is at least 4 thousand million years old?

**5** Give two reasons why we think new rocks are forming. Explain your answers.

**6** How can studies of rocks today tell us about what happened millions of years ago?

🔍 fold mountains GCSE

# Continental drift

**We are learning to:**
> understand what is meant by continental drift
> understand why and how the seafloor is spreading
> understand why scientists may be sceptical about an explanation
> understand how scientific explanations may be tested

## Are the oceans getting bigger?

The Atlantic Ocean is getting bigger. The UK is moving further away from America every year. It is happening very slowly – about 1 cm per year – which is about as fast as fingernails grow.

## What continental drift means

Continents are huge land masses on Earth. They include Europe, Africa, Asia, North and South America, Antarctica and Australia. A scientist called Wegener suggested that the continents were joined to each other in the past, and drifted apart over millions of years.

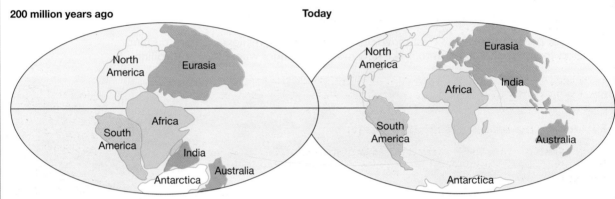

**FIGURE 1**: How the continents may have moved in the past.

Alfred Wegener published his hypothesis of **continental drift** in 1915. His ideas were:

> The shapes of the continents looked as if they could interlock.

> Similar fossils and rock types were found on continents separated by oceans.

> Mountain chains made from similar rocks appeared on the edges of different continents.

Wegener also thought that the forces causing the continents to move caused land to fold and deform into great mountain ranges along their edges.

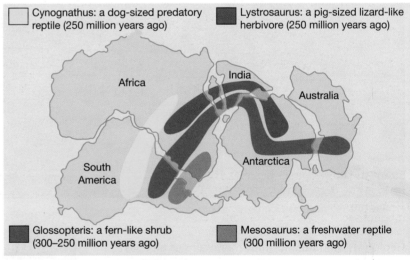

**FIGURE 2**: Fossil records provided evidence for continental drift.

### QUESTIONS

1. Look at Figure 1. Write down two pairs of continents that seem to fit next to each other.

2. Which fossils have been found just in Africa and South America? Which fossil has been found on all the land masses shown in Figure 2?

continental drift    Wegener

# P1 The Earth in the Universe

## Did scientists accept the theory of continental drift?

Geologists study rocks and the Earth, but Wegener was not a geologist. Geologists believed that continents were in fixed positions, so if Wegener was right, many of their previous ideas were wrong. Also:

> Wegener could not explain *how* the continents moved.
> The movement of the continents was too small to detect.
> Data about rocks and fossils was not available for all continents.
> There were simpler explanations for some of Wegener's ideas.

Wegener's ideas were so unexpected, scientists were sceptical. More evidence was needed to back up Wegener's hypothesis.

### QUESTIONS

**3** Why did geologists find it hard to believe Wegener's hypothesis?

**4** Explain whether you think they were right to be sceptical.

## The spreading seafloor

The Earth's core is extremely hot and heats rocks in the mantle. Some parts become hotter than others. This causes **convection**, which moves the rocks in the mantle. Sections of the Earth's crust above the mantle are forced apart, making the seafloor spread. This process is called **seafloor spreading**.

Every year, seafloors spread by several centimetres. We do not see a gap appear because fresh rock from the mantle comes to the surface to fill the space. This rock forms underwater mountains called **oceanic ridges**.

### Patterns in the seafloor (Higher tier only)

Over millions of years, the Earth's magnetic field changes its direction. Every time fresh rock comes to the surface, it is magnetised in the direction of the Earth's field at that time.

**Watch out!**
The widening of the seafloor means that some (not all) continents are moving apart.

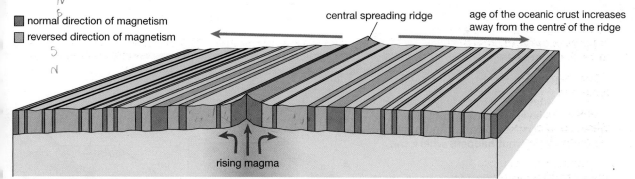

**FIGURE 3**: Changing magnetic field direction in the oceanic crust.

Scientists studying the magnetism of rocks in the oceanic ridges found that they contained strips of rock magnetised in opposite directions. This is evidence that the seabed is constantly forming in strips over millions of years.

The hypothesis of continental drift explained the fossil and rock records better than previous explanations. It could not explain how and why the continents moved. It was not correct but it was not abandoned until scientists developed the theory of plate tectonics, which explained the data better.

### Did you know?

Volcanoes can erupt underwater as well as on land. The oceanic ridges contain chains of underwater volcanoes.

### QUESTIONS

**5** What is the evidence that the seabed is spreading?

**6** Explain how the magnetism of the seabed supports the hypothesis of continental drift.

seafloor spreading

# Tectonic plates

## How is the Earth's surface like a jigsaw?

The Earth's outer surface, the crust, is made in sections called **tectonic plates** that fit together like a jigsaw puzzle.

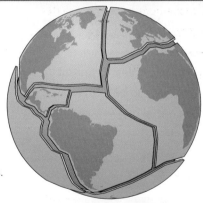

FIGURE 1: Tectonic plates.

**We are learning to:**
> understand what happens at tectonic plate boundaries
> understand why tectonic plates move
> understand that developing explanations needs creative thought
> understand how scientific explanations can be tested

## Tectonic plate boundaries

The solid tectonic plates float on top of semi-solid rocks below the crust. The place where two plates meet is called a **plate boundary**. Earthquakes, volcanoes and new mountains are found at plate boundaries.

**Watch out!**
Tectonic plates are not fixed shapes. Some plates shrink if they are forced back into the mantle. Fresh material coming to the surface from volcanoes makes some plates grow.

**Did you know?**
Mount Everest is the world's tallest mountain. Yet millions of years ago, it was not a mountain at all but under the ocean. As two tectonic plates in the Earth's crust moved together, the rocks at the boundary were pushed upwards and this created the mountain.

### QUESTION

1 Describe the structure of the Earth's crust.

## Tectonic plate movements (Higher tier only)

Tectonic plates move very slowly (a few centimetres a year), either pulling apart, sliding past each other or pushing together. Figure 2 shows what happens where the plates meet.

**A** Volcanoes occur where plates move apart making a gap in the Earth's crust. **Magma** (liquid rock) is forced through cracks in the surface and piles up, forming volcanic mountains.

**B** Mountains form when two tectonic plates move together and one plate is forced under the other. Volcanoes may occur.

**C Fold mountains** form where tectonic plates made from similar rock densities push together. This lifts and folds the land.

**D** Most earthquakes occur where the tectonic plates suddenly slide past each other, releasing energy in a sudden jerk.

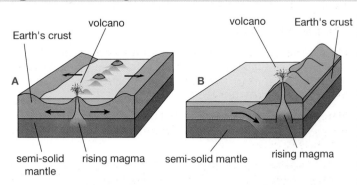
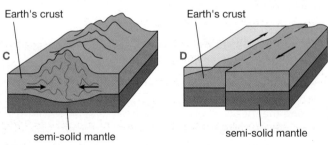

FIGURE 2: What happens at plate boundaries.

forming mountains

## The rock cycle

The **rock cycle** depends on the movement of tectonic plates.

> Mountains are weathered and fragments of rock break off.

> Rock fragments are transported away and deposited on the seabed as sediments.

> At plate boundaries, movement of the tectonic plates can force one plate beneath the other. Where an oceanic plate is forced below the land, the sediments are dragged down as well.

> As the plate moves deeper into the mantle the rock melts and becomes magma.

> Pressure can cause magma to rise. This magma either solidifies into rock beneath the land surface or escapes as a volcanic eruption to form new mountains.

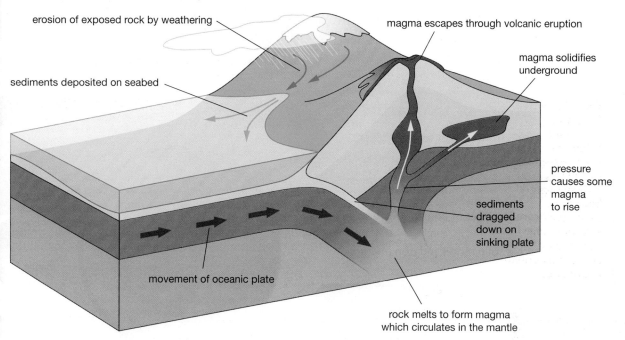

**FIGURE 3**: New rock from old in the rock cycle.

## The plate tectonics theory

**Plate tectonics** is a fairly new theory, developed in the 1960s. It explains data such as the location of mountains, volcanoes and earthquakes, the shape of continents and the ages of mountains. It took many years between recording the data and thinking up a creative explanation to match these findings.

Plate tectonics explains things that are not obviously related, such as why the direction of magnetism varies in rocks of different ages. It can also be used to predict where earthquakes are likely to occur.

### QUESTIONS

**2** In what ways can tectonic plates move?

**3** Explain why plotting volcanoes, mountains and earthquakes on a map shows where the boundaries of tectonic plates are.

**4** Describe how tectonic plate movement causes new mountains to form.

**5** Explain how the rock cycle and moving tectonic plates are linked.

**6** Explain why the theory of tectonic plates is now widely accepted.

tectonic plates theory GCSE

# Earthquake waves

## Are there earthquakes in the UK?

You may think you are safe from earthquakes. In fact every couple of years the UK has an earthquake strong enough for people to notice. In 2007, an earthquake in Folkestone shook many buildings. In parts of the town the tremor caused extensive damage to chimneys and walls, and five streets had to be evacuated.

**We are learning to:**
> understand that earthquakes produce waves in the Earth
> understand how these waves tell us about the Earth's structure
> describe the Earth's structure

**FIGURE 1**: Damage in Folkestone when the earth shook on 28 April 2007 at 8.18 am.

## The structure of the Earth

Our Earth is made from several separate layers.

> The **crust** is a layer of solid rock about 30 km thick.

> The **mantle** is a layer of semi-solid rock about 2900 km thick.

> The outer core is a layer of liquid nickel and iron about 2200 km thick.

> The inner core is solid nickel and iron about 1250 km thick.

### QUESTION

1  Draw a bar chart comparing the thickness of each layer in the Earth. Use a scale of 1 cm for 500 km.

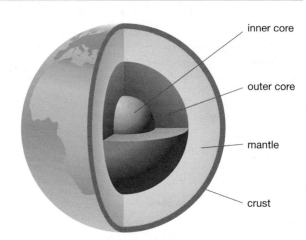

**FIGURE 2**: The structure of the Earth.

## Earthquake waves

The sudden movement of tectonic plates in an earthquake produces shock waves, known as **seismic waves**. These waves travel through the Earth. There are two main types of seismic wave.

> **P-waves** travel through solids and liquids.

> **S-waves** only travel through solids.

### Did you know?

There are several hundred earthquakes every day all around the world. Luckily, only a few of these cause damage.

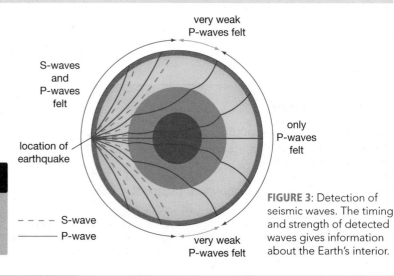

**FIGURE 3**: Detection of seismic waves. The timing and strength of detected waves gives information about the Earth's interior.

224  seismic wave GCSE    structure of Earth

Instruments on the Earth's surface can detect the waves after an earthquake. Studies of seismic data received by instruments in different parts of the world show that:

> P-waves arrive at detectors before S-waves. This means that P-waves travel faster than S-waves.

> In some places P-waves arrive but S-waves do not. This suggests that some of the interior of the Earth is liquid.

> The speed of a wave depends on the path it takes through the Earth. This suggests that the Earth is made from layers of different materials.

## QUESTIONS

**2** Which type(s) of seismic wave can travel through solids?

**3** How do we know that P-waves travel faster than S-waves?

**4** Explain why S-waves are not detected on the opposite side of the Earth from an earthquake.

## Seismic waves and the Earth's structure

Scientists analyse the speeds of seismic waves by measuring the time it takes the different waves from an earthquake to reach detectors all over the world. They use what they know about how waves travel in different materials to explain their observations.

> Seismic waves follow curved paths in the core and mantle. As seismic waves travel faster in denser materials, this tells us the Earth's density increases with depth.

> There is a **shadow zone** where very few P-waves are detected. This is because the P-waves change direction abruptly at the boundary between the mantle and the core. P-waves travel more slowly in the core, so the mantle and core must have different densities.

> No S-waves are detected on the opposite side of the Earth to an earthquake. As S-waves cannot pass through a liquid, this suggests that part of the Earth's core is a liquid.

**Watch out!**
The inside of Earth is not all liquid – the inner core at the centre is solid.

## QUESTIONS

**5a** Write down one difference and one similarity between longitudinal waves and transverse waves.

**b** Can transverse seismic waves travel through solids and liquids, or just solids?

**6** Seismic waves travel in different ways through different parts of the Earth. Explain how their detection around the Earth's surface allows us to learn about the structure of the Earth.

### Different types of wave

P-waves are **longitudinal** waves. The vibration moves backwards and forwards along the direction the wave travels, squashing and stretching the material. P-waves are sometimes called pressure, push or primary waves. Sound waves are also longitudinal waves.

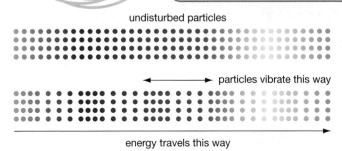

**FIGURE 4**: P-waves are longitudinal waves.

S-waves are **transverse** waves. The vibration moves at right angles to the direction the wave travels in. S-waves are sometimes called secondary or shake waves. Light is also a transverse wave.

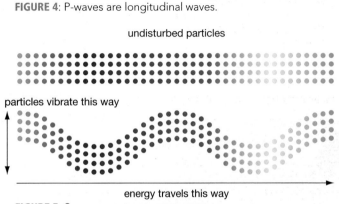

**FIGURE 5**: S-waves are transverse waves.

P-wave and S-wave

# What a wave is

**We are learning to:**
> understand what a wave is
> draw and use diagrams of waves
> calculate the speed of a wave
> understand what is meant by amplitude and wavelength

## How are ocean waves like space waves?

Massive stars can trigger disturbances through the clouds of gas surrounding them. Radiation from the stars spreads like a very fast-moving wind, making ripples in the clouds of gas. In the same way, the wind blowing over the surface of the ocean creates water waves.

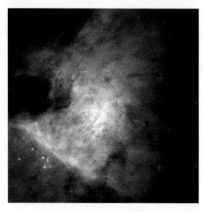

FIGURE 1: Making waves is much the same, whether on Earth or in space.

## Waves

When you drop a stone in a puddle, there is a splash when the stone hits the water. Ripples spread outwards on the water's surface. The ripples carry energy from the impact of the stone. As each ripple passes, the water returns to where it was before.

These ripples are a type of water wave. A **wave** is a series of disturbances that carry energy in the direction of the wave, without transferring matter.

For wave motion, we can use the equation:

$$\text{distance} = \text{wave speed} \times \text{time}$$

This can be rearranged to find out how fast a wave travels:

$$\text{wave speed (m/s)} = \frac{\text{distance travelled (m)}}{\text{time taken (s)}}$$

For example, if the ripple travels 0.5 m in 5 s, then we can calculate its speed.

$$\text{wave speed} = \frac{\text{distance}}{\text{time}}$$

$$\text{wave speed} = \frac{0.5\,\text{m}}{5\,\text{s}} = 0.1\,\text{m/s}$$

FIGURE 2: A wave travels out from the source, carrying energy with it.

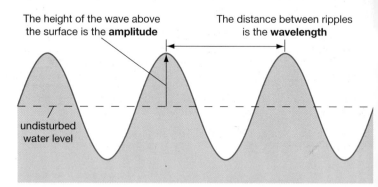

FIGURE 3: Amplitude and wavelength of a wave.

### Did you know?

The tallest waves recorded in the ocean were taller than 10-storey buildings. The longest ocean waves are Rossby waves with wavelengths of hundreds of kilometres.

## QUESTIONS

1. What is a wave?
2. Draw a diagram showing a wave. Label the wavelength and amplitude.
3. What is the speed of a wave that travels 1 m in 5 seconds?

amplitude    wavelength    wave GCSE

P1 The Earth in the Universe

## Seeing waves

Vibrations cause regular cycles of disturbances. This is how waves are created. You can see and feel the vibrations causing sound waves. For example, the strings of a guitar vibrate while a note is being played, and the voice box in your throat vibrates while you speak.

An **oscilloscope** is a machine that displays waves on a screen. A grid on the screen lets you compare the wavelength and amplitude of waves.

> A sound is louder if it has a larger amplitude.

> A sound is higher pitched if it has a shorter wavelength.

### QUESTIONS

**4** Look at Figure 4.

  **a** Which diagram shows the larger amplitude? Which diagram shows a louder sound wave?

  **b** Explain how loudness depends on amplitude.

**5a** Which diagram shows the longer wavelength? Which diagram shows a lower pitched note?

  **b** Explain how pitch depends on wavelength.

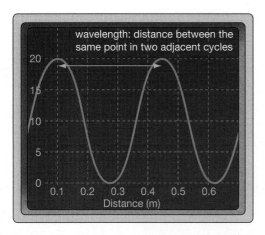 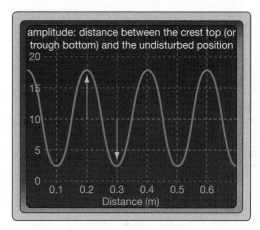

**FIGURE 4**: Two different sound waves on the same oscilloscope grid.

## Measuring waves

The scale on an oscilloscope is used to measure a wavelength or amplitude. Wavelengths are measured in metres, but you do not need to know the units of amplitude. The wavelength and amplitude of the left-hand wave in Figure 4 can be found like this:

> Each horizontal square is 0.1 m. The wavelength is 3.5 squares. So the wavelength is 0.35 m (0.1 m × 3.5 squares).

> Each vertical square is 5 units. The amplitude is 2 squares. So the amplitude is 10 units (5 units × 2 squares).

**Watch out!**
Measure a whole wavelength from one point in a cycle to the same point in the next cycle, for example from peak to peak. Don't measure half the wavelength by mistake.

**Watch out!**
Measure amplitude from the undisturbed level (midpoint) to the maximum displacement. Don't double it by mistake.

### QUESTIONS

**6** What is the wavelength and amplitude of the right-hand wave in Figure 4?

**7** Draw and label a diagram to the correct scale showing a wave with a wavelength of 4 cm. On your diagram, show the amplitude as 2 cm. Include three complete waves.

oscilloscope

# The wave equation

**We are learning to:**
> understand what is meant by wave frequency
> use the wave equation

## What makes the best surfing beach?

One of the most popular activities in Cornwall in the summer is surfing. One of the first skills to master is judging the speed of an incoming wave and then paddling just fast enough to match it. When the wave catches the board it's then a case of getting to your feet. Experienced surfers can sense the frequency of the waves and are ready for the next one that breaks.

FIGURE 1

## Wave frequency and speed

Imagine watching the waves in a swimming pool when the wave machine is on. You can count how many waves arrive in a fixed time. You can also estimate the distance between each wave crest. If you multiply these two numbers together, you get the distance the wave has travelled in that time – the wave's speed of approach. This idea works with all waves.

> The number of waves passing a point every second is called the **frequency** of the waves. This is the same as the number of waves produced by a vibration every second. Frequency is measured in **hertz**: 1 hertz (Hz) equals 1 wave (or 1 vibration) per second.

> The length of each wave (for example from crest to crest) is the wavelength.

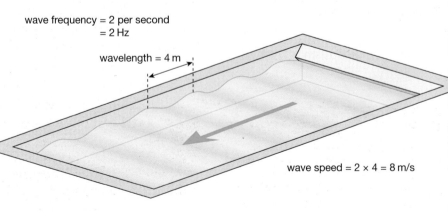

wave frequency = 2 per second = 2 Hz

wavelength = 4 m

wave speed = 2 × 4 = 8 m/s

**FIGURE 2:** Calculating wave speed

wave equation GCSE

The speed is given by:

wave speed = frequency × wavelength
(metres per second, m/s)   (hertz, Hz)   (metres, m)

This is called the **wave equation**.

All waves obey the wave equation. For example, the frequency of a particular sound wave is 440 Hz and its wavelength is 0.75 m. Using the wave equation, we can calculate its speed.

speed = frequency × wavelength = 440 × 0.75 = 330 m/s

## QUESTIONS

**1** The frequency of a sound wave is 400 Hz. How many waves are produced each second?

**2** If 20 waves pass a point in 5 seconds, what is their frequency?

**3** What is the speed of a wave that has a frequency of 7500 Hz and a wavelength of 0.2 m?

## Using the wave equation

On page 226 you learned that, for a wave:

$$\text{wave speed (m/s)} = \frac{\text{distance travelled (m)}}{\text{time taken (s)}}$$

Now, you have learned how wave speed can be calculated using the wave equation:

wave speed (m/s) = frequency (Hz) × wavelength (m)

Which equation should you use in a calculation?

> Read the question to find out what information you are told.

> Check what you need to calculate.

One consequence of the wave equation is that if the speed of a wave stays constant, then changing the frequency will change the wavelength.

All sound waves, for example, travel at the same speed in air. High-pitched sounds have higher frequencies than low-pitched sounds. This means that the wavelength of a high sound must be shorter than the wavelength of a low sound for the speed to be the same. The higher the frequency, the shorter the wavelength. The wavelength is **inversely proportional** to the frequency. This is true for all waves.

**Watch out!**
Use the equation that matches the information you know. In a Higher tier question, you may need to rearrange the equation.

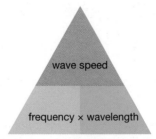

FIGURE 3: Cover up what you want to find out. Use the quantities you know to work out the answer.

## QUESTIONS

**4** Sound waves travel at a speed of 330 m/s in air. Describe how the wavelength of different notes changes as the frequency changes from 2000 Hz to 200 Hz.

**5** Calculate the wavelength of a sound wave with a frequency of 165 Hz (speed of sound in air is 330 m/s).

**6** Calculate the frequency of a sound wave with a wavelength of 1.2 m (speed of sound in air is 330 m/s).

**7** Light travels as a wave. All colours of light travel at the same speed in air, but red light has a longer wavelength than blue light. Is the frequency of blue light greater or smaller than that of red light? Explain your answer.

### Did you know?

Sound travels fastest in solids (like metal) and slowest in gases. Light travels fastest in empty space, and slowest through solids (like glass).

# P1 Checklist

## To achieve your forecast grade in the exam you'll need to revise

Use this checklist to see what you can do now. Refer back to pages 204–229 if you're not sure.

Look across the rows to see how you could progress – **_bold italic_** means Higher tier only.

Remember you'll need to be able to use these ideas in various ways, such as:
> interpreting pictures, diagrams and graphs
> applying ideas to new situations
> explaining ethical implications
> suggesting some benefits and risks to society
> drawing conclusions from evidence you've been given.

Look at pages 312–318 for more information about exams and how you'll be assessed.

| To aim for a grade E | To aim for a grade C | To aim for a grade A |
|---|---|---|
| recall the names, relative sizes and motion of different bodies in the solar system and the wider Universe; recall that the Sun is just one of the thousands of millions of stars in the Milky Way galaxy; recall that the Universe contains thousands of millions of galaxies which each contain thousands of millions of stars | | |
| recall that light travels through space at 300 000 km/s; recall that a light-year is the distance travelled by light in a year | use appropriate units, including the light-year, for describing distances in the Universe | |
| describe how we use the radiation from distant stars and galaxies to detect them | understand how we can learn about distant stars and galaxies using their radiation, although light pollution and atmospheric conditions interfere with these observations | understand why the finite speed of light means we see distant objects in the Universe as they were in the past; understand why our observations of distant objects may be unreliable |
| recall that nuclear fusion is when two nuclei join together, forming a new element; understand that nuclear fusion of hydrogen is the source of the Sun's energy | understand how nuclear fusion forms all the different elements in stars and provides the stars' energy | |
| describe how we can measure distances to stars by comparing their brightness | describe how the relative brightness of stars and stellar parallax help us to measure the distances to stars | understand methods of measuring distances to stars, and explain some problems with making and interpreting these measurements |

| To aim for a grade E | To aim for a grade C | To aim for a grade A |
|---|---|---|
| understand that the Universe began about 14 thousand million years ago, and that distant galaxies are moving away from us | understand that the observed movement of distant galaxies away from us gives an approximate age of the Universe | *understand that redshift tells us that more distant galaxies move away faster, so space itself may be expanding* |
| understand that the Universe may be different in the future; understand there are some difficulties in predicting the future of the Universe | understand that problems in measuring distances to and motion of distant objects mean that the future of the Universe cannot be accurately predicted | understand why problems measuring the distances to and motion of distant objects and *the mass of the Universe* cause uncertainty in predicting its ultimate fate |
| recall that the oldest rocks on Earth are about 4 thousand million years old | understand that the Earth is older than its oldest rocks | explain the implications of the Earth being older than its oldest rocks |
| describe some rock processes taking place today and what they suggest about the past | explain how rocks and rock processes seen today provide evidence for past changes | understand and compare how rocks and rock processes seen today provide evidence for past changes |
| recall that continental land masses are moving very slowly and that this was described by Wegener in his theory of continental drift | explain how Wegener's theory of continental drift was developed and modified; understand that heating of the core causes convection in the mantle, and seafloor spreading | understand implications of Wegener's theory of continental drift and why geologists at first rejected it; *understand the causes of seafloor spreading, and what the magnetisation of seafloor rocks can tell us* |
| understand that earthquakes, volcanoes and mountain building generally occur at the edges of tectonic plates | | *understand how moving tectonic plates cause earthquakes, volcanoes and mountain building* |
| understand that earthquakes produce P-waves and S-waves which can be detected; draw and label a diagram of the Earth's interior | describe the features of P-waves and S-waves, and how they give evidence for the Earth's structure | understand how differences in P-waves *(longitudinal waves)* and S-waves *(transverse waves)* give evidence about Earth's structure |
| recall that waves are disturbances that transfer energy; use the terms wavelength, frequency and amplitude; draw and interpret diagrams showing amplitude and wavelength; use equations involving wave speed | | explain how waves transfer energy; understand the difference between a transverse and longitudinal wave; use *and rearrange* the wave equation |

# Exam-style questions

## Foundation level

**1**

**AO1 a** Scientists use light-years to measure distances in the Milky Way. Which of these statements explain why they use light-years? There are two correct explanations. [2]
  A Light-years measure much larger distances than km.
  B Distances are extremely large in the Milky Way.
  C A light-year measures the distance light travels in a year.
  D Light from planets takes time to travel through the galaxy.

**AO2 b** Astronomers find out about distant objects by studying the radiation they emit. Suggest two reasons why is it easier for astronomers to study stars in a neighbouring galaxy than the planets orbiting these stars. [2]

[Total: 4]

## Foundation/Higher level

**AO1 2** One of the largest land-based telescopes in the world is in Hawaii. Scientists studying galaxies using the telescope have found much observational evidence about different galaxies in the Universe. Some statements below suggest why they may choose to stop collecting evidence. Which statement is most likely to be the correct explanation? [1]
  A Better data could be collected using a space telescope.
  B Scientists working on other research projects also need access to the telescope.
  C The telescope can only be used at night.
  D The scientists have enough data to confirm current theories, so there is no need to collect more data.

[Total: 1]

**3** Light from two distant galaxies has been analysed and displayed as a set of coloured lines called a spectrum. The spectrum of light from each galaxy is shown below, with the spectrum obtained from the same light on Earth. The arrow above the lines indicates how the colour of the different lines changes.

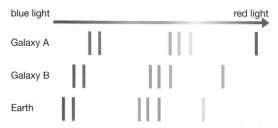

**AO2 a** Write these sentences correctly using information from the diagram. [3]

Galaxy A is moving *towards/away from* the Earth.
Galaxy B is moving *towards/away from* the Earth at a *faster/slower* speed than Galaxy A.

**AO3 b** How do the spectra shown here support the idea of an expanding Universe? [2]

[Total: 5]

**AO1 4** Wegener's theory of continental drift was a completely new idea when it was first suggested. It has now proved to be wrong in several ways and has been replaced. Explain why the theory was still an important step in our understanding of processes taking place on Earth. The quality of written communication will be assessed in your answer. [6]

[Total: 6]

**AO1 5** In the box below, the list on the left gives three different processes caused by moving tectonic plates. The list on the right describes different ways the tectonic plates move together. Match the process with the movement that causes it. [3]

| A | earthquakes | 1 | two plates move apart |
|---|---|---|---|
| B | mid-ocean ridges | 2 | two plates move together |
| C | fold mountains | 3 | two plates slide past each other |

[Total: 3]

## Higher level

**AO2 6** Sound travels at 1500 m/s in water. An echo sounder on a boat measures the water depth. It sends a sound wave to the sea bed and measures how long the signal takes to travel there and back.

**a** The signal takes 4 seconds to return to the boat. Use this information to calculate how deep the water is. [3]

**b** The frequency of the sound wave used by the echo sounder is 20 000 Hz. Calculate the wavelength of the sound wave. [3]

[Total: 6]

**AO1 7** Scientists have obtained evidence that the Universe is expanding from spectra of distant galaxies. Explain why, from this and other evidence, scientists can be more certain about the beginning of the Universe than they can be about its future. The quality of written communication will be assessed in your answer. [6]

[Total: 6]

AO1 recall the science   AO2 apply your knowledge   AO3 evaluate and analyse the evidence

P1 The Earth in the Universe

 **Worked example**

**AO1 a** There are many different elements on Earth, and scientists think some of these were formed in stars. Explain how scientists think elements formed in stars. [2]

*Elements formed in stars because of nuclear fusion.* ✔
*Nuclei of elements like hydrogen fuse together to form heavier elements in stars.* ✔

**AO2 b** Scientists found some heavier elements than they expect in young stars. Here are comments that two people made.

> **Professor Guy**
> I will collect more data myself looking at a different set of stars. I want to know which other stars also contain heavier elements. Maybe these elements only form in older stars, but that doesn't explain why I am seeing them in young stars. I wonder if they are present in all stars including the very first stars that formed?

> **Professor Jay**
> I can't imagine my results are correct. I will take some more measurements of the same stars and check my equipment carefully. If I get the same results again, I'm going to speak to a different group of scientists to find out why my readings are wrong. I may have to take the readings again using a different set of equipment.

Explain which scientist is developing a hypothesis. [2]

*Professor Guy.* ✔

**AO1 c** Suggest two reasons why it is a good idea for scientists to share the results of their experiments with other scientists. [2]

*Combining data from several places may provide enough data to prove or disprove a hypothesis.* ✔ *They may be able to identify mistakes they are making when taking results.* ✔

**AO1 d** Professor Guy did not suggest the correct explanation. Write down why scientists think heavier elements are found in young stars. [2]

*They think the heavy elements came from the remains of older stars that no longer exist* ✔ *after they exploded in a supernova that spread particles through space and eventually formed new stars.* ✔

## How to raise your grade

Take note of the comments from examiners – these will help you to improve your grade.

Always use and explain the correct scientific terms. This answer explains the term fusion as well as linking the answer to the processes in stars. It gains full marks.

Only 1 mark awarded here. Remember to answer the question – there is a mark available for naming the scientist and another mark for showing you recognise a hypothesis. A hypothesis is not just deduced from data, but is developed to explain data. Professor Guy has done this by thinking of a creative way to explain this data.

Two correct reasons have been given. Use the information from the question to help with the answer. These scientists had different reasons for sharing their data.

This answer is correct. It covers the two linked ideas in the question – how heavier elements are spread in space and why they are found in young stars.

233

# P2 Radiation and life

## What you should already know...

### Light is a form of radiation

Light travels much faster than sound.

Light travels in straight lines.

Light is reflected by shiny surfaces.

Light may be refracted when it passes from one material into another.

Radiation is emitted from sources. For example, a luminous object is a light source – it emits (gives off) light.

Radiation transfers energy.

Radiation does not need a material to transfer the energy through. For example light can travel through a vacuum.

 Name two properties of light.

### Radiation from the Sun is our ultimate energy supply

The Sun is the main source of energy for the Earth.

The Earth absorbs some of the radiation emitted by the Sun and is warmed by it.

Plants absorb light energy from the Sun and, through photosynthesis, change carbon dioxide and water to glucose and oxygen.

Fossil fuels store energy that originated from the Sun.

We burn fossil fuels for heat, for transport and to produce electricity. Carbon dioxide is given off into the environment.

 What happens to the temperature of the surface of the Earth when it absorbs radiation from the Sun?

## In P2 you will find out about...

> a family of radiations called the electromagnetic spectrum

> the energy of electromagnetic radiation being transferred in 'packets' called photons

> what affects the intensity of a beam of electromagnetic radiation

> what is meant by ionising radiation

> the properties of different types of radiation and the effects when it is absorbed

> the health risks posed by some types of radiation

> how ozone in our atmosphere is important for protecting us from the Sun's ultraviolet radiation

> how greenhouse gases such as carbon dioxide affect the amount of radiation absorbed by the Earth

> what is meant by global warming and what its effects may be

> how electromagnetic waves are used for communication

> the difference between analogue and digital signals

> the advantages of digital signals

# Electromagnetic radiation

## Can we cut with light?

Lasers are delicate enough to carry out delicate eye surgery, but powerful enough to slice through sheet metal several centimetres thick. They produce very narrow, intense beams of light that can be guided incredibly precisely. Whether it is eye tissue or steel, the material in the path of the beam melts and vaporises.

**We are learning to:**
> describe how electromagnetic radiation behaves
> describe the electromagnetic spectrum
> understand that electromagnetic waves of different frequency carry different amounts of energy

**FIGURE 1:** An industrial laser can cut through a thick steel block.

## How light behaves

Light is a form of **electromagnetic radiation**. Sources of light include things that glow such as the Sun, light bulbs and fires. These sources emit (produce) radiation that travels outwards in all directions. When the light meets an object in its path, the way it behaves depends on the material the object is made from.

Objects have different colours because they absorb and reflect different colours, or frequencies, of the light that lands on them. Light-coloured objects reflect most of the light that falls on them, whereas dull, black objects absorb most or all of the light.

We can only see objects that emit or reflect light. Light that reaches our eyes from objects is absorbed by special cells in the eye. Our eyes are examples of detectors because they respond to the light. For something to detect light, a change must happen when light is absorbed. Many objects absorb light but are not detectors.

**Watch out!**
Objects that just reflect light are not sources of light. Sources of light must *produce* the light.

### QUESTIONS

1 Which of these is a source of light? a mirror, a star, a lit candle, the Moon

2 Put these materials in order with the material that transmits the most light first:

cardboard, lens from sunglasses, tracing paper, glass from a windscreen

3 Describe how you can see the words on this page. Use some of these words in your answer: source, detector, transmit, reflect, absorb

Almost all light passes through glass and air. Light is **transmitted** well through **transparent** materials.

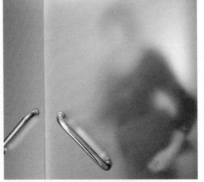

Only some light passes through clouded glass and coloured filters. Light is partly **absorbed** and partly transmitted by **translucent** materials.

Shiny objects **reflect** most of the light that falls on them.

**FIGURE 2**

electromagnetic radiation

# The electromagnetic spectrum

Light belongs to the **electromagnetic spectrum,** a family of electromagnetic waves that travel at a speed of 300 000 km/s through space, which is a **vacuum.**

The behaviour of electromagnetic radiation depends on the frequency of the waves. Waves with higher frequencies carry more energy. Electromagnetic waves are grouped in ranges of frequency. The types of electromagnetic radiation are:

> radio waves (lowest frequency and lowest energy)

> microwaves

> infrared

> visible light (red light has the lowest frequency and violet light has the highest frequency)

> ultraviolet

> X-rays

> gamma rays (highest frequency and highest energy).

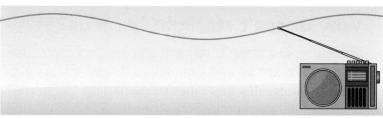

Radio waves do not harm us because they have a low frequency and carry little energy.

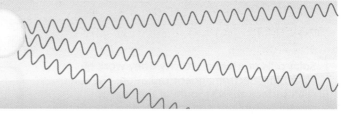

Ultraviolet radiation from the Sun causes sunburn and skin cancer as it has a much higher frequency and carries much more energy.
FIGURE 3

## QUESTION

**4** Why are gamma rays potentially harmful to humans?

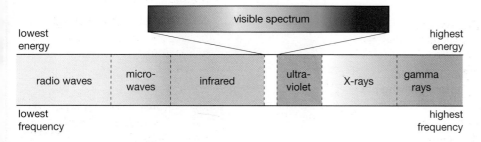

FIGURE 4: The electromagnetic spectrum.

# Light photons

In the 20th century, scientists developed a new model for light. They found evidence that not only did it behave as a wave, but it also in some ways behaved as a stream of particles. These 'particles' of light are packets of energy called **photons**. Photons have no mass. Light – and all other types of electromagnetic radiation – can be thought of as a stream of photons.

The energy carried by each photon depends on the frequency of the electromagnetic radiation. The higher the frequency, the more energy each photon transfers when it is absorbed.

## QUESTIONS

**5** Explain what a photon is.

**6** Photons of blue light carry more energy than photons of red light. Explain which colour of light has the higher frequency.

### Did you know?

Lasers produce a very pure colour of light because all the photons they produce have exactly the same frequency.

electromagnetic spectrum GCSE    photon GCSE

# Radiation intensity

**You will find out:**
- understand how energy is transferred by electromagnetic waves
- understand what radiation intensity means
- understand what affects intensity

## Why doesn't starlight harm us?

The Sun is our closest star, and its radiation can cause skin cancer, heat stroke and even kill us. Massive stars in our galaxy emit much greater quantities of damaging electromagnetic radiation which eventually reaches the Earth. This does not harm us, because the intensity of the radiation is so small after it has travelled many light-years to reach the Earth.

## Beams of electromagnetic radiation

**Solar cells** produce electricity. They work by absorbing electromagnetic radiation. The solar cell's surface absorbs some of the energy carried by sunlight and transfers it to electrical energy. The amount of energy absorbed by the solar cell depends on the strength (or **intensity**) of the radiation arriving at its surface.

The intensity of a beam of radiation is a measure of the energy transferred each second. Electromagnetic radiation transfers energy in 'packets' of energy called **photons**.

The energy arriving per second at a surface from the beam of radiation depends on:

> the number of photons arriving per second

> the energy transferred by the individual photons.

In winter, sunlight is less intense than in summer. There are fewer photons arriving per second on the surface of a solar cell, so the solar cell absorbs less energy.

The energy carried by individual photons depends on the type of radiation. Photons of ultraviolet radiation have a higher frequency and more energy than photons of infrared radiation.

**FIGURE 1:** Solar cells on this calculator absorb energy from the Sun.

**FIGURE 2:** Solar cells are only effective in intense sunlight.

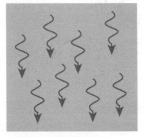

Red light carries less energy than blue light, as its photons carry less energy.

**FIGURE 3:** The energy carried by electromagnetic radiation can be thought of as a stream of photons.

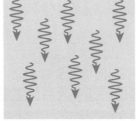

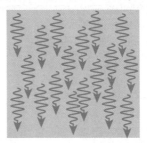

The beam is more intense when it contains more photons.

### QUESTIONS

1 Why doesn't a solar cell absorb energy at night? Use 'photons' in your answer.

2 What affects the amount of energy transferred by an electromagnetic wave?

# P2 Radiation and life

## Distance and radiation intensity

The intensity of electromagnetic radiation gets less as you move further away from the source. The intensity of the Sun's radiation decreases further away from it because it is spread over a larger area. This is why planets more distant from the Sun than the Earth are much cooler than our planet. The Earth receives more of the Sun's electromagnetic radiation because the intensity of the radiation is greater closer to the Sun.

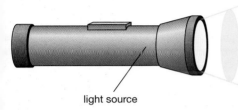

FIGURE 4: As you move away from the source, the energy is increasingly spread out.

### QUESTIONS

**3** Give three reasons why less radiation is absorbed 10 m from a lit candle compared with the radiation absorbed 5 m from a blue spotlight.

**4** Cygni is a star much brighter than the Sun, about 11 light-years away from Earth. Why is the Sun's light more intense on Earth than the light from Cygni?

### Did you know?

Pupils in our eyes change size to match the intensity of light. In bright light, the pupils become small and in dim light they get larger.

## Defining intensity (Higher tier only)

The complete definition of the intensity of a beam of electromagnetic radiation is the energy it transfers *every second per square metre of surface*:

intensity = energy transferred per second per m$^2$

If you move a torch closer to a surface, its light spreads over a smaller area and becomes brighter. The total energy from the torch each second is the same, but the intensity of the light on the surface increases. Move the torch away again and the intensity of light on the surface decreases. This happens because as the distance doubles, the area the energy spreads over increases four-fold, and so the intensity decreases four-fold.

When you shine a torch into a pool of water, the beam of light does not reach very far. The intensity falls to almost zero in a short distance. The light energy is absorbed by molecules in the water as the light travels through it. Some materials absorb electromagnetic radiation more than others. Light does not travel as far in water as it does in air. The type of electromagnetic radiation also affects how well it is absorbed. X-rays travel easily through paper although light does not.

**Watch out!**
The total energy given out by a source can remain constant, but the intensity of the radiation depends on the distance away from the source.

### QUESTIONS

**5** How does the intensity of light from a spotlight change when the area it spreads over becomes three times bigger?

**6** How does the intensity of a beam of light change when viewed from a distance three times greater than before?

intensity electromagnetic radiation

# Ionisation

**We are learning to:**
- understand what is meant by ionising radiation
- understand how ionisation occurs
- understand that exposure to ionising radiation can damage living cells

## How can you purify water in a bottle?

Dirty water kills millions of people every year. Chemicals like chlorine and iodine kill the microorganisms that cause illness, but these chemicals taste nasty. A new method uses a very fine filter inside a bottle to remove sediments. The microorganisms remaining in the water are killed using an ultraviolet bulb that gives off rays which damage their cells.

## What is ionisation?

Everything is made up from the basic building blocks of matter: atoms, molecules and ions. These basic building blocks are formed from even smaller particles, which include **electrons**. The building blocks are tiny – if you lined up one million atoms, the line would be less than a millimetre long.

> **Atoms** are the smallest particles of an element. Pure elements like copper contain just one type of atom.

FIGURE 1: All atoms in a piece of copper are the same.

> **Molecules** are the smallest part of a substance, and are formed from more than one atom joined together. Compounds are formed from two or more types of atoms.

> **Ions** are charged particles, formed when atoms or molecules are broken into smaller pieces. When an ion forms, either electrons are knocked out of an atom or molecule, or electrons join onto an atom or molecule. Either way, charged particles are formed because electrons have a negative charge. This is called **ionisation**.

All ions have a charge. Positively charged ions have lost electrons, and negatively charged ions have gained electrons.

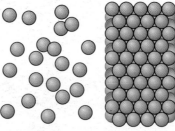

FIGURE 2: When an atom of sodium loses an electron it becomes a sodium ion.

### QUESTIONS

1. Write down one difference between ions and atoms, and one difference between ions and molecules.

2. What charge does an ion have in these cases?
   a. When it has lost an electron.
   b. When it has gained an electron.

## How electromagnetic radiation causes ionisation

When an atom absorbs a gamma ray photon, the photon may have enough energy to knock an electron out of the atom. A molecule that absorbs a gamma ray photon may break into bits, each bit being an ion. Gamma rays are a type of **ionising radiation**. Other types of ionising radiation include X-rays and high-frequency ultraviolet radiation.

The photons of gamma rays have the highest frequency and so the greatest energy compared to the rest of the electromagnetic spectrum.

Ionisation can also happen when a molecule absorbs X-rays and high-frequency ultraviolet radiation, as long as each photon has enough energy to knock an electron out of the molecule.

electromagnetic spectrum    ionising radiation

# P2 Radiation and life

Ionisation does not happen with visible light, microwaves, infrared or radio waves because the photon energy of each of these types of radiation is too small to knock electrons out of any molecules or atoms.

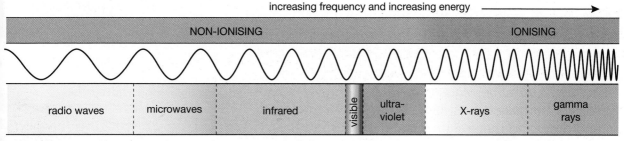

FIGURE 3: High frequency electromagnetic waves are ionising.

## QUESTIONS

3 Put these statements describing the ionisation process in the correct order, choosing the correct words.
 A positive/negative ion is formed.
 An electron/neutron is knocked out of the atom.
 An atom absorbs an X-ray/microwave photon.

4 Explain why an intense beam of radio waves cannot cause ionisation but a much fainter beam of gamma radiation can.

### Watch out!

Each photon must have enough energy to ionise a molecule. You cannot add the energy from two photons together (so visible light, microwaves, infrared and radio waves never cause ionisation).

# How ionisation is harmful

If molecules in our cells are ionised, the processes that occur within the cells change.

> Ultraviolet rays from sunlight can damage skin cells and excess exposure can eventually lead to skin cancer.

> Too much exposure to X-rays can cause cancers to develop.

> Gamma rays can severely damage cells, causing cancers and even cell death.

The damage to cells is greater when we are exposed for longer periods of time, or to a higher intensity of radiation.

## Damage to cells (Higher tier only)

The ions produced by ionising radiation take part in chemical reactions. Because they are charged particles, these reactions are different from the reactions involving atoms and molecules. In our body cells, processes can go wrong. Damage to the DNA in our cells results in **mutations** and some of these are the starting points for cancerous growths.

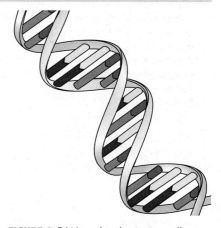

FIGURE 4: DNA molecules in our cells are complex. Ionisation alters how they instruct our bodies to function.

## QUESTIONS

5 Why is ionisation a problem in the human body?

6 Explain which form of ionising electromagnetic radiation is most harmful to humans.

### Did you know?

Our cells are being damaged all the time by low levels of ionising radiation from our surroundings. Most damaged cells either repair themselves or are replaced by healthy cells.

electromagnetic spectrum    ionising radiation

# Effects of ionising radiation

**We are learning to:**
- understand the risks and uses of X-rays and gamma rays
- understand precautions to take with X-rays and gamma rays
- understand how people perceive and accept risks

## Where are accidents most likely?

For most people most of the time, home feels like the safest place you could be. In fact, it is full of risks that we often forget are there. Over half a million children are admitted to hospital every year in the UK after accidents at home. We are so used to the everyday risks of burns and cuts, falls and trips, electrocution and drowning that we underestimate the dangers.

## Gamma rays and X-rays

**Gamma rays** have the most energy of all electromagnetic radiation as they have the highest frequency. They are capable of ionising (knocking electrons out of) molecules of the material they pass through.

**Radioactive** materials emit gamma rays as well as other ionising radiation. Gamma rays penetrate easily into the human body, causing ionisation in cells. This changes the molecules in the cells, which can change the reactions taking place inside them. Over time, living cells may die or become cancerous because molecules in the cells are damaged.

**X-rays** are slightly less energetic than gamma rays, but can still damage cells by ionising the molecules inside them. Bones absorb X-rays. This is why we can use X-ray shadow pictures to see if a bone has been damaged.

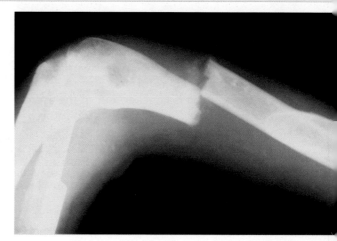

FIGURE 1: This X-ray of a broken arm shows that the risks of not taking an X-ray outweigh the slight risk caused by having the X-ray taken.

### Did you know?
We are naturally exposed to ionising radiation. The extra radiation from a chest X-ray is equivalent to a few days of natural exposure. A spinal X-ray is equivalent to about a year's exposure.

### QUESTIONS

1. Why can cells inside the body be damaged by gamma rays?

2. Suggest why gamma rays pass through bones but X-rays do not.

## Protection from ionising radiation

X-rays are used in many different ways. In hospitals, X-rays provide useful information about a patient's bones and tissues. In airports, X-rays are used to inspect objects in a passenger's luggage without opening the case. The X-rays pass through the soft case and clothes, but are absorbed differently by metal objects and plastic objects. Detectors produce an image of what is inside the bag.

People who work with X-rays need protecting from small doses of radiation over months or years. To reduce the damage to their cells, they leave the room while an X-ray is taken, or stand behind a special barrier that absorbs X-rays. Very energetic ionising radiation is best absorbed by heavy, dense materials such as thick lead and concrete.

X-rays gamma rays uses risks GCSE

# P2 Radiation and life

> Barriers built from lead or concrete protect people from exposure to X-rays and gamma rays.

> Radioactive materials are stored in lead containers.

> When an X-ray is taken, the rest of the patient's body is shielded with a lead apron to reduce damage to healthy cells nearby.

**Radiographers** prepare and analyse X-ray images, as well as treating patients with radiation as needed.

## Assessing risks

People are more willing to accept risks if:

> they choose the risk rather than having it imposed on them

> the effects are short-term rather than long lasting.

Radiation workers increase their risk of exposure to X-rays and gamma rays. They can choose whether to accept this risk. The risk can be controlled using dosimeters. These monitor a person's exposure to X-rays and gamma rays over time. Problems can be spotted before too much harm is done.

People who have mobile phone masts built near their homes have exposure to microwaves imposed on them. They tend to be less willing to accept any level of risk, and are more likely to overstate possible risks.

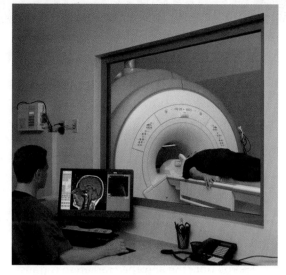

**FIGURE 2:** The glass screen contains lead to protect radiographers from ionising radiation.

### QUESTIONS

**3** How do lead and concrete reduce the risks from gamma and X-rays?

**4** Why are X-rays and gamma rays useful?

**5** Why are people more willing to take antibiotics for a week for a minor infection, but less willing to take statins long term, which reduce the risk of heart attack? Both drugs have some side effects.

**Watch out!**
An increased risk of accident or illness just means it is more likely people will have an accident or become ill. It does not mean that the illness or accident will definitely happen. If the risk of a problem is low, doubling the risk is still a low risk.

## Radiation all around us

We are surrounded by ionising radiation all the time. Most of our exposure comes from radioactive radon gas seeping naturally from rocks and building materials in our surroundings, from the food we eat and from space (cosmic rays). The risk from radiation to health is very low, but people tend to overstate it. This is because:

> radiation is unfamiliar to us

> it is invisible and hard to detect without special equipment

> its effects can be long lasting, for example causing cancer.

### QUESTION

**6** It is much safer to travel in a plane than travel the same distance in a car. Explain why many people feel the risk of flying is greater than the risk of driving.

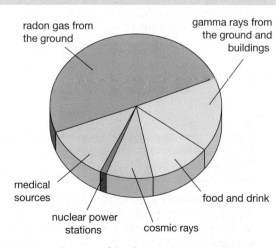

**FIGURE 3:** Sources of 'background radiation'.

Q X-rays gamma rays uses risks GCSE

# Microwaves

## Why do eggs explode in the microwave?

People have been badly injured after microwaving whole eggs. The egg absorbs so much energy, it reaches a far higher temperatures than when boiled in water. Water inside the egg changes to steam, causing the egg to explode, either in the oven or when it has been taken out. If an egg needs to be cooked whole in a microwave, it *must* be pierced first.

> **We are learning to:**
> - understand that the heating effect of radiation can damage cells
> - understand how microwaves can be useful but why some people have concerns
> - evaluate how scientists investigate factors that may be causing an effect
> - understand how risks may be assessed

## Radiation and heating

When a substance **absorbs** radiation it warms up because energy from the absorbed radiation is transferred to thermal (heat) energy. If living cells absorb radiation, this heating effect can damage the cells. More energy is absorbed if:

> the radiation is very intense

> the radiation is absorbed for a longer time.

### QUESTION

1. How can radiation damage living cells?

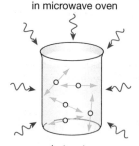

FIGURE 1: Absorbed radiation has a heating effect.

## Using microwave ovens

Energy from **microwave** radiation is absorbed by food in a microwave oven. Water molecules in the food strongly absorb microwaves. The energy from the microwaves is transferred to thermal energy (the water molecules move and vibrate faster). This heat cooks the food. Settings on the microwave oven alter how long the food is cooked for, and how intense the beams of microwaves are. Larger portions of food need a longer cooking time, or a higher setting.

Microwaves reflect off metal sheets and mesh. The oven is surrounded by a metal case, and the door has a metal screen. This ensures that microwaves stay inside the oven and do not affect people nearby.

FIGURE 2: Metal sheets and mesh shield the user from microwaves from the oven.

244  effect of microwaves on cells    uses of microwaves

## P2 Radiation and life

### QUESTIONS

**2** What two things affect how much energy is absorbed from microwave radiation?

**3** Why is it important that microwave ovens have shielding?

### Did you know?

The greatest risk from food cooked in microwave ovens is when food is not cooked or heated properly. Some parts of the food cook more quickly than others.

## Protection from microwaves

The risks to living cells from microwaves are very low because microwaves have a low frequency and carry small amounts of energy. Some people wonder if a long exposure to low levels of radiation can still be damaging to health.

> Mobile phones use low levels of microwave radiation. However, people use mobile phones for many years and can use them for several hours a day.

> Mobile phone masts communicate with neighbouring masts and phones using low intensity microwave radiation. Although the mast constantly emits microwave radiation, the radiation intensity is extremely small at the base of the mast.

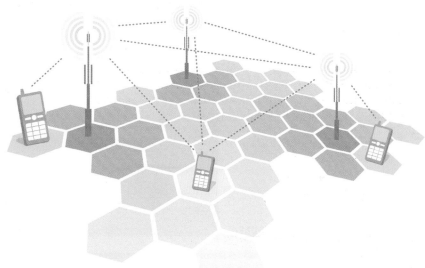

**FIGURE 3:** Calls from mobile phones pass through many masts before linking up with the person receiving the call.

Studies carried out to assess if mobile phone use increases cancer have used different groups (or samples) of people. These included:

> small groups, matched in as many ways as possible, but with different patterns of phone usage

> very large groups across many countries, chosen at random.

No clear evidence of increased or decreased risk has been found. As more studies are carried out, scientists can be more confident that this conclusion is right.

These studies are important for governments and public bodies to assess how much risk is acceptable. Mobile phone users who do not live near masts may feel a higher risk is acceptable. People living near the masts have to put up with greater risks for the same benefit.

### Watch out!

When researching controversial topics like mobile phone risks, choose sources of information carefully. Check facts from several reputable sites to avoid **bias**.

### QUESTIONS

**4** What factors affect the risks from mobile phone usage?

**5** Explain why scientists are increasingly confident that the risks from mobile phone usage are small.

**6** Discuss whether the government should ban mobile phone masts from locations near schools.

effect of microwaves on cells    uses of microwaves

# Ozone

**We are learning to:**
- understand how ozone protects living organisms
- understand how to decide whether to accept different risks

## Can your old fridge cause sunburn?

Fridges used to be cooled using chemicals called CFCs. These were safe and stable chemicals. No-one realised CFC molecules damaged the ozone layer in the atmosphere. This meant that more ultraviolet from the Sun reached certain places on Earth, increasing the risk of sunburn and skin cancer. Now CFCs are banned.

FIGURE 1

## What ozone is

**Ozone** is a form of oxygen found in the outer layers of our atmosphere. Its molecules contain three oxygen atoms. Ozone is very effective at absorbing ultraviolet radiation from the Sun. This means that less ultraviolet radiation reaches Earth. Most ozone is found about 32 km above the Earth's surface.

Ultraviolet radiation causes sunburn and can lead to skin cancer and eye damage. The ozone layer protects living organisms, especially animals, from the effects of too much ultraviolet radiation.

**Watch out!**

Ozone in the upper atmosphere protects us from ultraviolet radiation. Ozone at ground level is harmful, causing breathing difficulties and sore throats.

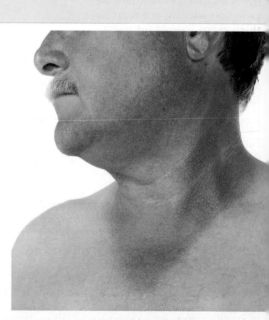

**FIGURE 2:** Sunburn is dangerous. Sunscreens and clothing can be used to absorb most of the ultraviolet radiation from the Sun.

### QUESTIONS

1. What type of radiation does ozone absorb?
2. How does ozone protect animals from sunburn and skin cancer?

ozone layer    ultraviolet radiation

## The risks from ultraviolet radiation

Everything we do carries a certain risk of accident or harm. Skin cancer is the most common type of cancer diagnosed in the UK. The risk of skin cancer increases with exposure to ultraviolet radiation, so outdoor workers are more likely to suffer from skin cancer. Nothing is risk free, even something as natural as enjoying a sunny day. Staying out of the sun is no solution – too little sunlight increases our risk of rickets, a preventable disease of the bones.

Sometimes new technologies cause new risks. Over 80 years ago, specialist chemicals containing fluorine, chlorine and bromine were invented with many uses including refrigeration and fire-fighting. About 40 years ago researchers realised that these chemicals were damaging the ozone layer. Particularly above the Antarctic, there was less ozone in the atmosphere so more ultraviolet radiation reached Earth.

We can assess the risk of skin cancer caused by depletion of the ozone layer. Studies of thousands of people living in affected areas over several years suggest that if the ozone layer thins by 1%, the risk of skin cancer increases by about 4%.

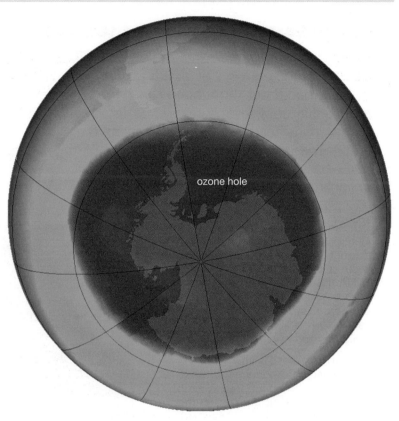

**FIGURE 3:** The blue colour in this image shows the regions in the atmosphere above the Antarctic where the concentration of ozone is low.

### QUESTIONS

**3** Why can't we reduce our risk of skin cancer to zero?

**4** How do scientists assess how the thinning ozone layer affects our risk of skin cancer?

### Did you know?

The amount of ozone in the atmosphere varies with the location on Earth, the type and amount of man-made chemicals in the atmosphere, the season and the conditions in the atmosphere.

## What happens in the ozone layer? (Higher tier only)

Ozone is produced when oxygen molecules absorb high energy ultraviolet radiation and react together. Ozone molecules may split into an oxygen molecule and a free oxygen atom if they absorb low energy ultraviolet radiation. In both these reactions, infrared radiation is emitted. Ozone also reacts with nitrogen, hydrogen, chlorine and bromine compounds.

The absorption of ultraviolet radiation in the ozone layer reduces the amount of ultraviolet radiation reaching the Earth's surface. It also causes chemical changes in this part of the atmosphere.

### QUESTIONS

**5** How does the energy of the ultraviolet radiation affect the chemical reactions that take place?

**6** What happens to the energy from the ultraviolet radiation when it is absorbed by a molecule of oxygen or ozone?

ozone layer    ultraviolet radiation

# The greenhouse effect

**We are learning to:**
> understand what the greenhouse effect is and its causes
> understand that scientists need good evidence to be sure that a factor causes an outcome

## Why doesn't it freeze at the equator?

When clouds block the Sun's rays on a sunny day, it feels colder. But the presence of clouds actually warms the Earth. If clouds had never surrounded Earth, the temperature, even at the equator, would be below freezing. The water vapour in the clouds helps to trap and re-radiate the energy that originally came from the Sun back to the Earth's surface.

## The greenhouse effect

The Sun emits different frequencies of electromagnetic radiation, which travel through space to reach Earth. Our Earth is surrounded by a layer of gases called the **atmosphere**. When the radiation from the Sun reaches the atmosphere, some frequencies pass through. The Earth's surface absorbs most of this radiation, and warms up.

The Earth's surface itself also emits electromagnetic radiation (infrared) as it warms up. When infrared radiation from the Earth's surface reaches the atmosphere:

> the atmosphere *absorbs* and *re-radiates* most of the radiation. Some of the re-radiated radiation reaches the Earth

> the atmosphere *reflects* some radiation back to Earth

> a small amount passes through the atmosphere to space.

The infrared radiation is absorbed by the Earth's surface and atmosphere, which both become hotter. The warming of the Earth's atmosphere is called the **greenhouse effect**. The Earth is about 30°C hotter than expected for its distance from the Sun. The greenhouse effect is one reason why Earth is the right temperature to sustain life.

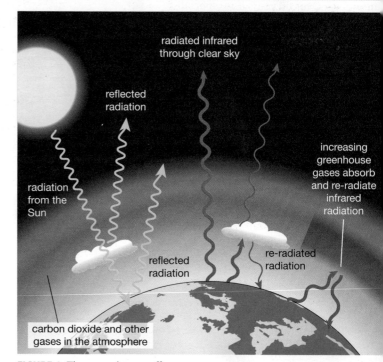

**FIGURE 1:** The greenhouse effect.

**Watch out!** The atmosphere does not form a solid barrier, so it does not act in exactly the same way as the glass in a greenhouse.

### Did you know?

The greenhouse effect happens on other planets too – Venus has the highest average temperature in the solar system even though Mercury is closer to the Sun.

### QUESTIONS

1 What is the greenhouse effect and how does it warm the Earth?

2 How does the atmosphere affect the amount of radiation **a** reaching and **b** leaving Earth?

greenhouse effect    greenhouse gases GCSE

## Temperature and electromagnetic radiation

All objects, large and small, emit some electromagnetic radiation at all temperatures. This radiation has a range of frequencies in varying amounts. The frequency emitted in the highest intensity is the **principal frequency**. The principal frequency of the emitted radiation increases as the temperature rises. This can be seen as a colour change for objects that are hot enough to emit visible light. A metal rod glows red when it is first heated, becoming yellow then white hot as it gets hotter. Hotter objects radiate higher frequencies of radiation than cooler objects.

The principal frequency of electromagnetic radiation from the Earth's surface is lower than the principal frequency emitted by the Sun. This is because the Earth is cooler than the Sun.

FIGURE 2: The hottest part of the metal is bright yellow.

### QUESTION

**3** Explain which of these emits the lowest frequency of radiation: an ice cube, a glass of cold water, a cup of hot water.

**4** Do greenhouse gases absorb more or less high-frequency radiation than infrared radiation?

## The cause of global warming?

The level of carbon dioxide in the atmosphere has been rising for more than 50 years. The global average temperature has also risen over this period. Scientists could see that the two effects were related. It took longer for many scientists to accept that rising carbon dioxide levels could *cause* the greenhouse effect.

A factor does not always cause the other even if the two are correlated. Once a plausible mechanism was suggested explaining how rising carbon dioxide levels could increase global temperatures, more scientists accepted that this was a possibility.

### Effects of different greenhouse gases (Higher tier only)

The atmosphere contains several greenhouse gases. Water vapour is a greenhouse gas, responsible for over two-thirds of the greenhouse effect. The activities of humans have very little effect on the amount of water vapour in the atmosphere. **Methane** (present in very small amounts) also adds to the greenhouse effect. Each molecule of methane contributes about 20 times more to the greenhouse effect than a molecule of carbon dioxide, although the amount of methane present is much less than carbon dioxide. Different amounts of these gases are present in the atmosphere at different times.

FIGURE 3: Water vapour in the atmosphere is the main cause of the greenhouse effect.

### QUESTIONS

**5** Explain why the amount of water vapour in the atmosphere varies.

**6** People grow rapidly in their teenage years. They also take more exams then than at any other time in their lives. Explain whether the two observations are linked, or whether one causes the other.

# Carbon cycling

## Where have you been before?

All the carbon in your body existed when the solar system formed billions of years ago. Carbon is constantly cycling through the land, sea, and atmosphere and even inside the Earth. You cannot tell if the carbon atoms in your cells have already been in stars, in coal or even in a worm.

**We are learning to:**
- describe the carbon cycle
- understand why the amount of carbon in the atmosphere is increasing
- understand that a correlation may not mean that one factor causes an outcome

## Recycling carbon

Carbon is found in all living things. Carbon in your body comes from the food you eat and the air you breathe. The carbon from carbon dioxide in the atmosphere is constantly being recycled through **photosynthesis** and **respiration**, as Figure 1 shows.

This diagram shows only some processes in the carbon cycle. Carbon from organisms is also released as carbon dioxide when they decompose. Carbon from some plants and animals that were alive millions of years ago is stored in **fossil fuels**. When fossil fuels are burned, carbon dioxide is released into the atmosphere.

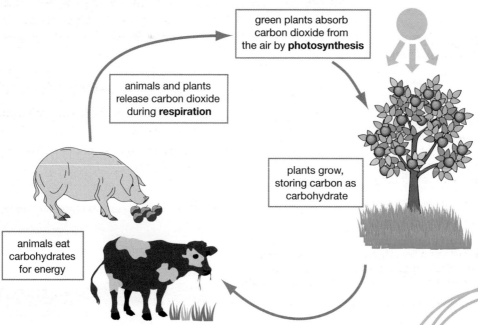

**FIGURE 1**: Part of the carbon cycle.

### Watch out!

Carbon dioxide does not have its own cycle, but carbon does. The carbon changes from carbon dioxide in the atmosphere to more complex compounds in living things.

 **QUESTIONS**

1. Which processes absorb carbon dioxide from the atmosphere?
2. Which processes release carbon dioxide to the atmosphere?

carbon cycle GCSE

## Carbon dioxide in the atmosphere

For thousands of years, the levels of carbon dioxide in the Earth's atmosphere stayed approximately constant. Carbon absorbed and stored by plants matched the amount released during rotting (decomposing) and respiration. Trees hundreds of years old in ancient forests stored large amounts of carbon. Over millions of years, fossil fuels formed from dead plants and animals that did not decompose, or decomposed very slowly.

About two hundred years ago, a big social change started to spread across the world. During the Industrial Revolution, machines powered by fossil fuels started manufacturing goods on a large scale in factories. Carbon locked away in the fossil fuels for millions of years was released to the atmosphere.

Forests were cut down or burned to clear the land and provide fuel. This left fewer trees to absorb carbon dioxide. Carbon dioxide was also released as the wood was burned or left to rot.

In the past two hundred years, the amount of carbon dioxide in the atmosphere has been steadily rising. This century, the amount of carbon dioxide released from **deforestation** and burning fossil fuels has increased greatly.

**FIGURE 2**: Deforestation and burning fossil fuels increases the amount of atmospheric carbon dioxide.

### Did you know?

We estimate how much carbon dioxide the atmosphere has contained in the past by studying air bubbles trapped in ice cores in the Antarctic.

### QUESTIONS

**3** Explain why more carbon dioxide is now being released to the atmosphere than is absorbed.

**4** Has the total amount of carbon on Earth and in the atmosphere changed? Explain your answer.

## Is there a link?

If there is a **correlation** between two things, then increasing one will increase (or decrease) the other. If there is a correlation between atmospheric carbon dioxide and burning fossil fuels, then:

> burning more fossil fuels will increase the amount of atmospheric carbon dioxide

> burning less fossil fuels will decrease the amount of atmospheric carbon dioxide.

There may be a correlation if one factor affects the chance of something happening, even if it does not cause it. For example, it is likely that a larger population will burn more fossil fuels.

Sometimes, two things increase at the same time but are not correlated. The levels of carbon dioxide and worldwide bicycle sales rose at the same time. These factors are not correlated, but both are affected by the increase in population.

### QUESTIONS

**5** How can you tell if two things are correlated?

**6** One sunny morning, it became warmer as the time increased from 6 am to 12 noon. Explain if there is a correlation between the time and warmth. Do any other factors affect the correlation?

# Global warming

**We are learning to:**
- understand some of the possible results of global warming
- understand how risks are assessed

## Why did the mammoths die?

Global warming and cooling have taken place in regular cycles over thousands of years. Twenty-one thousand years ago, rising temperatures allowed forests to spread over prairies and grasslands. This may have killed off the mammoths.

FIGURE 1

## What is global warming?

**Global warming** is the increase in average temperature worldwide caused by the greenhouse effect. The greenhouse effect occurs naturally. But scientists are concerned that recent increases in greenhouse gas emissions will cause global warming to continue and at a possibly increasing rate.

If global warming continues, the climate will change, with some places becoming hotter, drier or damper. It may become impossible to grow certain crops in these regions.

The ice at the poles may start melting because of global warming. Sea levels may rise if water from melted ice on land (glaciers) flows into the sea. It may also rise as the water in the ocean expands as it warms. Rising sea levels will leave low lying islands vulnerable to flooding.

Global warming may increase the amount of severe weather in places. It may cause more severe or frequent heat waves, storms, hurricanes, floods, rain or snow.

FIGURE 2: Sea levels could rise when water from melting glaciers reaches them.

**Watch out!** Climate and weather are not the same. Climate is the long-term weather trend over years. Weather events are short term and have many causes.

### Did you know?
The rise in sea levels is measured using satellites to take measurements every few days, accurate to several centimetres. This makes it possible to calculate any underlying trend.

## QUESTIONS

1. How are global warming and greenhouse gases linked?
2. Write down three effects of global warming.
3. Explain two reasons why sea levels may rise.

🔍 global warming GCSE

## What should we do?

We are not sure what the scale of global warming will be. If we do nothing now, we may be able to adapt successfully, but large changes will be very difficult to adapt to. Governments making decisions about global warming must consider how likely it is that global warming is taking place because of human activity. They must also consider how serious its effects may be.

Most governments now believe we have enough evidence that global warming is taking place and we should take steps to reduce it. We do not know for certain how serious the risks and benefits of global warming will be. It is important to consider the effects of global warming on different individuals and groups, and the steps taken to reduce its impact.

> **QUESTIONS**
>
> **4** Why have governments decided to take steps to reduce global warming?
>
> **5** Reducing global warming costs money. Many governments have started charging 'green taxes' on some activities and products to cover these costs. How should they persuade people that this tax is worth paying?

## Causes of extreme weather (Higher tier only)

More severe weather can happen when the atmosphere is warmer. A larger temperature difference between the atmosphere and the oceans may increase **convection** – the transfer of energy through the atmosphere by circulating masses of air at different temperatures. Warm air holds more water vapour than cooler air, so the amount of water vapour in the atmosphere increases. In the right conditions it is released as heavy rain or snow. This **hypothesis** is still being debated and the processes taking place are not yet fully accepted or understood.

### Computer climate models

Climate scientists rely on computer models to make forecasts. The computer model is tested using past data to confirm which assumptions should be included in the model. These assumptions include changes caused by human activities such as increased burning of fossils fuels. Once the test model runs, scientists check if its predictions match what actually did happen.

Predictions from computer models of global temperature increase and climate change are increasingly accurate. This enables climate scientists to be confident that human activities included in their models are one of the principal causes of climate change. While scientists agree that climate change is occurring, some point out that human activity is only partly responsible. There are many factors that affect the world's climate, and natural changes take place over thousands of years. The computer models are simpler than real life, and assumptions need to be made when designing them.

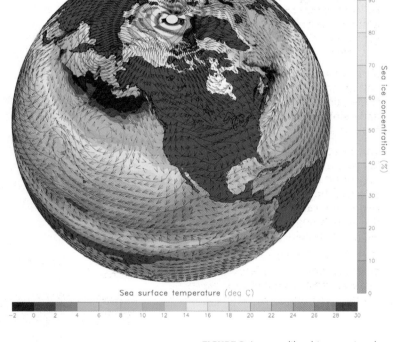

**FIGURE 3**: Images like this contain a large amount of information about surface temperatures, sea ice concentrations and sea level pressures.

> **QUESTION**
>
> **6** Explain the benefits and weaknesses of using computer climate models.

computer climate models

# Electromagnetic waves for communication

**We are learning to:**
> understand why some types of electromagnetic radiation are used for communication
> understand how information is carried by electromagnetic waves

## Keeping in touch from space

The first astronauts who orbited the Moon communicated with Earth using radio waves. In each orbit, radio contact was lost for 45 minutes while the spacecraft was on the far side of the Moon. While on the Moon, the astronauts used radio waves to speak to each other as sound waves cannot travel in space.

**FIGURE 1**: In December 1968, on the first manned mission to orbit the Moon, James Lovell was able to speak to his mother.

## Waves that can carry information

Conversations, music, pictures and other information can be carried from place to place using electromagnetic waves. The main types of radiation used for communication are radio waves, microwaves, infrared radiation and visible light.

### QUESTION

1 List the types of radiation that are used for communication.

## Using waves for communicating

Radio waves transmit terrestrial TV and radio programmes from a transmitter to a receiver. The waves are not strongly absorbed by air so they can travel many kilometres. Radio waves can spread around hills and buildings, and reflect off layers in the atmosphere.

Microwaves have a shorter wavelength than radio waves, and are not absorbed much in air. Narrow beams of microwaves travel many kilometres through the atmosphere and communicate with satellites. We use microwaves for satellite TV broadcasts, satnav and mobile phone connections.

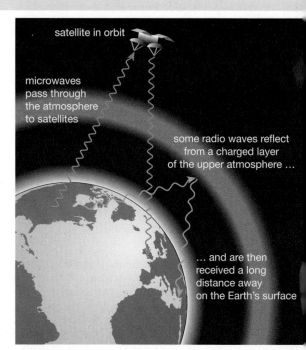

**FIGURE 2**: Microwaves can pass through the atmosphere to satellites, but longer wavelength radio waves are reflected.

254  electromagnetic communication GCSE    analogue broadcast GCSE

**P2 Radiation and life**

Wii controllers and TV remotes use infrared radiation, which only travels short distances in air, and cannot pass through walls.

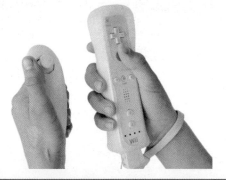

Very narrow glass fibres called **optical fibres** carry infrared and light signals long distances. Infrared radiation and visible light are not absorbed much in glass, repeatedly reflecting off the sides of the glass fibre. Optical fibres are used in telephone, internet and TV cables. More than one signal can pass through an optical fibre at the same time.

### QUESTIONS

**2** Why are optical fibres used in communications?

**3** Explain why infrared radiation is not used to communicate with satellites.

## How waves carry the information

You tune your radio to a certain frequency to listen to your favourite radio station. This frequency is the frequency of the carrier wave.

> A **carrier wave** is a radio wave which carries information from the broadcasting station to your radio.

> Information containing the programme's sounds or images is added to the carrier wave. The carrier wave is **modulated** (changed).

> This creates a **signal** that is transmitted to your radio set.

> Your radio can separate out the modulation from the carrier, so you can hear the programme.

In **analogue** broadcasting, the signal added to the carrier wave can vary continuously. This is an **analogue signal**. The frequency of the carrier wave and its amplitude can have any value.

Carrier wave (no programme is broadcast)

Sound signal from the programme to be broadcast

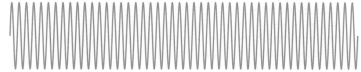

Carrier wave modulated by sound signal

**FIGURE 3**: Modulating a carrier wave.

**Watch out!** You don't hear radio waves. Your radio detects the signal and converts it into sound waves.

### QUESTIONS

**4** Explain which part of the transmitted radio wave changes if you change radio station.

**5** Why do different radio stations use different frequencies?

**6** Figure 3 shows a modulated wave transmitted at one frequency. If the same programme were transmitted at a lower frequency, how would the modulated wave different?

### Did you know?

The first sound signal was transmitted across the Atlantic in 1907.

electromagnetic communication GCSE    analogue broadcast GCSE

# Preparing for assessment: Evaluating and analysing evidence

*To achieve a good grade in science, you not only have to know and understand scientific ideas, but you need to be able to apply them to other situations and to analyse evidence. These tasks will support you in developing these skills.*

## The beginnings of satellite TV

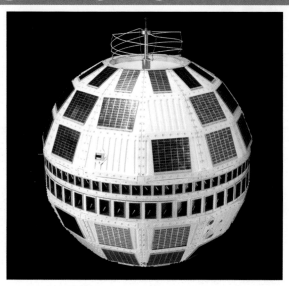

In 1962, television was about to enter a whole new world – that of programmes being transmitted from one continent to another. The plan was to use three huge transmitter and receiver sets, one on the east coast of America, one in France and one at Goonhilly, in Cornwall. The one in America could beam signals up to a satellite in orbit above the Earth, which would then beam them down to England and France, and vice versa. Telstar was the satellite built to do this job. It was spherical and about one metre in diameter. It had a mass of 77 kg and was powered by solar cells.

The satellite was launched into low orbit in April 1962. It orbited the Earth once every 2 hours 37 minutes. It didn't transmit directly to people's homes, but to the TV company which edited and re-transmitted to viewers. It could only connect the USA with Britain and France for about 20 minutes of its orbit, and then vanished for over 2 hours.

The huge aerial at Goonhilly (called Arthur, after the legendary king) had to pan across the sky and then move back, ready for the next pass. Its parabolic reflector was over 26 metres across and weighed over 1000 tonnes.

In fact, Telstar did not last long. The transistors in its circuits were damaged by radiation released by weapons tests and although engineers managed to get it working again, it wasn't to be for long. Within a year of launching, it was out of use, but it had shown the way for global links.

TV signals are still relayed across the Atlantic (and elsewhere) but the satellites used these days are all geostationary ones. They are in orbit much further out than Telstar, and orbit the Earth at the same rate that the Earth rotates. They are therefore always in the same place relative to a point on the surface of the Earth.

Satellites can now broadcast direct to people's homes. Arthur has been consigned to history. Most of the North Atlantic telecommunications traffic, however, now goes through low-cost submarine cables.

## Task 1

> How was Telstar powered? Why was it designed this way?

> Why did the aerial at Goonhilly have to be so large compared with modern satellite dishes?

## Task 2

Telstar was in a much lower orbit than modern TV satellites and orbited the Earth more quickly.

> How did that give it an advantage?

> In what way was this a significant disadvantage?

## Task 3

> Draw and label a diagram to explain why it isn't possible to just beam signals directly from the USA to the UK.

> Explain why it was difficult to pick up the signal from the satellite. Why do you think the TV pictures were of rather poor quality by modern standards?

## Task 4

> The signal to Telstar was broadcast as a microwave. How fast would it be travelling?

## Task 5

> Why would satellites like Telstar not be suitable for providing satellite TV directly to people's homes? You should be able to think of several reasons.

## Maximise your grade

These sentences show what you need to include in your work to achieve each grade. Use them to improve your work and be more successful.

For grade E, your answers should show that you can:
> describe how a source emits electromagnetic radiation which can be reflected, transmitted, and detected by a receiver
> describe how intensity of radiation varies with distance from the source
> understand that higher quality sound or images use more information

For grades D, C, in addition show that you can:
> understand that electromagnetic radiation affects a receiver when it is absorbed
> understand that electromagnetic radiation spreads over an increasing surface area and is partially absorbed further from the source
> explain the implications of higher quality sound or images using more information

For grades B, A, in addition show that you can:
> explain how the properties of waves determine the transmission and reception of signals
> explain the need for higher transmission rates to improve signal quality

# Digital signals

**We are learning to:**
- know what digital signals are
- understand how digital signals are transmitted
- describe advantages of digital signals

## How is a nerve impulse like a digital signal?

Nerve cells in your body are either turned on or off. When they are turned on, impulses travel through to your brain. When they are off, no impulse is passing through. In the same way, a digital signal is either on or off. The information about an image, for example, can be stored or sent digitally as a series of data with the value 0 or 1. These numbers can then be used to re-create the original picture.

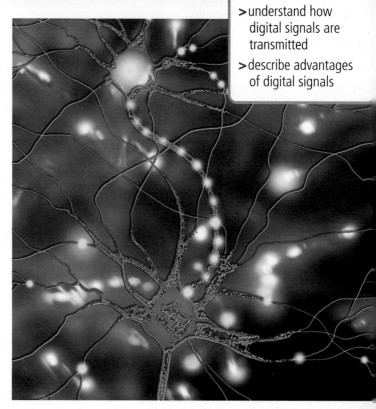

FIGURE 1: Nerve cells carry electrical signals or impulses.

## What digital signals are

A signal is used to carry information from a transmitter to a receiver. A **digital signal** is a type of signal that can take one of a small number of fixed (discrete) values, usually two. The values change in discrete steps, and cannot take any values in-between.

Many TV and radio programmes are transmitted using digital signals. Sounds and pictures can be converted into digital signals that are streams of just two values, 0 and 1.

You can use a lamp to create a simple digital signal. The lamp is either on or off. Turning the lamp on represents a signal value of 1, and turning the lamp off represents a signal value of 0.

This is different from analogue signals, which can vary continuously. A lamp with a dimmer switch can be set to any value of brightness to give an analogue signal.

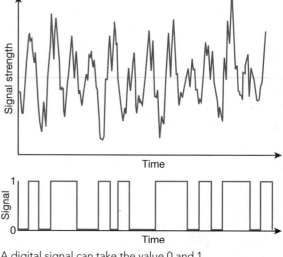

An analogue signal can take any value.

A digital signal can take the value 0 and 1.

FIGURE 2

### QUESTIONS

1. How are digital signals different from analogue signals?
2. How could you create a digital signal using a buzzer?

digital signal GCSE   analogue signal GCSE

# Transmitting digital signals

When information is to be broadcast the height of the analogue signal is measured regularly. Each measurement is coded as a string of numbers, either 0 or 1. An example is:

The codes for successive measurements are joined together in one long string. This string of numbers can be transmitted using an electromagnetic carrier wave. The carrier wave is turned on or off, creating short bursts of waves called **pulses**. The value of the signal is 1 if there is a pulse, and it is 0 if there is no pulse. The signal above would change into a wave transmitted like this:

off, off, pulse, off, pulse, pulse, off, pulse, off

Once the digital signal is received, the series of pulses is **decoded** inside the radio, TV or other digital device. This produces a copy of the original sound wave or picture that the user can hear or see.

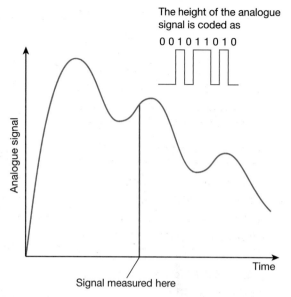

**FIGURE 3**: The height of the analogue signal is measured. Its value is turned into a digital signal at regular intervals.

## QUESTIONS

**3** How does the carrier wave change when the digital signal changes?

**4** Why is a decoder needed?

# Advantages of digital signals

There are several advantages of sending information using digital signals:

> Different types of information can be sent at the same time (such as a picture inside a written document).

> Digital signals can be sent more accurately than analogue signals.

> Digital information can be stored and processed by computers.

> Unwanted information mixed in with the original signal, called **noise**, can be removed more easily and the original signal recovered. Noise can be caused by interference with signals from other equipment or the pick-up of a random signal.

## QUESTIONS

**5** Why is noise a problem in communication?

**6** How can a decoder tell if part of a digital signal is caused by noise?

## Why noise is less of a problem with digital signals (Higher tier only)

Digital signals are not affected by noise as badly as analogue signals. During transmission, all signals may be amplified (made stronger). They are all decoded by the devices that receive them.

Analogue signals can take any value so when they are amplified or decoded, it is not possible to tell which part of the signal is real and which part is noise. Digital signals can only have certain values, so the decoders and amplifiers assume signals close to a certain value have that value. The decoder ignores parts of the signal that are not close to expected values.

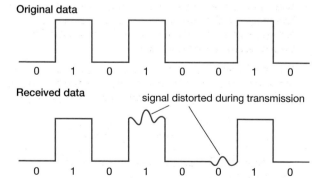

**FIGURE 4**: A digital signal can still be 'read' if it is distorted with noise.

# Storing digital information

**We are learning to:**
> understand how digital information is stored
> understand the advantages of using digital information

## Has your memory improved?

In 1969 when the first astronauts landed on the Moon, the computers in the rocket and at mission control used less memory between them than a simple memory stick contains today.

### Storing images

The strings of 0s and 1s that make up digital information are called **binary digits** or 'bits'. To store text like that on this page, letters are converted to binary digits. Each letter is represented by a string of eight digits, which is called a **byte**. The letter A is represented as 01000001.

All digital information is converted into a string of binary values, measured in bytes: each eight digits make 1 byte (1 B) of information.

To store an image digitally, the area of the image is divided into lots of tiny boxes. The colour and brightness of each box are represented by a long series of binary digits.

To produce an image, the values used to store the image are assigned to tiny areas on the screen called **pixels**. The quality of a picture improves as more pixels are used. Each pixel then represents a smaller area of the original picture. Because there are more areas, more information is needed, so more data must be stored for a higher quality image. The number of pixels that form an image is called the **resolution**. High resolution images use more pixels than low resolution images for the same size of image.

Good quality digital cameras produce high resolution images.

The individual pixels can be seen on a low resolution image.

**FIGURE 1**

### QUESTIONS

1. What do bytes measure?
2. How many bytes does the word 'digital' contain?
3. How is the quality of a picture affected by the amount of data stored?

### Storing sound

When creating a digital signal from an analogue sound signal, information about the sound signal is collected (or 'sampled') at regular time intervals (see Figure 3 on page 259). To transmit or store a higher quality sound image, the signal must be sampled more often. More information about the sound is then stored and a higher quality transmission or recording is achieved.

In general, the more information that is stored (the larger the number of bytes), the better the quality when the stored data is converted back to sound.

**Watch out!**
The capacity of data storage devices is quoted in kilobytes or kB (about 1000 bytes), megabytes or MB (about a million bytes) and gigabytes or GB (about a thousand million bytes).

### QUESTIONS

4. Why is it possible to store better quality images and sound files now than a few years ago?

260    Q byte GCSE    digital information GCSE

# P2 Radiation and life

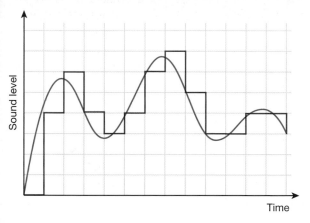 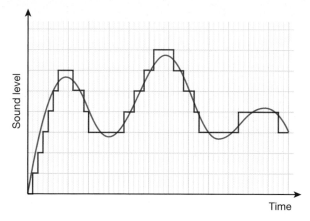

**FIGURE 2**: If a sound signal is sampled more often, the quality improves.

## Advantages of digital information

Computers store and process digital information. They can store and process any information that can be changed into digital code, such as sound, video, text and images.

Digital information is versatile: it can be processed in many different ways. A digital file stored on a digital camera, for example, can be transferred to a memory stick or a computer, it can be e-mailed, uploaded onto the internet, or sent by Bluetooth to mobile phones, and still stay in its original form. Bluetooth transmits signals wirelessly over short distances between telephones, computers and other equipment.

### Did you know?

All symbols, as well as letters, are described using digital values in 'ASCII' tables.

| Letter | ASCII code | Binary | Letter | ASCII code | Binary |
|---|---|---|---|---|---|
| a | 097 | 01100001 | A | 065 | 01000001 |
| b | 098 | 01100010 | B | 066 | 01000010 |
| c | 099 | 01100011 | C | 067 | 01000011 |
| d | 100 | 01100100 | D | 068 | 01000100 |
| e | 101 | 01100101 | E | 069 | 01000101 |
| f | 101 | 01100110 | F | 070 | 01000110 |
| g | 103 | 01100111 | G | 071 | 01000111 |
| h | 104 | 01101000 | H | 072 | 01001000 |
| i | 105 | 01101001 | I | 073 | 01001001 |
| j | 106 | 01101010 | J | 074 | 01001010 |
| k | 107 | 01101011 | K | 075 | 01001011 |
| l | 108 | 01101100 | L | 076 | 01001100 |
| m | 109 | 01101101 | M | 077 | 01001101 |
| n | 110 | 01101110 | N | 078 | 01001110 |
| o | 111 | 01101111 | O | 079 | 01001111 |
| p | 112 | 01110000 | P | 080 | 01010000 |
| q | 113 | 01110001 | Q | 081 | 01010001 |
| r | 114 | 01110010 | R | 082 | 01010010 |
| s | 115 | 01110011 | S | 083 | 01010011 |
| t | 116 | 01110100 | T | 084 | 01010100 |
| u | 117 | 01110101 | U | 085 | 01010101 |
| v | 118 | 01110110 | V | 086 | 01010110 |

**FIGURE 3**: The same digital file can be processed by all these devices.

## QUESTIONS

**5** Explain the advantages of storing data as digital information.

**6** Why is it important that the same files can be used by different computers?

🔍 byte GCSE   digital information GCSE

# P2 Checklist

## To achieve your forecast grade in the exam you'll need to revise

Use this checklist to see what you can do *now*. Refer back to pages 236–261 if you're not sure.

Look across the rows to see how you could progress – **bold italic** means Higher tier only.

Remember you'll need to be able to *use* these ideas in various ways, such as:
> interpreting pictures, diagrams and graphs
> applying ideas to new situations
> explaining ethical implications
> suggesting some benefits and risks to society
> drawing conclusions from evidence you've been given.

Look at pages 312–318 for more information about exams and how you'll be assessed.

| To aim for a grade E | To aim for a grade C | To aim for a grade A |
|---|---|---|
| understand that a source emits electromagnetic radiation which is reflected, transmitted or absorbed by materials, and which affects a detector when it is absorbed; list the electromagnetic radiations in order of frequency, and recall their speed through space | | |
| understand that energy from electromagnetic radiation is transferred by photons, and that higher frequency photons transfer more energy | understand that the energy transferred by electromagnetic radiation depends on the frequency and number of photons arriving; and that electromagnetic radiation is less intense further from the source | understand that the intensity of electromagnetic radiation is the energy arriving *per square metre* per second; and that electromagnetic radiation spreads over *an increasing surface area and is partially absorbed further from the source* |
| understand that absorbed electromagnetic radiation can heat and damage living cells | relate the heating effect of radiation to the intensity of the radiation and its duration; understand that water molecules strongly absorb microwave energy | |
| understand that some people worry about health risks from low intensity microwave radiation; understand that evidence for the health risk from microwaves is disputed | explain why evidence for the health risk from microwaves is disputed | |
| understand that high-energy ultraviolet radiation, X-rays and gamma rays are ionising radiation; understand that exposure to ionising radiation can damage living cells | understand that photons of ionising electromagnetic radiations have enough energy to remove electrons from atoms or molecules when absorbed by substances | ***understand that ionised molecules can take part in chemical reactions*** |
| recall that sunscreen and clothing absorb ultraviolet radiation; understand that the ozone layer absorbs ultraviolet radiation from the Sun, protecting living organisms | | ***understand that chemical changes occur in the atmosphere when ozone absorbs ultraviolet radiation*** |

| To aim for a grade E | To aim for a grade C | To aim for a grade A |
|---|---|---|
| understand that lead and concrete absorb X-rays; describe how X-rays can produce shadow pictures | apply understanding of the behaviour of X-rays to explain how images are produced | |
| recall that some radiation from the Sun passes through the Earth's atmosphere, warming the Earth's surface | understand that all objects emit electromagnetic radiation, with a principal frequency that increases with temperature | understand that radiation emitted by the Earth *has a lower principal frequency than radiation from the Sun, and that this radiation* is absorbed or reflected back by some gases in the atmosphere |
| recall that carbon dioxide is a greenhouse gas present in the Earth's atmosphere; explain the causes of the greenhouse effect | | *recall that greenhouse gases include methane and water vapour* |
| interpret diagrams representing the carbon cycle; recall that over the last 200 years the carbon dioxide in the atmosphere has been steadily increasing | use the carbon cycle to explain why the amount of carbon dioxide in the atmosphere was constant, and that it is now increasing due mainly to burning fossil fuels and deforestation | *understand that computer climate models provide evidence that human activities are causing global warming* |
| understand that global warming causes climate change and describe some of these effects | | *explain that increased convection and more water vapour in the warmer atmosphere can cause more extreme weather* |
| recall that information can be superimposed onto an electromagnetic carrier wave to create a signal that can be transmitted | explain and compare different ways in which electromagnetic radiation can transmit information | |
| recall the features of an analogue signal and a digital signal; recall some advantages of digital signals over analogue signals; recall that a digital signal is transmitted as pulses of an electromagnetic wave | understand the advantages of digital signals over analogue signals; understand that digital information is carried as pulses of an electromagnetic carrier wave, which are decoded when received | *explain why digital signals are less prone to noise than analogue signals;* understand how pulses of an electromagnetic carrier wave are created, used to carry digital information, and decoded |
| recall that higher quality sound or images use more digital information | | |

# Exam-style questions

## Foundation level

**AO1 1 a** Ultraviolet radiation from the Sun makes our skin tan. Sarah works outside and has developed a suntan. Which one of these reasons explains why Sarah has developed a tan? [2]
A Ultraviolet radiation is absorbed by Sarah's skin.
B Sarah has not used sun cream.
C The Sun emits ultraviolet radiation.

**b** Annie works inside and has not tanned even though she sits by a window. Which one of these reasons explain why Annie has not tanned? [1]
A Annie's skin is not sensitive to ultraviolet radiation.
B Ultraviolet radiation is not transmitted through the window.
C The window transmits ultraviolet radiation.
[Total: 2]

**2** As a fire cools down, light from the burning logs changes from being bright and yellow, to being a faint red glow.

**AO2 a** Describe how the number and type of photons from the fire change as the fire cools down. [2]

**AO1 b** A person further away from the fire feels cooler than a person closer to the fire. Explain this. [2]
[Total: 4]

## Foundation/Higher level

**AO1 3 a** Gamma rays and X-rays are two types of electromagnetic radiation that can cause ionisation. Explain why only some types of electromagnetic radiation cause ionisation. [2]

**b** Jodie works in the X-ray department of a hospital. She wears a badge to monitor her exposure to X-rays. Which **three** of these statements explain why Jodie's exposure to X-rays should be monitored? [3]
A Exposure to X-rays can ionise molecules in cells.
B Wearing the badge reduces Jodie's exposure to X-rays.
C There is no safe limit for exposure to X-rays.
D There is more risk of cell damage with higher exposure.
E Monitoring gives a warning if levels of exposure are getting too high.
[Total: 5]

**AO1 4** Ozone is a molecule that is found in the atmosphere. Describe how ozone in the atmosphere can protect us from ultraviolet radiation. [3]
[Total: 3]

**AO2 / AO3 5** Ahmed's teacher told him that in the past 50 years carbon dioxide levels in the atmosphere have been rising. His geography teacher told him that in the past 20 years there has been large-scale deforestation in the world. Ahmed thinks there is a correlation between rising carbon dioxide levels and deforestation. Explain whether a such correlation means deforestation caused rising carbon dioxide levels. The quality of your written communication will be judged in your answer. [6]
[Total: 6]

## Higher level

**AO1 6** Research on global warming indicates that warmer weather can increase the chances of severe storms. Say whether each of the following statements is true or false. [3]
A Warmer air holds less water than cooler air.
B There is more convection between the ocean and atmosphere if there is a bigger temperature difference.
C Severe storms in one year are proof that the climate is changing.
[Total: 3]

**7** Stars and other objects in space emit all frequencies of electromagnetic radiation.

**AO2 a** Explain whether a space telescope will detect the same electromagnetic frequencies from stars as a ground-based telescope. [2]

**AO3 b** Explain how scientists might use this argument to support the case for investment in a new space telescope. [2]
[Total: 4]

**AO2 8** Scientists are carrying out a study to investigate whether humans can live permanently on the Moon. The Moon has virtually no atmosphere, which affects the levels of radiation on the surface. A large percentage of radiation falling on the Moon from the Sun reaches the Moon's surface, compared with the proportion reaching the Earth's surface. Humans living on the Moon would need to live in a fully enclosed biodome.

Use this information to explain why the intensity of X-rays reaching the Moon's surface is greater than that on Earth, and how precautions against exposure to X-rays for volunteers living in the biodome should compare with precautions taken for radiation workers on Earth. The quality of your written communication will be assessed in your answer. [6]
[Total: 6]

AO1 recall the science   AO2 apply your knowledge   AO3 evaluate and analyse the evidence

# P2 Radiation and life

## Worked example

**AO1 a** Match each type of radiation with its medical use. [2]

- gamma radiation → killing cancer cells
- visible light → sealing blood vessels
- X-rays → making detailed images of teeth

> Correct. In questions like these, make the connections that you are sure of first.

**b** A mother refused to allow her son to have an X-ray after he hurt his back. The doctor thinks that he may have fractured a vertebra, but his mother does not want him exposed to radiation. The doctor said 'It is true that he will have a small additional exposure to radiation, but this is less than he would receive from the surroundings in a year. The risk of not diagnosing the problem is much greater.' The boy's mother thinks doctors can treat him without using X-rays.

**AO1 i)** Suggest a risk that the mother may be concerned about. [1]

*The exposure to radiation increases the risk of cancer.* ✓

**AO2 ii)** Suggest a specific risk that the doctor may be concerned about. [1]

*The risk of not diagnosing the problem.*

> Part i is correct. The answer to ii gives no more information than is given in the question. A suitable answer to gain a mark might be: 'The doctor is concerned that the boy's bones may heal in the wrong position/the damage could worsen/cause further problems without diagnosis and treatment'.

**AO1** **c** Discuss whether or not the doctor should treat the boy without
**AO3** first using X-rays. The quality of written communication will be assessed in your answer. [6]

*If he has an X-ray the extra risk of his exposure to it is small, as it is weak compared with what he gets from background radiation. A broken vertebra is serious and the X-ray will show where the damage is and how bad it is. Finding exactly where the damage is and how bad it is will mean the treatment will be more effective.*

### How to raise your grade

Take note of the comments from examiners – these will help you to improve your grade.

> The answer indicates the advantage of having an X-ray, balanced by the small extra risk presented by the X-ray exposure over and above background radiation. Information is relevant and presented in a structured way. However there is no discussion of the other point of view – whether treatment could go ahead without an X-ray. The answer uses appropriate scientific terms, such as X-rays, exposure and background radiation. From a banded mark scheme such as that shown on page 317, this answer would gain 4 of the 6 marks available.

# P3 Sustainable energy

## What you should already know...

### Electricity is a vital energy supply for our lifestyle

 Write down six major effects of a week without electricity in the UK.

### Sources of energy are needed to produce electricity

Energy cannot be created, but is transferred from one form to another

When a fuel is used, energy is released in the form of heat.

We get a lot of the energy we need from fossil fuels: coal, oil and gas.

Fossil fuels are burnt, and energy is released in a chemical reaction. Carbon dioxide and pollutants are produced.

Fossil fuels and nuclear fuels are primary energy sources used to generate electricity in power stations.

Alternative sources of energy for electricity generation include moving water, wind, the Sun (solar energy) and biomass (plant material).

 Name two different types of fuel, and two different renewable energy sources.

### Magnets can give rise to forces

A magnetic field is a region of space around a magnet pole, in which magnetic materials experience a force.

 Name two devices that use magnets.

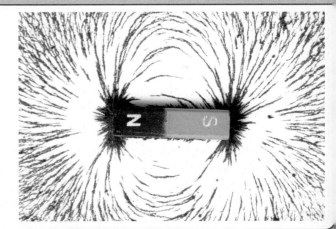

# In P3 you will find out about...

> what is meant by energy transfer, and how to show energy transfers in diagrams

> how energy is transferred in electrical devices, and what is meant by the power of a device

> the importance of the efficiency of energy transfer

> how a generator produces electricity by spinning a magnet near a coil of wire

> how in power stations a primary energy source is used to spin turbines which spin the generator

> how the turbines are made to spin in different types of power station

> how electricity is transported to where it is needed

> the cost of electricity

> the advantages and disadvantages of different types of power stations, considering their cost, their efficiency, their waste and other environmental impacts

> the problem of dealing with radioactive waste from nuclear power stations

> the need to reduce our demand for energy

# Energy sources

**We are learning to:**
> understand which energy sources we use to generate electricity
> appreciate why fossil-fuel power stations contribute to global warming
> understand how impacts of human activity can be reduced

## Where are the hot spots?

Lights from cities all over the world are visible from space. Figure 1 shows where most of the population lives in Europe – in cities, relying on electricity even at night. Lighting takes up a large proportion of worldwide energy demand, and light pollution has a big impact on wildlife. Reducing our use of electricity for lighting would reduce our impact on the environment.

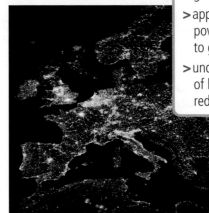

FIGURE 1

## Primary and secondary energy sources

**Primary energy sources** are those that are used in the form in which they are found. They may be processed, or refined, but their energy has not been transferred from other energy stores. Examples of primary energy sources include:

> **fossil fuels** (coal, oil, gas)
> **nuclear fuels** (uranium, plutonium)
> **biofuels** (plant and animal material used as fuels)
> wind
> waves
> radiation from the Sun.

**Secondary energy sources** are produced from primary energy sources. Electricity is a secondary energy source because it is generated from primary energy sources. Energy stored in fossil fuels, for example, is transferred to electrical energy.

### QUESTIONS

1 What is the difference between a primary energy source and a secondary energy source?

## Our changing use of energy

During the 1800s, our use of energy sources began to increase. Factories started making things in large quantities.

By the 1880s, it was possible to generate and distribute electricity on a large scale. Our use of energy sources then increased dramatically, especially in the 20th century. A bigger percentage of an increasing world population used electricity in their daily lives. To cope with these growing energy needs, more and more fossil fuels were extracted and burned.

FIGURE 2: Mass production in the late 19th century.

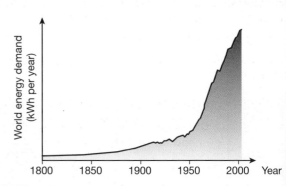

FIGURE 3: How our use of energy has increased in the last 200 years.

Figure 4 shows how the proportion of energy supplied from different energy sources has changed over time:

> In 1850, coal provided about 10% of our energy needs, with 90% coming from wood.

> Coal provided about three-quarters of our energy needs in 1910.

> In 1970, coal only provided 20% of our energy with nearly 50% coming from oil.

> Nuclear energy started providing significant amounts of energy from 1970 onwards.

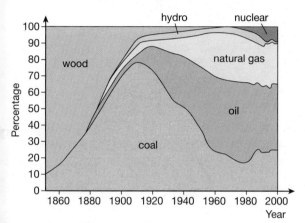

FIGURE 4: How our choice of energy sources has changed since 1850.

Fossil fuels take millions of years to form, and we are now using them much quicker than they can be replaced. Stores of fossil fuels are becoming harder to find and more expensive to extract. At some point, it will not be economic to use fossil fuels in power stations because they will become too expensive. Reserves of fossil fuels may become too small to be used for generating large amounts of power.

## QUESTIONS

**2** Why are we using more types of energy sources now compared with 200 years ago?

**3** Why will our supply of fossil fuels run out?

**Watch out!**
It is hard to predict just when our fossil fuels will run out because we change the pattern of use when they become harder to find and more expensive to use.

## The problems with fossil fuel use

Burning fossil fuels produces carbon dioxide, which is a **greenhouse gas**. As more greenhouse gases are produced worldwide, global temperatures have gradually risen. It is believed that this **global warming** causes climate change and erratic weather patterns bringing, for example, droughts in some places and floods in others.

There are other environmental effects caused by using fossil fuels:

> damage to local areas from coal mining

> production of large quantities of flue ash, which contains small amounts of toxic elements such as heavy metals and traces of radioactive elements

> oil spills when drilling for oil and transporting it

> production of sulfur dioxide, which contributes to acid rain.

### Did you know?
New designs for coal-fired power stations include the option of capturing the carbon dioxide emitted, to be stored in disused North Sea gas fields.

Recognising the damage we are causing and designing better methods of extracting and transporting fuels can reduce the impact of these activities. Safety features, more efficient extraction techniques and proper treatments of waste will all help. Whatever energy source is used to generate electricity on a large scale, there are environmental effects. You will learn more about these on pages 287 and 290.

## QUESTIONS

**4** How do fossil-fuel power stations contribute to global warming?

**5** What damage is caused to the environment when:

**a** extracting fossil fuels

**b** burning fossil fuels

**c** disposing of waste after fossil fuels are burned?

**6** We have used fossil fuels for centuries. Why is this use causing so many problems now?

# Power

**We are learning to:**
> understand what power is
> calculate the energy transferred by equipment
> use the correct units of energy, time and power

## How often do you recharge your batteries?

Remote controlled cars use batteries up very quickly, especially when climbing slopes. They need a larger, heavier battery so they can be used for a reasonable amount of time before the battery needs recharging. Batteries come in different sizes to cope with the varied energy demands of battery-operated devices.

**FIGURE 1:** Remote controlled cars use energy from batteries.

## Measuring energy transferred

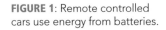

When you listen to music on a radio, electrical energy is being transferred. The amount of energy transferred each second is the radio's **power**.

If the radio is on for a period of time, the energy transferred is its power multiplied by the time:

energy transferred = power × time

The units used in this equation depend on the equipment and how long it is switched on for.

Power is measured in **watts** (W) if energy is measured in joules (J) and time is measured in seconds (s).

For example, if the power of an analogue radio was 2 watts, each second it would transfer 2 joules of energy. An 8 W digital radio would be more powerful, transferring 8 joules every second.

Power is measured in **kilowatts** (kW) if energy is measured in kilojoules (kJ) and time is measured in seconds (s). A kilojoule is 1000 joules.

For example, a kettle transfers 1200 joules (1.2 kilojoules) every second. One kilowatt is the same as 1000 watts, so its power is 1.2 kilowatts (kW).

The amount of energy transferred is measured in **kilowatt-hours (kWh)** if the power is measured in kilowatts and time is measured in hours.

For example, if the power of a fridge is 50 W, or 0.05 kW, every day it uses 0.05 kW × 24 hours = 1.2 kilowatt-hours.

**Watch out!**
A **joule** is very small amount of energy, equal to the energy used by a 100 W bulb in one-hundredth of a second. Larger units are needed for calculations involving electrical appliances.

### QUESTIONS

**1 a** How much energy is transferred by the 2 W analogue radio in 1 minute (60 s)?

**b** How much energy is transferred by the 8 W digital radio in 1 minute?

**2** How much energy is transferred by the 0.05 kW fridge in one year?

electrical power GCSE

# P3 Sustainable energy

## Calculating power

When a kettle is turned on, an electric current flows through the circuit transferring energy from the power supply to the kettle. The power of an electrical device like a kettle is calculated using:

power (watts, W) = voltage (volts, V) × current (amperes or 'amps', A)

> **Voltage** is supplied by the mains supply or a battery. A higher voltage means more energy is transferred. The voltage of the mains supply in the UK is 230 V; the voltage of many batteries is much lower at 1.5 V.

> **Current** is the flow of electricity through a circuit. It is measured in **amperes**, shortened to amps (A). A large current means more electricity flows in the circuit, so the current can carry more energy.

For example, the power of a kettle may be 2.3 kW, the mains voltage is 230 V, and the current in the kettle element 10 A.

2300 W = 230 V × 10 A

**Watch out!** Power can be calculated using two equations. Check what information you are given to help you choose the right equation, and check the units.

### QUESTIONS

**3** What is the power of a microwave oven in which the current is 4 A and the voltage supplied is 230 V?

**4** What is the power of a radio if it uses four 1.5 V batteries of total voltage 6 V, and the current is 0.8 A?

## Rate of transfer

Power measures how quickly energy is transferred by a device. This is the **rate** at which it transfers energy. The rate of energy transfer is quicker at higher powers. The power of a device is usually given as a **rating**.

A microwave oven may have a power rating of 1000 W. It transfers energy at a rate of 1000 joules per second, or 1.0 kJ per second. What current allows it to transfer energy at this rate?

power = voltage × current

current = $\frac{power}{voltage}$ = $\frac{1000}{230}$ = 4.3 A

```
PROline
MOD.: ST44
                    2450MHz
230V ~ 50Hz   MICROWAVE INPUT POWER    : 1550 W
              MICROWAVE ENERGY OUTPUT :  950 W

SERIAL NO.    81000138
MADE IN KOREA                              CE
```

**FIGURE 2**. Instructions for heating food using a microwave oven give different times for different oven ratings. A more powerful microwave oven heats food more quickly.

### QUESTIONS

**5** In America, voltage is supplied at 110 V. It is supplied at 230 V in the UK. Calculate the current used by a 1100 W hairdryer in America and in the UK. Use your answers to explain why it may not be safe to use the same hairdryer in America.

**6** Two kettles use the same amount of energy to heat the same amount of water to boiling. Explain why the time taken to heat the water may be different for the two kettles.

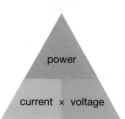

**FIGURE 3**: Cover up what you want to find out.

### Did you know?

All equipment that plugs into the mains has a plate giving you information about the power, voltage and current it uses.

# Buying electricity

**We are learning to:**
> understand how electric current transfers energy
> calculate the cost of electricity used
> use appropriate units of energy

## Buy new or make do?

If you want to save money, change the light bulbs you use. Energy-efficient bulbs can be on five times longer than normal bulbs and still use the same amount of energy. Many new electrical appliances use less electricity. Remember the savings are only worthwhile for equipment using lots of electricity and if they are being replaced anyway.

**FIGURE 1:** Newer is often cheaper to run.

## Transferring energy

An electric current transfers energy from one part of a circuit to another. Whether electricity is generated in a power station or a battery, energy is transferred as the current flows round the circuit. A **component** (or device) is a part of the circuit that transfers electrical energy into a different useful form. For example:

> a *bulb* transfers electrical energy to the environment in the form of *light*, and also heat

> a *motor* in a food mixer transfers electrical energy into the *kinetic energy* of the blades and food mix, and also heat and sound to the environment

Whenever an energy transfer takes place, the component heats up by a certain amount. The wires and cables of a circuit also heat up slightly as current flows through them. This energy transfer is not always wanted because it transfers some energy to warm the surroundings – this is not useful. Other unwanted energy transfers result in noise and vibration. After several energy transfers, all the energy from the power supply spreads to the environment.

### QUESTIONS

**1** What transfers electrical energy in a circuit?

**2** Why does the energy eventually spread to the surroundings?

# P3 Sustainable energy

## Measuring the energy used

All homes are fitted with an electricity meter, which records the amount of electricity used. It is not practical to measure the large amount of electrical energy used at home in joules, or even kilojoules, so we use kilowatt-hours instead. One kilowatt-hour is equal to 3.6 million joules – the energy used by a 100 watt bulb in 10 hours. Kilowatt-hours are called Units by electricity supply companies.

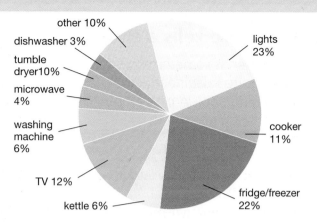

**FIGURE 2**: Electricity use in a typical home (not including room heating and hot water).

### QUESTIONS

**3** Why are different units of energy needed?

**4** What unit of energy is used in electricity meters?

## Buying electricity

Meter readings help us to work out how much our electricity bill will be. The electricity meter shows how many Units of electricity have been used.

> The number of Units used since the last reading is the current reading minus the previous reading.

> The cost of Units used is price per Unit × number of Units used.

June        September

**FIGURE 3**: Electrical energy used: 34183 − 33463 = 720 Units

It is useful to work out the cost of using a single piece of equipment:

the amount of electricity supplied in kilowatt-hours is the power (in kilowatts) × time it is on for (in hours).

Always make sure you use power in kilowatts, and time in hours.

For example, Emma used the following equipment:

> a 2 kW electric fire switched on for 5 hours uses 10 kWh (2 kW × 5 hours)

> a 100 W bulb switched on for 10 hours uses 1 kWh (0.1 kW × 10 hours)

> a 2 kW kettle switched on for 15 minutes uses 0.5 kW (2 kW × 0.25 hours)

In all, Emma used 11.5 kWh (or Units). Each Unit cost 16p, so this cost Emma 184p, or £1.84.

**Watch out!**
Remember that a kilowatt is 1000 W and an hour is 60 minutes:
– divide by 1000 to change watts into kilowatts
– divide by 60 to change minutes to hours.

**Watch out!**
Remember to check if you have calculated the cost in pence or pounds.

### QUESTIONS

**5** Joe's meter reading is 11567. Three months ago, it read 10567.

  **a** How many units (kilowatt-hours) has he used in three months?

  **b** If each Unit costs 16p, how much does this cost in pence?

  **c** How much does this cost in pounds?

**6** Calculate the Units used and the total cost when this equipment is used:

  **a** a 1.5 kW boiler on for 5 hours

  **b** a 1 kW microwave oven on for 15 minutes

  **c** two 60 W lamps on for 3 hours each.

Each Unit costs 16p.

kilowatt-hour GCSE

# Energy diagrams

**We are learning to:**
> interpret and process data on energy use
> interpret and construct Sankey diagrams

## A picture tells a thousand words

How often is something described in terms of 'as big as ten Olympic swimming pools'? It is hard to imagine this size – especially if you have never seen or been in an Olympic swimming pool. A picture would make the scale much easier to imagine.

## Presenting data

Information about energy use can be provided in different ways.

> A pie chart shows proportions. The one in Figure 1 shows how the total energy used in the UK is shared between different purposes. Figure 2 compares the proportions of electricity used by different appliances in the home.

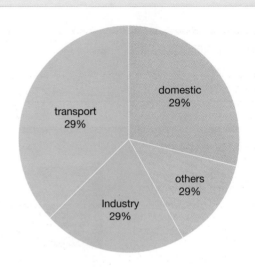

**FIGURE 1**: Comparing energy use in the UK.

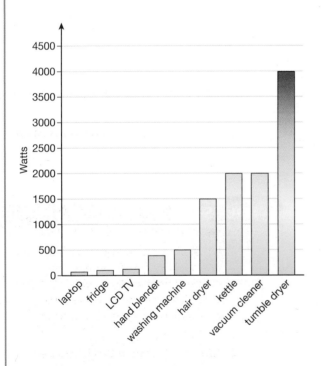

**FIGURE 2**: Comparing the powers of some appliances.

> A bar chart is also used for comparisons, but can show actual amounts. The one in Figure 2 compares the powers of different pieces of household electrical equipment.

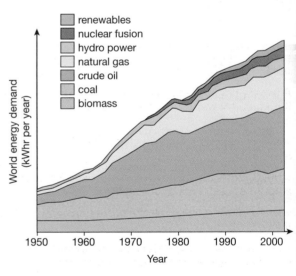

**FIGURE 3**: Change in use of energy since 1950.

> The type of graph shown in Figure 3 can also be very useful for comparing data. It shows how the overall world energy demand has grown, as well as showing the trends for individual energy sources.

UK energy use GCSE

## QUESTIONS

1. What proportion of energy is used for transport in the UK?
2. Look back at Figure 2 on page 273. From this data, which are the two biggest uses of electricity in the home?
3. **a** From the bar chart in Figure 2 on the opposite page, which appliances use less than 1000 joules per second?
   **b** One of these appliances is one of the biggest users of electrical energy in the home. Explain why.

**Watch out!**
Pie charts show proportions – the total should come to 100%.

## Sankey diagrams

A **Sankey diagram** shows how energy supplied to a device is transferred into different forms of energy. It is assumed that *energy is conserved* so the **energy input** and the **energy output** match. Figure 4 shows that in a filament light bulb 100 J of electrical energy changes into 5 J of light energy and 95 J of heat energy.

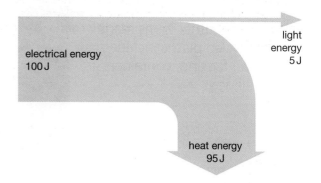

FIGURE 4: Sankey diagram for a filament light bulb.

### QUESTIONS

4. For the Sankey diagram in Figure 4, write down the total input energy and the total output energy. Explain what you notice about the two values.

## Drawing Sankey diagrams

A Sankey diagram shows the energy transfers in a device. It is an arrow which splits to show how the output energy is transferred:

> energy input is shown on the left side and energy output is shown on the right

> the thickness of the arrows is in proportion to the amount of each type of energy transferred

> the total thickness of all the output arrows equals the thickness of the input arrow

> *all* output energy transfers are shown, including non-useful energy transferred to the environment

> arrows pointing downwards show energy transferred to the environment.

### QUESTIONS

5. **a** Draw a Sankey diagram to show the energy transfers in an energy-efficient light bulb, in which 100 J of electrical energy transfers into 25 J of light energy and 75 J of heat energy.
   **b** Comment on how this compares with Figure 4.

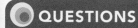
sankey diagram GCSE

# Efficiency

**We are learning to:**
- calculate efficiency for electrical devices and power stations
- suggest ways of reducing use of energy
- understand how the impact of human activity on the environment can be reduced

## Were your clothes once plastic bottles?

As the world's growing population draws on the Earth's limited resources of raw materials and energy, and waste is an increasing problem, scientists are finding more and more ways of recycling materials. Roughly 25 plastic bottles can make a fleece top – other items made from recycled plastic include furniture, fencing, containers and walkways.

**FIGURE 1**: Synthetic clothing, such as these fleece jackets, can be made from recycled plastic bottles.

## What is efficiency?

Efficient equipment and processes do not waste much energy. An efficient device uses less energy to do the same job as a similar but less efficient device.

We measure **efficiency** using:

$$\text{efficiency} = \frac{\text{energy usefully transferred}}{\text{total energy supplied}}$$

An energy-efficient bulb uses 100 J of electricity to provide 25 J light energy, so its efficiency is 0.25.

Sometimes, efficiency is written as a percentage – the efficiency of this light bulb is 25%.

**Watch out!**
Efficiency is always less than 1, or less than 100%.
It has no units.

### QUESTIONS

1. From Figure 4 on the previous page (the Sankey diagram), what is the efficiency of a filament light bulb?

2. What is the efficiency of a fridge that uses 250 J of electricity for every 200 J of useful output energy? Give your answer as a decimal fraction.

3. For every 30 kJ of electricity generated in a coal-fired power station, 100 kJ of energy is supplied from the coal. Give its efficiency as a percentage.

energy efficiency in the home GCSE

# Reducing our energy use

We can reduce our energy use by using more efficient equipment, or by using equipment differently. In the home, possible changes include:

> Using energy-efficient lighting.

> Buying appliances such as freezers and fridges that have high energy-efficiency ratings. The efficiency of an A++ rated fridge can be up to 50% higher than an A-rated fridge.

> Turning lights and heating off when they are not needed.

> Using equipment more efficiently – for example heating only the amount of water needed in a kettle, heating small quantities of food in a microwave oven rather than an oven, not using a tumble dryer for a very small load of clothes.

The contribution of any particular individual to national energy consumption will be small – there are about 24 million households in the UK. But all the little savings add up – national policies could have a big impact. These include passing laws or providing incentives so that:

> companies improve the efficiency of the equipment they manufacture, such as boilers and cars

> house-builders make new homes that are better insulated and more energy efficient

> new power stations use more energy-efficient methods of generating electricity

> existing power stations are altered to improve their efficiency – the efficiency of coal-fired power stations ranges from 45% to about 28%.

FIGURE 2: An energy-efficient bulb uses five times less energy compared with a filament bulb to provide the same light.

## QUESTIONS

**4** Write down three ways you could reduce your energy usage on a school day.

**5** Explain why it is important for companies to produce goods that are more efficient.

# Reducing our impact on the environment

Increases in population put demands on the environment. As the quality of life improves in developed countries, populations use more resources. People in less well developed countries would like to enjoy better living standards and to live longer through improved medical care and better access to sufficient clean water and food. To reduce the impact of human activities on the environment, scientists have helped to develop methods of recycling resources such as metals, glass and plastics. Renewable energy resources are being developed to generate electricity and to provide sustainable methods of transport. This way, people living in developing countries can access improved living standards that people in developed countries take for granted.

FIGURE 3: Recycling reduces waste, the use of raw materials and energy.

## QUESTIONS

**6** Describe three steps that scientists have taken to reduce the impact of an increasing population.

**7** Explain the benefits that can come from scientists developing ways of recycling more materials.

# Preparing for assessment: Applying your knowledge

*To achieve a good grade in science, you not only have to know and understand scientific ideas, but you need to be able to apply them to other situations and to analyse evidence. These tasks will support you in developing these skills.*

## ✸ Getting up steam

In the 18th century, Britain was on the verge of an Industrial Revolution that would bring about huge changes. People in their thousands would move from living in the country and working on the land to working in factories or mines, in rapidly expanding towns and cities. New technologies would enable the large-scale production of textiles, building materials and iron. Although flowing water was often used as a source of energy, it was the development of the steam engine that made large factories and deep mines possible on a scale never previously seen.

Two of the people behind the development of steam power were Matthew Boulton and James Watt. Engines of their design powered machines, pumped water, raised and lowered workers and lifted material from the bottom of mineshafts.

A Boulton & Watt engine needed a supply of steam from a boiler. The boiler burned a fuel such as coal to heat water and turn it into steam. The steam went into the engine and pushed a piston along a cylinder. (Imagine blowing into a syringe and making the plunger move.) The piston was linked to a wheel which then turned (rather like when your leg pushing down on a bicycle pedal makes the wheels turn). This rotating movement could be linked to, for example, a water pump, or a cable down a mineshaft.

The engines were huge, often standing far taller than the men who worked them. The engine houses were hot and noisy; stoking the boilers was hard work.

These engines didn't move very quickly (they turned far slower than a car engine, for example) but they were large and powerful. They were much more efficient than the crude atmospheric engines in use before, but not very efficient by today's standards. Compared with the energy content of the fuel, only about 5% reached the machines being driven by the engine. In comparison, a gas turbine in a modern power station will run at about 60% efficiency.

## P3 Applying your knowledge

### Task 1

> What do you think it would be like to be near a Boulton & Watt engine? Describe the sight, sounds and smells.

> What kinds of useful work might they do in a mine?

> In which wasteful ways can you imagine energy being transferred in these engines?

### Task 2

> Why do you think coal was selected as the fuel for the boilers to produce steam?

> What energy transfers would be taking place in the boiler?

### Task 3

> Most of the energy ends up being transferred to the environment. How does this happen?

> Thinking about the useful and wasted outputs and the efficiency of the engine, sketch a Sankey diagram to show the energy transfers.

### Task 4

> Tell the story of the transfer of useful energy, starting with the fuel in the boiler and finishing with a mine cage of ore being raised up.

> Why did factory and mine owners use Boulton & Watt engines, as they were so inefficient?

> The engines were particularly popular in areas like Cornwall, which had a lot of copper, lead and tin mines but no coal mines. Can you explain why?

### Maximise your grade

These sentences show what you need to include in your work to achieve each grade. Use them to improve your work and be more successful.

**E**
For grade E, your answers should show that you can:
> describe advantages and disadvantages of different energy sources
> comment on the significance of data relating to energy use and waste
> draw and interpret simple Sankey diagrams
> discuss methods used to reduce energy use

**C**
For grades D, C, in addition show that you can:
> interpret and process data on energy sources, use and waste
> draw and interpret more complex Sankey diagrams
> discuss qualitatively and quantitatively the effectiveness of methods used to reduce energy use in the work place

**A**
For grades B, A, in addition show that you can:
> draw and interpret Sankey diagrams including an analysis of the efficiency of energy transfers
> discuss qualitatively and quantitatively the effectiveness of methods used to reduce energy use nationally

# Generators

**We are learning to:**
> understand how generators produce electricity
> understand the need for fuel for electricity generation
> understand that power generation needs to conform to standards

## Why do power station workers need to watch TV?

During large televised sporting events like the World Cup and the Olympics, people use advert breaks to put the kettle on and heat snacks. The demand for electricity increases so suddenly that there is a risk of power cuts. Some power stations must generate extra electricity at this time. Workers check the progress of the match to plan ahead.

## How generators produce electricity

Nearly 200 years ago, Michael Faraday realised that moving a magnet near a circuit caused an electric current to flow in the circuit. The current flowed only when the **magnetic field** was changing – that is, when the magnet was moving.

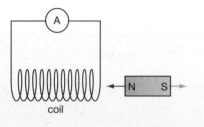

**FIGURE 1**: When the magnet moves into or out of the coil, a current flows in the coil.

Power stations use this idea to produce mains electricity. The magnet and wire coil are called an electric **generator**. An **electromagnet** constantly spins inside the coil. An electromagnet is used because its magnetism is much stronger than a permanent magnet.

**FIGURE 2**: This is the coil that Faraday used to discover how to generate electricity.

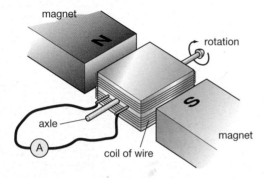

**FIGURE 3**: Power stations generate electricity using the same idea as this simple generator – here the coil spins and the magnets are stationary.

**Watch out!**
A generator does not create electrical energy – it transforms kinetic energy into electrical energy.

## QUESTIONS

1. Where is mains electricity produced?
2. Describe the two parts of a generator.
3. Why are electromagnets the best type of magnet to use in power stations?

# P3 Sustainable energy

## Providing the energy for electricity production

A power station generator has two main sections:

> a stationary coil made of hundreds of turns of wire – the stator

> a large rotating electromagnet – the rotor.

The rotor fits inside the hollow stator and is made to spin by a **turbine**.

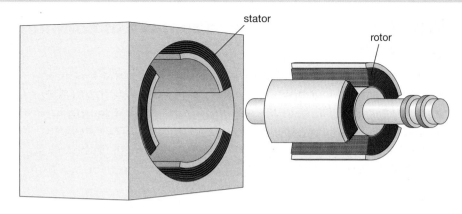

FIGURE 4: Parts of a generator.

When the electromagnet spins, it produces a voltage across the coil of wire. If the electromagnet spins faster, a larger voltage is produced. When the generator is connected to an outside circuit, a current flows in the circuit.

Primary fuels such as fossil fuels, nuclear fuels or biofuels, or other primary energy sources such as wind or moving water, provide the energy to turn the turbines, which make the rotor spin. When a larger current is needed, the rotor is made to spin faster, or more generators can be used. More primary fuel is needed every second for this to happen.

### QUESTIONS

**4** Why are primary fuels used in power stations?

**5** Why are more primary fuels needed when there is a bigger demand for electricity?

### Did you know?

Equipment in the home is designed to work at a voltage of 230 V. Equipment can be damaged if faults in the supply mean that the voltage is supplied at lower or higher values.

## Regulating electricity production

Generators in power stations supply their electricity to a national network of cables (the National Grid). For our electricity supply to be safe, reliable and consistent across this network, all power stations have to conform to the same set of standards. As new methods of generating electricity are developed, regulations are needed to ensure that these standards are being met, and that safety standards are acceptable. New developments in power station technology include:

> ways of reducing pollution levels

> more efficient power stations

> new ways of generating electricity.

### QUESTIONS

**6** Explain the advantages of having the same regulations applying to all power stations.

FIGURE 5: New developments in power generation must conform to government regulations.

Faraday GCSE

# How power stations work

**We are learning to:**
- understand how primary fuels are used to produce electricity in power stations
- describe the structure and components of different types of power stations

## What does a power station have in common with a steam engine?

Trains designed nearly 200 years ago used the same technology to turn the wheels as we use now to generate electricity. Steam was produced in boilers and this drove pistons in the engines – this made the wheels turn. In power stations, jets of steam force turbines to spin and this generates electricity. In both cases, energy from moving steam forces something else to move that makes the machinery work.

## How a generator is made to spin

A **turbine** is basically a set of blades that are linked to a generator by an axle. The blades spin when water, steam, gas or wind are aimed at it. Fossil-fuel power stations and nuclear power stations use jets of steam to spin turbines. The turbine axle is connected to the electromagnet in the generator, which is made to spin inside the coil, generating electricity.

### Producing the steam

When fossil fuels are burned, or when fission reactions take place in a nuclear reactor, heat is produced. This heats water in pipes, which produce jets of steam to turn the turbine.

### QUESTIONS

1. Why do some power stations produce steam?
2. What is the job of a turbine?

**FIGURE 1**: Turbine blades.

## The structure of power stations

In a coal-fired power station, coal is burned in the furnace. Heat from the burning coal heats water in a set of pipes. Steam is produced and this is used to spin turbines. This is shown as a block diagram in Figure 3.

**FIGURE 2**: A coal-fired power station.

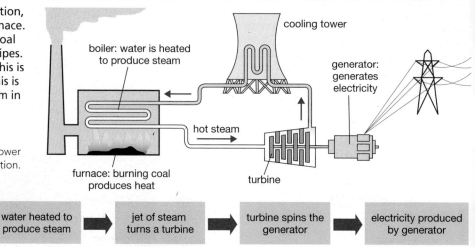

**FIGURE 3**: Processes in a coal-fired power station. Oil-fired power stations are very similar – oil is burned instead of coal to produce heat. They are called 'thermal' power stations.

coal-fired power GCSE, gas-fired power GCSE, nuclear power GCSE

# P3 Sustainable energy

Gas-fired power stations have two sets of turbines. When the gas is burned, one set of turbines spins when jets of hot burning gases flow over them. These hot gases are used to change water into steam, which spins a second set of turbines. Gas-fired power stations are much more efficient than coal or oil-fired power stations.

Turbines are forced to spin in a different way in hydroelectric power stations. Water is trapped behind a dam. When gates are opened, water rushes down pipes and through the turbines at the base of the dam. This generates electricity very quickly.

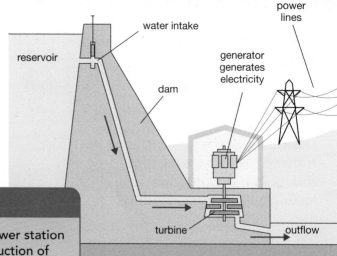

FIGURE 4: A hydroelectric power station.

## QUESTIONS

**3** Write down these parts of the coal-fired power station in the order that they are involved in the production of electricity: furnace, generator, set of pipes, turbine.

**4** Write down one difference in the way a turbine is spun in a hydroelectric power station compared with a coal-fired power station.

**5** Draw a block diagram like Figure 3 showing what happens in a hydroelectric power station.

### Did you know?

The world's first hydroelectric power station was built on the Niagara Falls in 1895.

## Nuclear power stations

Figure 5 shows how electricity is produced using steam in a nuclear power station.

The **reactor** contains nuclear fuel rods, which produce heat. Control rods are raised or lowered to control the amount of heat produced. Water circulating through pipes leaving the reactor is put under high pressure which means it can be heated to high temperatures. A second set of pipes filled with water surrounds these heated pipes. Heat is transferred to water inside the second set of pipes, changing it into steam.

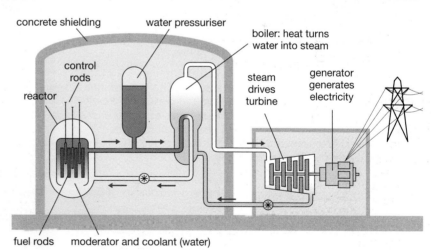

FIGURE 5: One type of nuclear power station.

### Watch out!
Fossil fuels are burnt to release heat. Nuclear fuels do not burn – a **nuclear fission** reaction takes place which releases heat.

## QUESTIONS

**6 a** Which components do coal-fired power stations and nuclear power stations have in common?

**b** Which components do nuclear power stations have that coal-fired power stations do not have?

hydroelectric power GCSE, power station block diagram GCSE

# Waste from power stations

**We are learning to:**
- understand that nuclear and other power stations produce radioactive waste
- understand the difference between irradiation and contamination, and their different risks
- understand how people may perceive different risks

## Could you do a risky job?

Some statistics give the ten riskiest industries, in order, as: commercial fishing, timber production, farming, steel and iron industry, recycling and rubbish disposal, flying, roofing, coal mining, merchant navy, flour milling. Although these are familiar, they are also dangerous jobs.

FIGURE 1

## Dealing with radioactive waste

Nuclear power stations use uranium and plutonium to generate electricity. After nuclear fuels have been used, the waste products are **radioactive**. This means that the waste emits **ionising radiation**, which may affect molecules and cells (see page 241), and can pose a serious health risk. Some of the waste can be recycled and reused in power stations, but most of it needs to be stored safely until the radioactivity dies away naturally. Different types of radioactive waste from nuclear power stations have different strengths, so will be treated and stored differently.

> 93% of the volume of waste does not need special handling because its radioactivity is so low.

> Less than 1% of the volume of waste contains 95% of the radioactivity from nuclear power stations. This is combined into solid glass (by heating it with melted glass and allowing it to cool). Then it is kept cool in large tanks of water and eventually stored underground in sealed stainless steel containers. After about 40 years, the radioactivity typically falls to a one-thousandth of its original amount.

**FIGURE 2**: 'High-level' radioactive waste is converted to glass for safety.

### QUESTIONS

1. What is meant by radioactive waste?
2. Why are different types of radioactive waste treated differently?

## Irradiation and contamination

If an object is in the path of ionising radiation, it will be **irradiated** – as soon as the object moves out of the path, it stops being irradiated. The object does *not* become radioactive. Thick shielding around a radioactive object stops it irradiating its surroundings. Radioactive waste is stored in specially built underground storage units because the ground provides shielding.

nuclear power, waste disposal GCSE

P3 Sustainable energy

Something is **contaminated** if radioactive gases, liquids or particles mix with it, making the material radioactive. Contamination is hard to remove when particles of both materials have mixed. Radioactive waste is turned into solid glass or mixed with concrete and contained in steel drums so it cannot contaminate its surroundings.

Contamination is a bigger health threat than irradiation because:

> it is hard to limit how far the contamination spreads and what it spreads to

> it is harder to clear up contamination than to prevent irradiation

> it leaves substances radioactive, increasing the time of exposure to radiation.

FIGURE 3: Casks for storing nuclear waste are lined with concrete to absorb the radioactivity.

## QUESTIONS

3 What is the difference between contamination and irradiation?

4 How is the risk of contamination and irradiation from nuclear waste reduced?

## Understanding risks

Ionising radiation cannot be seen and its effects are not immediately obvious, so people tend overestimate the risk it causes. We underestimate familiar risks such as health risks from acid rain, smog and other pollution from fossil fuels. It was recently estimated that 50,000 people die early each year in the UK as a result of pollution from fossil fuels.

We tend to trust familiar technology and to worry about unfamiliar technology. Coal-fired power stations do not have to limit their radioactive emissions but nuclear power stations do. Coal contains traces of uranium, so the area surrounding coal-fired power stations is several times more radioactive than the area surrounding nuclear power stations. It is still much lower than radioactivity from natural causes and is not believed to be a health risk.

Statistics provide information that can be used to compare different risks. Using statistics helps us to select the risks that cause problems, and which need to be controlled, rather than worrying about risks that are not a problem.

FIGURE 4: Risks from coal-fired stations may be greater than those from nuclear power stations but they are viewed differently.

### Did you know?

Our greatest exposure to radioactivity is from natural sources like rocks and building materials. Lifestyle choices such as smoking, diet and just where we live have a much bigger impact on our risk of developing cancer than any exposure to radiation. A recent study showed that people who work with radiation are less likely to die from cancer than members of the wider population.

## QUESTIONS

5 Give an example of a risk that we overestimate.

6 Why is it important to estimate risks correctly?

irradiation contamination GCSE

# Renewable energy sources

**We are learning to:**
> understand that some renewable sources of energy drive a turbine directly
> understand that while technology can improve the quality of life, unintended impacts may be undesirable

## The renewable revolution?

Samso is a small island that is part of Denmark. In 1997, it began converting to renewable energy sources. Now 100% of its electricity and 75% of its heating comes from renewable energy. This includes wind power and solar power.

**FIGURE 1**: An offshore wind farm in Denmark.

## What do we mean by renewable?

Renewable energy sources will not run out. About 7% of the UK's electricity comes from renewable energy sources. Some schemes involving spinning turbines directly including:

> **hydroelectric** schemes
> **wind turbines**.

**Wave technology** is still being developed for widespread commercial use.

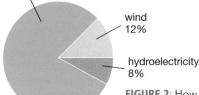

**FIGURE 2**: How the UK's renewable energy was supplied in 2009.

### QUESTIONS

1. How are renewable energy resources different from non-renewable resources?
2. Write down two renewable energy sources that spin turbines directly.

## Renewable resources: pros and cons

Hydroelectric schemes use dams to trap river water in a reservoir. When water is released, it spins the turbines directly (see page 283).

Across the world, hydroelectric schemes are providing large amounts of electricity. In the UK, hydroelectricity tops up surges of demand for electricity. When demand for electricity is low, water is pumped back up to the reservoir.

However large areas are flooded, and downstream vegetation on the sides of the river die and rot, releasing greenhouse gases. Although hydroelectricity can seem cheap to run, it is costly to build and the dams must be maintained.

**FIGURE 3**: Hydroelectric power stations release water as electricity is required.

renewable energy sources GCSE, sustainable energy GCSE

## P3 Sustainable energy

Wind turbines use a set of blades to trap the energy from wind and spin a turbine directly. Wind turbines are tall and are built in windy locations both on land and offshore. The amount of electricity generated depends on:

> how many wind turbines are connected to and contributing to the **National Grid**

> whether it is windy enough for wind turbines to spin effectively, but not so stormy that wind turbines have to shut down.

One problem is that winds are unreliable. Demand for electricity continues even if the weather conditions are wrong, so thermal power stations must be available too. One future solution may be to store surplus energy in giant batteries.

Wave power is useful on remote islands that cannot use other methods to generate electricity. The waves spin turbines directly, using structures moving in the water or on the shoreline. It is not easy to capture the energy from waves, so this method is not yet used to generate electricity widely. The technology is still being developed.

**FIGURE 4**: This was the world's first commercial-scale wave power station. It's on the coast of Islay, a Scottish island. Waves move in and out of a partly submerged concrete chamber containing a turbine.

### QUESTIONS

**3** Write down one benefit and one problem caused by hydroelectric schemes.

**4** Electricity generation by wind turbines is unreliable. Write down one possible solution.

**5** Why is wave power only used in a small number of places?

### Did you know?

The amount of renewable energy that a country has access to depends on its geography. Almost all of Norway's energy comes from hydroelectricity; and two-thirds of Iceland's energy is from geothermal sources (underground heat).

## Unintended consequences

A reliable electricity supply is fundamental to our way of life, but it has had an impact on the environment. The benefits of using renewable energy sources should help to reduce these impacts. However, there are unintended consequences – especially when renewable energy sources are used widely. Careful planning can reduce these effects:

> the impact of noise and visual pollution is reduced by building offshore wind farms

> careful siting and use of hydroelectric dams reduces the impact of flooding and disruption to normal river-flow patterns.

We need to weigh up the benefits against the costs when evaluating renewable power.

**Watch out!**
The financial costs of any energy source include the costs of structures, fuels and maintenance, as well as clean up costs.

### QUESTIONS

**6** Explain whether you think the benefits of using:

**a** wind turbines

**b** hydroelectric schemes

outweigh the costs and any undesirable impacts they may have.

hydroelectric schemes GCSE, offshore farms GCSE

# The National Grid

## Why can birds sit safely on electric cables?

Electric cables carry electricity from power stations to towns and cities. Signs near the pylons warn of dangerously high voltages, and yet birds sit safely on these cables. This is because when they land on the cables they are not connected to the ground – there is no complete circuit through their bodies, so no electric current flows through them.

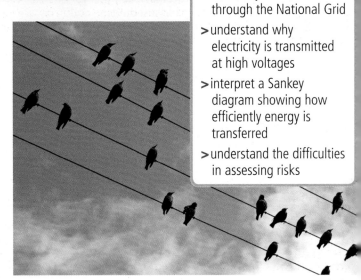

FIGURE 1: Blissfully unaware.

**We are learning to:**
> understand that electricity is transmitted through the National Grid
> understand why electricity is transmitted at high voltages
> interpret a Sankey diagram showing how efficiently energy is transferred
> understand the difficulties in assessing risks

## Getting our electricity

The National Grid is a network of electric cables that supply electricity throughout the UK. Its job is to carry electricity from power stations to all users. Electricity is a convenient form of energy because it can be transmitted over long distances. When it is used it does not pollute and can be used in many ways.

### QUESTIONS

1. What does the National Grid do?
2. Why is electricity convenient?

## Voltages in the National Grid

Energy is wasted at all stages in the generation and transmission of electricity. When power cables carry an electric current they warm up. Over hundreds of miles, a significant amount of energy is transferred to the surrounding air. Transmitting electricity using a very small current reduces this energy loss, but a very high voltage is needed for enough electrical energy to be transmitted. This can be up to 400 000 V. The voltage supplied to homes is 230 V, which is much lower for safety reasons. Very high voltages are reduced in substations before they reach homes.

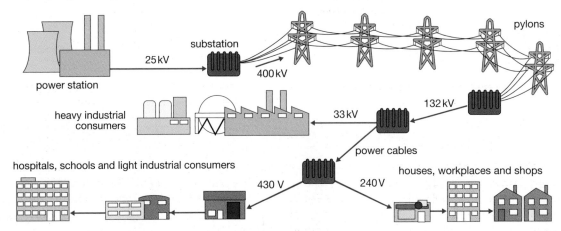

FIGURE 2: The National Grid.

Electricity is supplied at 230V in the UK because this is large enough to supply useful amounts of energy. However, this voltage can electrocute people if equipment is faulty or misused. Some countries use lower voltages, but equipment overheating may be more likely because the current is larger if the same amount of energy is supplied.

### Did you know?

Some power cables are routed underground to protect special landscapes, public spaces or river crossings. This is 17 times more expensive than using overhead cables, and there is more disruption to the landscape.

### QUESTIONS

**3** How are energy losses reduced in power cables?

**4** Explain whether or not 230V is acceptable as our electricity supply voltage.

## Energy transfers in supplying electricity

Energy from fuels and other sources is transformed into electricity in power stations. The energy wasted at each stage of generation can be shown in a Sankey diagram. Figure 3 shows several stages of energy changes, starting with 100 kJ of energy in the fuel and ending with the electricity delivered to the customer.

The efficiency of electricity generation (including the efficiency of the National Grid) is the proportion of the energy in the original fuel that is finally delivered to customers as electricity. In Figure 3 it is:

$$\frac{33\,kJ}{100\,kJ} = 0.33 \text{ (or 33\%)}$$

The efficiency of a particular power station does not take into account the energy wasted in the transmission system. In Figure 3, the useful energy is delivered to the customers (33 kJ) plus the 5 kJ used in the power station itself – that's 38 kJ, so the power station has an efficiency of 0.38 (or 38%).

### Watch out!

Check what the question is asking for when interpreting a Sankey diagram. It may only want calculations for one stage.

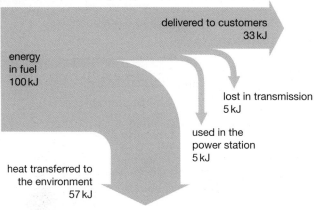

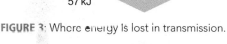

FIGURE 3: Where energy is lost in transmission.

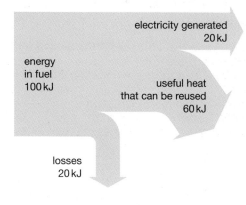

FIGURE 4: Energy transfer in a combined heat and power gas turbine plant.

### QUESTIONS

**5** Explain why a Sankey diagram for one power station does not show the true proportion of electricity reaching the consumer.

**6** Look at the Sankey diagram for the 'combined heat and power' gas turbine plant in Figure 4. The heat and the electricity are both useful forms of energy. What is the useful energy output of this power station? What is the efficiency of this power station?

# Choosing the best energy source

## Why windmills at sea?

Britain now generates more wind power offshore than any other country in the world. We are building large wind farms offshore like that shown on page 281 because there is more wind offshore. This means that the turbines can generate more electricity for longer periods of time. There is also less public objection to their siting offshore.

**We are learning to:**
> state the advantages and disadvantages of different energy sources
> interpret and evaluate information about energy sources

## Advantages and disadvantages of different energy sources

The advantages of different energy sources are:
> fossil fuels, nuclear power, hydroelectricity and biofuels generate large amounts of electricity when needed
> nuclear power and energy from water, waves and the Sun do not produce greenhouse gases
> energy from water, wind and the Sun are free once the power stations are built
> renewable energy sources will not run out.

All energy sources have disadvantages:
> wind, waves and solar power are weather dependent, producing small amounts of energy
> coal mining and drilling for oil and gas are dangerous occupations, causing deaths every year
> hydroelectricity and tidal barrages need suitable sites.

Some disadvantages affect the environment:
> fossil fuels and biofuels produce greenhouse gases
> extracting and mining large quantities of fossil fuels damages the landscape, causing pollution and oil spills
> nuclear power creates radioactive waste
> hydroelectricity and tidal barrages flood large areas, disrupting normal river flow
> large-scale wind farms cause visual and noise pollution.

### QUESTIONS

1 Which energy source is renewable but produces greenhouse gases?

2 Why is it a good idea to reduce the use of fossil fuels to generate electricity?

**FIGURE 1**: This hydroelectric scheme in China is the largest in the world. What problems might its construction have caused?

## Choosing an energy source

Our choice of energy source depends on:
> the impact of the energy source on the environment
> how expensive it is to build and run power stations
> the quantity and type of waste products produced
> whether or not it produces carbon dioxide.

🔍 three gorges dam GCSE

# P3 Sustainable energy

Many power stations use fossil fuels or nuclear power to produce large amounts of electricity reliably and cost effectively. Small and remote islands often use small-scale renewable energy sources to reduce the environmental impact and waste produced. But renewable energy is expensive and less reliable.

Our choice of electricity generation needs to balance the benefits with the costs and other disadvantages. If we choose to use new technology to generate electricity, we may have more power cuts if the technology is unreliable and much larger electricity bills to cover development costs. These can affect vulnerable people badly.

The graph in Figure 2 shows how the cost of generating each kilowatt-hour of electricity varies with different energy sources.

The graph in Figure 3 compares how efficiently different energy sources transform energy into electricity.

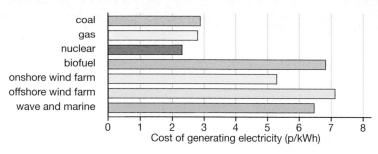

**FIGURE 2**: Cost of generating electricity (in pence per kWh).

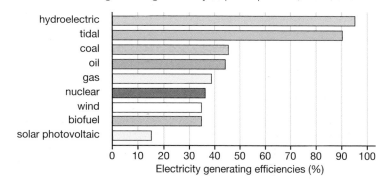

**FIGURE 3**: Electricity generation efficiency.

**Watch out!**
You may be given data in different forms. Check carefully what you are being asked to do with the data.

## QUESTIONS

**3** Why is nuclear power unsuitable for a small remote island?

**4** Use the data in Figure 2 to suggest why fossil fuels are used to generate electricity, despite producing greenhouse gases.

**5** Use the data in Figure 3 to explain why large-scale hydroelectric schemes are used, even though they are expensive to build and flood the local environment.

## Comparing power stations

Different types of power station produce different amounts of electrical energy. The power output of power stations is stated in megawatts (MW) – one megawatt is one million watts and is the energy in joules produced every second. Nuclear power stations produce about 1000 MW, compared with fossil fuel power stations at 500–2000 MW. The power output of wind farms varies with the weather. Wind farms are designed to produce up to 300 MW, but most produce only 10 to 50 per cent of their maximum power.

The setting-up cost of a power station can be huge, so the lifetime of a power station is another important consideration. The lifetime of nuclear power stations was originally planned to be around 30–40 years, but this has been extended as new technology has developed. Coal-fired power stations are also designed to last 40 or more years, and hydroelectric power stations are designed to last for up to 100 years. Wind turbines have an expected life of 20 years, but many individual turbines can be built for the cost of one large power station.

## QUESTIONS

**6** 'Wind turbines are as cost effective as a coal-fired power station in the long run.' Discuss this statement.

### Did you know?

The output of the Three-Gorges Dam hydroelectric power station in China, shown in Figure 1, is 22 500 MW – equivalent to 18 nuclear power stations or to burning 40 million tons of coal a year. Its reservoir is 370 miles long.

# Dealing with future energy demand

**We are learning to:**
> discuss the effectiveness of ways of reducing energy demand
> understand that some decisions have ethical implications
> know that there are different ways to assess the ethics of a decision
> understand that we will continue to need a mix of energy sources

## Will new houses help save the planet?

New houses are better insulated with energy-saving ideas built into their design. We now know it is possible to make houses that are so energy efficient that they are cheap to run, and do not contribute to carbon dioxide emissions overall. As older houses need replacing, newer more efficient homes will gradually slow down the increase in our energy demands.

**FIGURE 1**: New housing has effective insulation and built-in solar panels.

## Can we reduce energy demands?

Changes must be made to reduce the world's energy demands. When people make changes at home, the effect is very small. However, small changes made by everyone add up – switching on lights only when needed and turning off equipment can make a big difference. Changes made by businesses have a bigger effect.

Improved efficiency of production and of vehicle engines reduces the fuel used and emissions released.

The UK produces only 2% of the world's carbon dioxide emissions, but global warming is a worldwide problem. All countries must make some changes to ensure that emissions fall across the world. Many industrialised countries have agreed to limit their carbon dioxide emissions, setting targets together.

### QUESTIONS

**1** Explain why all countries should have a target to reduce their carbon dioxide emissions.

**2** Describe how businesses can have a big impact on energy demand.

## How can we reduce our emissions?

Energy demand is increasing, causing emissions of greenhouse gases to increase. Ideally we would produce our energy from sources that don't produce emissions, but it is unlikely that these can meet all our energy needs. As the world's population increases, so does the demand for energy. To reduce, or even stabilise our emissions, we need to significantly reduce the energy demand per person.

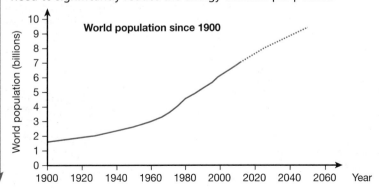

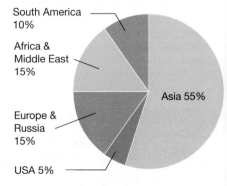

**FIGURE 2**: Population growth and distribution.

## P3 Sustainable energy

People in different countries disagree about who should reduce emissions and how this should be achieved. Emissions from many developing countries are increasing as they become more industrialised. They often rely on older, dirtier technology because it is cheaper. Individuals use more energy as their lifestyle improves. Reducing emissions from these countries may unfairly limit access to healthcare and education. Running schools and hospitals uses electricity and other energy-hungry resources.

Wealthy industrialised countries increasingly use cleaner technology, but current pollution levels are due to them having used less efficient technology in the past. Such countries often import goods made elsewhere, and they should take responsibility for the carbon emissions these create. There is a real risk that imposing the same $CO_2$ emission targets for everyone will result in industrialised countries consuming as much as before, while developing countries make sacrifices to cover the emissions caused by the production of exported goods.

If all countries reduced carbon dioxide limits by the same amount, the impact of the greenhouse effect would be reduced, benefiting most people. However, this would be achieved by developing countries sacrificing their increased standards of living. Any international decision on emissions targets that affects economic and social development in some countries is controversial.

**FIGURE 3**: Industrialisation in China has enabled people there to have a better standard of living. It doesn't seem right to make them cut their emissions, especially when a lot of the goods they produce are exported for our consumption.

### QUESTIONS

**3** Explain why carbon dioxide emissions should be controlled by developing countries and industrialised countries together.

**4** China, the US and India are the three countries that emit the most carbon dioxide. Why isn't it fair to simply make the biggest carbon dioxide producers reduce emissions the most?

### Did you know?

Nationally, China emits more carbon dioxide than the US. On average though, a person in America produces more than four times as much carbon dioxide as a person in China.

## How can we achieve a secure energy supply?

Even if we manage to reduce our energy demands, we will still need vast amounts of electricity to sustain our lifestyles.

We must reliably generate enough electricity now and in the future to avoid power cuts and to keep electricity affordable. We need enough power stations, and constant access to sources of energy. Many power stations are reaching the end of their planned lives and will need to be replaced. Our supplies of fossil fuels are running low, and even though we can import fossil fuels from elsewhere the cost of this would be controlled by other countries. In any case, to reduce carbon dioxide emissions, we must depend less on fossil fuels. It is likely that a mix of cleaner thermal power stations and nuclear power stations along with power generation using renewable energy, such as hydroelectricity and wind, will provide our future energy needs. Until renewable energy technology develops further (providing cheap, reliable energy), fossil-fuel power stations will continue to provide a significant amount of our electricity.

### QUESTIONS

**5** Analyse the benefits of using several different energy sources to generate the UK's electricity.

**6** Explain why decisions made now about our future energy supplies will have a big impact in the future.

### Watch out!

These problems have more than one answer. You will get credit for explaining your reasoning.

🔍 developing countries energy demand GCSE

# P3 Checklist

## To achieve your forecast grade in the exam you'll need to revise

Use this checklist to see what you can do *now*. Refer back to pages 268–293 if you're not sure.

Look across the rows to see how you could progress – **bold italic** means Higher tier only.

Remember you'll need to be able to *use* these ideas in various ways, such as:
> interpreting pictures, diagrams and graphs
> applying ideas to new situations
> explaining ethical implications
> suggesting some benefits and risks to society
> drawing conclusions from evidence you've been given.

Look at pages 312–318 for more information about exams and how you'll be assessed.

| To aim for a grade E | To aim for a grade C | To aim for a grade A |
|---|---|---|
| understand that power stations that burn fossil fuels produce carbon dioxide; understand that increasing demand for energy raises issues about the availability of energy sources and their environmental effects | | |
| understand that an electric current passing through a component transfers energy to the component and/or to the environment | | |
| recall that the power in watts is the energy transferred each second; use 'power = voltage × current' to calculate how quickly an electrical device transfers energy | | ***recall that power is the rate of energy transfer;*** use ***and rearrange*** 'power = voltage × current' in calculations involving rate of electrical energy transfer |
| use 'energy transferred = power × time' to calculate the electrical energy transferred in joules or kilowatt-hours | | use ***and rearrange*** 'energy transferred = power × time' in calculations involving electrical energy transfer |
| understand that a domestic electricity meter measures energy use in kilowatt-hours | calculate the cost of energy supplied by electricity | |
| interpret information about energy use; understand that a more efficient appliance transfers more of the energy supplied to a useful outcome | use 'efficiency = (energy usefully transferred ÷ total energy supplied) × 100%' to calculate the efficiency of an electrical device or power station | use ***and rearrange*** 'efficiency = (energy usefully transferred ÷ total energy supplied) × 100%' in calculations |

| To aim for a grade E | To aim for a grade C | To aim for a grade A |
|---|---|---|
| interpret and construct simple Sankey diagrams showing energy transfer | interpret and construct Sankey diagrams for various contexts including electricity generation and distribution, and use them to calculate efficiency of transfer | |
| recall that mains electricity is produced by generators; understand that a generator produces a voltage across a coil of wire by spinning a magnet near it | explain how the voltage produced, and current supplied, by a generator can be increased; understand that a generator uses more primary fuel per second when it supplies a bigger current | |
| understand that electricity is convenient because it is easily transmitted over long distances and has many uses | | |
| understand that in many power stations a primary energy source heats water, producing steam which drives a turbine coupled to an electrical generator; label a block diagram of the basic components of hydroelectric, thermal and nuclear power stations; understand that some renewable energy sources drive the turbine directly | | |
| recall that nuclear power stations produce radioactive waste and that this emits ionising radiation; describe some of the hazards of ionising radiation | understand that radioactive waste emits ionising radiation; explain the difference between contamination and irradiation | explain why contamination by a radioactive material is more dangerous than a short period of irradiation |
| recall that electricity is distributed through the National Grid and that the mains supply voltage to our homes is 230 volts | explain how the distribution of electricity through the National Grid at high voltages reduces energy losses | |
| discuss methods of reducing energy use at home and at work | discuss qualitatively and quantitatively the effectiveness of methods of reducing energy demand in a national context | |
| describe advantages and disadvantages of different energy sources | understand how different factors affect the choice of energy source for a given situation | *understand that to ensure a security of electricity supply nationally, we need a mix of energy sources* |
| interpret information about energy sources for generating electricity and apply it to different situations | interpret and evaluate information about different energy sources for generating electricity, considering efficiency, economic costs and environmental impact | interpret and evaluate information about different energy sources for generating electricity, *also considering power output and lifetime* |

# Exam-style questions

## Foundation level

**1** AO1, AO2

**a** Calculate the power of a computer that transfers 18 kJ per minute. [3]

**b** State and explain whether the computer uses a larger current than a microwave oven of power 1000 W. [2]

[Total: 5]

**2** AO1, AO2 Ashley used a 3kW tumble dryer for 90 minutes. The cost of electricity is 16p per kilowatt-hour.

**a** How much energy did Ashley use in kWh? (Use the equation: energy = power × time.) [2]

**b** How much did this energy cost? [2]

**c** Describe what eventually happened to this energy. [1]

[Total: 5]

## Foundation/Higher level

**3** AO1

| 1 Water is trapped behind a dam. | |
|---|---|
| A Moving water spins turbines. | B Electricity is generated. |
| C Water is released through pipes. | D Moving turbines make a generator spin. |

Use your knowledge of power stations to number statements A to D correctly, so that they could be made into a flowchart with statement 1. [3]

[Total: 3]

**4** AO1, AO3 An electrical shop claims that the efficiency of one new model of fridge is 20% better than the efficiency of a similar model.

**a** Explain what is meant by efficiency. [1]

**b** Which one of the following is a good reason for someone to use a more efficient fridge? [1]

*It costs less to buy the fridge / It will mean lower electricity bills / It stays cleaner / It creates less noise*

**c** Alice has a ten-year old fridge that is still working but is not as energy-efficient as new models. She would like to change it because she is worried about the effect on the environment. Which two of the following are factors affecting the environment that she should consider in making her decision? [2]

A The reduction in her electricity bill if she uses a new fridge
B Additional greenhouse gas emissions from manufacturing a new fridge
C The cost of buying a new fridge
D Additional greenhouse gas emissions from greater electricity usage of an inefficient fridge

[Total: 4]

**5** AO1 Complete these sentences by choosing the correct word.

A Radon gas is a *radioactive / reactive* gas that increases the risk of lung cancer.
B The lungs of a person breathing in radon gas are *irradiated / contaminated* by radon.
C Surgical equipment is irradiated to kill microorganisms so it *is / is not* radioactive afterwards.
D High level nuclear waste is crushed and melted into solid glass or concrete blocks to reduce the risk of *irradiation / contamination*. [4]

[Total: 4]

## Higher level

**6** AO2, AO3 A new power station is needed for a city in southern England. The city has good road and rail links. It is about 30 miles inland, beside a river, in flat countryside. The choice is between a wind farm, a coal-fired power station or a nuclear power station.

Explain which power station would be most suitable.

The quality of your written communication will be assessed in your answer. [6]

[Total: 6]

**7** AO2, AO3 Matt has measured the monthly electricity consumption of his family. They do not use electricity for heating.

Jan: 700 kWh   Feb: 500 kWh   Mar: 460 kWh
April: 380 kWh   May: 300 kWh   June: 170 kWh
July: 170 kWh   August: 170 kWh   Sept: 260 kWh
Oct: 400 kWh   Nov: 500 kWh   Dec: 710 kWh

**a** Draw a bar chart showing the family's monthly energy consumption. [3]

**b** Explain one reason why their demand for electricity changes through the year. [1]

**c** Their neighbours live in the same style and size of house but their electricity bill is lower. Give **three** possible reasons why their bill is lower. [3]

[Total: 7]

AO1 recall the science    AO2 apply your knowledge    AO3 evaluate and analyse the evidence

# P3 Sustainable energy

## Worked example

Peta took readings from her electricity meter.

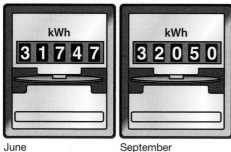

June  September

**How to raise your grade**

Take note of the comments from examiners – these will help you to improve your grade.

**AO1 a** The joule and kilowatt-hour are both units of energy. Explain which is the more useful when measuring the energy used in the home. [2]

*Kilowatt-hour* ✔ *because it measures the larger amounts of energy we use in the home.* ✔

This answer identifies the correct unit and explains why it is used: 2 marks.

**AO2 b** How many kWh of electricity did Peta use in this three-month period? [2]

*32050 – 31747* ✔ *= 313 kWh* ✗

This answer gets 1 mark for showing the correct working even though the final answer is wrong. Always show your working.

**AO2 c** Peta used more electricity in the same period last year. Since then she has started using low-energy light bulbs all over the house and has not used her tumble dryer as much.

**i)** Explain why switching to low-energy light bulbs has made a significant difference to Peta's electricity bill. [2]

*Low-energy light bulbs are more energy-efficient than filament bulbs – they use a smaller amount of electricity to produce the same amount of light.* ✔ *She has used them all over the house and it's likely that several lights are on for several hours each day.* ✔

This shows an understanding that low-energy bulbs use less electricity and that their number and likely length of usage are significant. This gets 2 marks.

**ii)** Why has reducing the use of the tumble dryer also made a significant difference? [2]

*Using it less means using less electricity.* ✔

This answer has recognised that less usage means less electricity used but hasn't referred to the fact that large appliances such as tumble dryers use a lot of electricity, so that even a reduction has a significant impact. This will be awarded 1 mark only.

297

# Carrying out practical work in GCSE Science

## Introduction

As part your GCSE Science course, you will develop practical skills and will carry out investigative work and practical data analysis.

Investigative work can be divided into several parts:

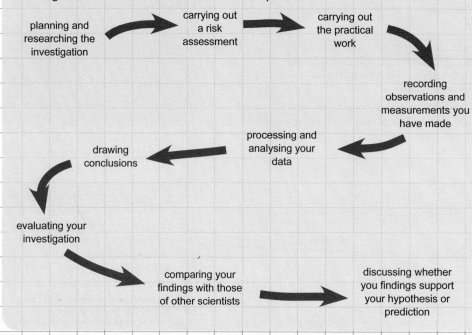

### ✳ Planning and researching your investigation

A scientific investigation usually begins with a scientist testing an idea, answering a question, or trying to solve a problem.

You first have to plan how you will carry out the investigation.

Your planning will involve testing a **hypothesis**. For example, you might observe during a fermentation with yeast investigation that beer or wine is produced faster at a higher temperature.

So your hypothesis might be 'as the temperature increases, the rate of fermentation increases'.

To formulate a hypothesis you may have to research some of the background science.

First of all, use your lesson notes and your textbook. The topic you've been given to investigate will relate to the science you've learnt in class.

Also make use of the Internet, but make sure that your Internet search is closely focused on the topic you're investigating.

✓ The search terms you use on the Internet are very important. 'Investigating fermentation' is a better search term than just 'fermentation', as it's more likely to provide links to websites that are more relevant to your investigation.

### Definition

A **hypothesis** is a possible explanation that someone suggests to explain some scientific observations.

### Tip

If you're formulating a hypothesis, it's important that it's testable. In other words, you must be able to test the hypothesis in the school lab.

### Tip

In the planning stage, scientific research is very important.

# Carrying out practical work in GCSE Science

✓ The information on websites also varies in its reliability. Free encyclopaedias often contain information that hasn't been written by experts. Some question and answer websites might appear to give you the exact answer to your question, but be aware that they may sometimes be incorrect.

✓ Most GCSE Science websites are more reliable, but if in doubt, use other information sources to verify the information.

As a result of your research, you may be able to extend your hypothesis and make a prediction that's based on science.

### Example 1

Investigation: Plan and research an investigation into the activity of enzymes

Your hypothesis might be 'When I increase the temperature, the rate of reaction increases'.

You may be able to add more detail: 'This is because as I increase the temperature, the frequency of collisions between the enzyme and the reactant increases'.

## Choosing a method and suitable apparatus

As part of your planning, you must choose a suitable way of carrying out the investigation.

You will have to choose suitable techniques, equipment and technology, if this is appropriate. How do you make this choice?

For most of the practical work you are likely to do, there will be a choice of techniques available. You must select the technique:

✓ that is most appropriate to the context of your investigation, and

✓ that will enable you to collect valid data, for example if you are measuring the effects of light intensity on photosynthesis, you may decide to use an LED (light-emitting diode) at different distances from the plant, rather than a light bulb. The light bulb produces more heat, and temperature is another independent variable in photosynthesis.

Your choice of equipment, too, will be influenced by measurements you need to make. For example:

✓ you might use a one-mark or graduated pipette to measure out the volume of liquid for a titration, but

✓ you may use a measuring cylinder or beaker when adding a volume of acid to a reaction mixture, so that the volume of acid is in excess to that required to dissolve, for example, the calcium carbonate.

> **Tip**
> Technology, such as data-logging and other measuring and monitoring techniques, for example heart sensors, may help you to carry out your experiment.

> **Tip**
> Carrying out a preliminary investigation, along with the necessary research, may help you to select the appropriate technique to use.

##  Variables

In your investigation, you will work with factors which may affect an outcome.

The factors you choose, or are given, to investigate the effect of are called input variables or **independent variables**.

What you choose to measure, as affected by the independent variable, is called the outcome variable or **dependent variable**.

##  Independent variables

In your practical work, you will be provided with an independent variable to test, or will have to choose one – or more – of these to test. Some examples are given in the table.

| Investigation | Possible independent variables to test |
|---|---|
| activity of yeast | > temperature<br>> sugar concentration |
| rate of a chemical reaction | > temperature<br>> concentration of reactants |
| stopping distance of a moving object | > speed of the object<br>> the surface on which it's moving |

Independent variables can be **discrete** or **continuous**.

> When you are testing the effect of different disinfectants on bacteria you are looking at discrete variables.

> When you are testing the effect of a range of concentrations of the same disinfectant on the growth of bacteria you are looking at continuous variables.

**Definition**
Variables that fall into a range of separate types are called **discrete variables**.

**Definition**
Variables that have a continuous range are called **continuous variables**.

**Definition**
The **range** defines the extent of the independent variables being tested.

### Range

When working with an independent variable, you need to choose an appropriate **range** over which to investigate the variable.

You need to decide:

✔ which treatments you will test, and/or
✔ the upper and lower limits of the independent variables to investigate, if the variable is continuous.

Once you have defined the range to be tested, you also need to decide the appropriate intervals at which you will make measurements.

The range you would test depends on:

✔ the nature of the test
✔ the context in which it is given
✔ practical considerations, and
✔ common sense.

## Carrying out practical work in GCSE Science

### Example 2

1 Investigation: Investigating the factors that affect how quickly a weak acid works to remove limescale from an appliance

You may have to decide on which acids to use from a range you're provided with. You would choose a weak acid, or weak acids, to test, rather than a strong acid, such as concentrated sulfuric acid. This is because of safety reasons, but also because the acid might damage the appliance you were trying to clean. You would then have to select a range of concentrations of your chosen weak acid to test.

2 Investigation: How speed affects the stopping distance of a trolley in the lab

The range of speeds you would choose would clearly depend on the speeds you could produce in the lab.

> **Tip**
> Again, it's often best to carry out a trial run or preliminary investigation, or carry out research, to determine the range to be investigated.

### Concentration

You might be trying to find out the best, or optimum, concentration of a disinfectant to prevent the growth of bacteria.

The 'best' concentration would be the lowest in a range that prevented the growth of the bacteria. Concentrations higher than this would be just wasting disinfectant.

If, in a preliminary test, no bacteria were killed by the concentration you used, you would have to increase it (or test another disinfectant). However, if there was no growth of bacteria in your preliminary test, you would have to lower the concentration range. A starting point might be to look at concentrations around those recommended by the manufacturer.

## Dependent variables

The dependent variable may be clear from the problem you're investigating, for example the stopping distance of moving objects. But you may have to make a choice.

> **Tip**
> The value of the **depend**ent variable is likely to **depend** on the value of the independent variable. This is a good way of remembering the definition of a dependent variable.

### Example 3

1 Investigation: Measuring the rate of photosynthesis in a plant

There are several ways in which you could measure the rate of photosynthesis in a plant. These include:

> counting the number of bubbles of oxygen produced in a minute by a water plant such as *Elodea* or *Cabomba*

> measuring the volume of oxygen produced over several days by a water plant such as *Elodea* or *Cabomba*

> monitoring the concentration of oxygen in a polythene bag enclosing a potted plant using an oxygen sensor

> measuring the colour change of hydrogencarbonate indicator containing algae embedded in gel.

## 2 Investigation: Measuring the rate of a chemical reaction

You could measure the rate of a chemical reaction in the following ways:

> the rate of formation of a product

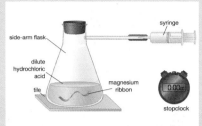

> the rate at which the reactant disappears

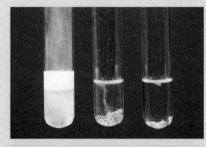

> a colour change

> a pH change.

### ✸ Control variables

The validity of your measurements depends on you measuring what you're supposed to be measuring.

Other variables that you're not investigating may also have an influence on your measurements. In most investigations, it's important that you investigate just one variable at a time. For a 'fair test', other variables, apart from the one you're testing at the time, must be kept constant. These are called **control variables**.

Some of these variables may be difficult to control. For example, in an ecology investigation in the field, factors such as varying weather conditions are impossible to control.

## Experimental controls

Experimental controls are often very important, particularly in biological investigations where you're testing the effect of a treatment.

> ### Example 4
> Investigation: The effect of disinfectants on the growth of bacteria
>
> If the bacteria don't grow, it could be because they have been killed by the disinfectant. But the bacteria in your investigation may have died for some other reason. Another factor may be involved. To test whether any effects were down to the disinfectant, you need to set up the same practical, but this time using distilled water in place of the disinfectant. The distilled water is your control. If the bacteria are inhibited by the disinfectant, but grow normally in the dish containing distilled water, it's reasonable to assume that the disinfectant inhibited their growth.

> **Definition**
> An **experimental control** is used to find out whether the effect you obtain is from the treatment, or whether you get the same result in the absence of the treatment.

## ✹ Assessing and managing risk

Before you begin any practical work, you must assess and minimise the possible risks involved.

Before you carry out an investigation, you must identify the possible **hazards**. These can be grouped into biological hazards, chemical hazards and physical hazards.

| Biological hazards include: | Chemical hazards can be grouped into: | Physical hazards include: |
|---|---|---|
| > microorganisms<br>> body fluids<br>> animals and plants. | > irritant and harmful substances<br>> toxic<br>> oxidising agents<br>> corrosive<br>> harmful to the environment. | > equipment<br>> objects<br>> radiation. |

Scientists use an international series of symbols so that investigators can identify hazards.

Hazards pose **risks** to the person carrying out the investigation.

A risk posed by concentrated sulfuric acid, for example, will be lower if you're adding one drop of it to a reaction mixture to make an ester, than if you're mixing a large volume of it with water.

When you use hazardous materials, chemicals or equipment in the laboratory, you must use them in such a way as to keep the risks to an absolute minimum. For example, one way is to wear eye protection when using hydrochloric acid.

> **Definition**
> A **hazard** is something that has the potential to cause harm. Even substances, organisms and equipment that we think of as being harmless, used in the wrong way may be hazardous.

Hazard symbols are used on chemical bottles so that hazards can be identified

> **Definition**
> The **risk** is the likelihood of a hazard to cause harm in the circumstances of its use.

# Carrying out practical work in GCSE Science

## ✸ Risk assessment

Before you begin an investigation, you must carry out a risk assessment. Your risk assessment must include:

✔ all relevant hazards (use the correct terms to describe each hazard, and make sure you include them all, even if you think they will pose minimal risk)
✔ risks associated with these hazards
✔ ways in which the risks can be minimised
✔ results of research into emergency procedures that you may have to take if something goes wrong.

You should also consider what to do at the end of the practical. For example, used agar plates should be left for a technician to sterilise; solutions of heavy metals should be collected in a bottle and disposed of safely.

### Tip
To make sure that your risk assessment is full and appropriate:

> remember that for a chemical reaction, the risk assessment should be carried out for the products and the reactants

> when using chemicals, make sure the hazard and ways of minimising risk match the concentration of the chemical you're using; many acids, for instance, while being corrosive in higher concentrations, are harmful or irritant at low concentrations.

## ✸ Collecting primary data

✔ You should make sure that observations, if appropriate, are recorded in detail. For example, it's worth recording the appearance of the potato chips in an osmosis practical, in addition to the measurements you make.
✔ Measurements should be recorded in tables. Have one ready so that you can record your readings as you carry out the practical work.
✔ Think about the dependent variable and define this carefully in your column headings.
✔ You should make sure that the table headings describe properly the type of measurements you've made, for example 'time taken for magnesium ribbon to dissolve'.
✔ It's also essential that you include units – your results are meaningless without these.
✔ The units should appear in the column head, and not be repeated in each row of the table.

### Definition
When you carry out an investigation, the data you collect are called **primary data.** The term 'data' is normally used to include your observations as well as measurements you might make.

## ✸ Repeatability and reproducibility of results

When making measurements, in most instances, it's essential that you carry out repeats.

|  | Test 1 | Test 2 | Test 3 |
|---|---|---|---|
|  |  |  |  |

These repeats are one way of checking your results. One set of results from your investigation may not reflect what truly happens. Carrying out repeats enables you to identify any results that don't fit.

Results will not be repeatable of course, if you allow the conditions the investigation is carried out in to change.

### Definition
A reading that is very different from the rest, is called an anomalous result, or **outlier**.

You need to make sure that you carry out sufficient repeats, but not too many. In a titration, for example, if you obtain two values that are within 0.1 cm³ of each other, carrying out any more will not improve the reliability of your results.

This is particularly important when scientists are carrying out scientific research and make new discoveries.

> **Definition**
> If, when you carry out the same experiment several times, and get the same, or very similar results, we say the results are **repeatable**.

## Processing data

### Calculating the mean
Using your repeat measurements you can calculate the arithmetical mean (or just 'mean') of these data. We often refer to the mean as the 'average.'

| Temperature, °C | Number of yeast cells/mm³ | | | Mean number of yeast cells/mm³ |
|---|---|---|---|---|
| | Test 1 | Test 2 | Test 3 | |
| 10 | 1000 | 1040 | 1200 | 1080 |
| 20 | 2400 | 2200 | 2300 | 2300 |
| 30 | 4600 | 5000 | 4800 | 4800 |
| 40 | 4800 | 5000 | 5200 | 5000 |
| 50 | 200 | 1200 | 700 | 700 |

You may also be required to use formulae when processing data.

### Significant figures
When calculating the mean, you should be aware of significant figures.

For example, for the set of data below:

| 18 | 13 | 17 | 15 | 14 | 16 | 15 | 14 | 13 | 18 |

The total for the data set is 153, and ten measurements have been made. The mean is 15, and not 15.3.

This is because each of the recorded values has two significant figures. The answer must therefore have two significant figures. An answer cannot have more significant figures than the number being multiplied or divided.

### Using your data
When calculating means (and displaying data), you should be careful to look out for any data that don't fit in with the general pattern.

It might be the consequence of an error made in measurement, but sometimes outliers are genuine results. If you think an outlier has been introduced by careless practical work, you should ignore it when calculating the mean. But you should examine possible reasons carefully before just leaving it out.

> **Definition**
> The **reproducibility** of data is the ability of the results of an investigation to be reproduced by someone else, who may be in a different lab, carrying out the same work.

> **Definition**
> The **mean** is calculated by adding together all the measurements, and dividing by the number of measurements.

> **Definition**
> **Significant figures** are the number of digits in a number based on the precision of your measurements.

## Displaying your data

Displaying your data – usually the mean values – makes it easy to pick out and show any patterns. And it also helps you to pick out any anomalous data.

It is likely that you will have recorded your results in tables, and you could also use additional tables to summarise your results. The most usual way of displaying data is to use graphs. The table will help you decide which type to use.

| Type of graph | When you would use the graph | Example |
|---|---|---|
| Bar charts or bar graph | Where one of the variables is discrete | 'The diameters of the clear zones where the growth of bacteria was inhibited by different types of disinfectant' |
| Line graph | Where independent and dependent variables are both continuous | 'The volume of carbon dioxide produced by a range of different concentrations of hydrochloric acid' |
| Scatter graph | To show an association between two (or more) variables | 'The association between length and breadth of a number of privet leaves'<br><br>In scatter graphs, the points are plotted, but not usually joined |

If it's possible from the data, join the points of a line graph using a straight line or, in some instances, a curve. In this way graphs can also help us to process data.

### Tip

Remember when drawing graphs, plot the independent variable on the x-axis, and the dependent variable on the y-axis.

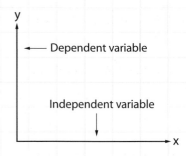

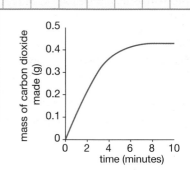

We can calculate the rate of production of carbon dioxide from the gradient of the graph

## Variation in data

Plotting a graph of just the means doesn't tell you anything about the spread of data that has been used to calculate the mean.

You can show the spread or range of the data on your graphs using error bars or range bars.

Range bars are very useful, but they don't show how the data are spread between the extreme values. It is important to have information about this range. It may affect the analysis you do of the data, and the conclusions you draw.

Scientists use a number of techniques to look at the spread of data. You could refer to the work that you've done in Maths to look at some of these techniques.

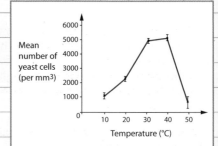

Range bars indicate the spread or range of values

Carrying out practical work in GCSE Science

 ## Conclusions from differences in data sets

When comparing two (or more) sets of data, we often compare the values of two sets of means.

### Example 5

Investigation: Comparing the effectiveness of two disinfectants

Two groups of students compared the effectiveness of two disinfectants, labelled A and B. Their results are shown in the table.

| Disinfectant | Diameter of zone of inhibition (clear zone), mm | | | | | | | | | | Mean dia. mm |
|---|---|---|---|---|---|---|---|---|---|---|---|
| | 1 | 2 | 3 | 4 | 5 | 6 | 7 | 8 | 9 | 10 | |
| A | 15 | 13 | 17 | 15 | 14 | 16 | 15 | 14 | 13 | 18 | 15 |
| B | 25 | 23 | 24 | 23 | 26 | 27 | 25 | 24 | 23 | 22 | 24 |

When the means are compared it appears that disinfectant B is more effective in inhibiting the growth of bacteria. But can we be sure? The differences might have resulted from the treatment of the bacteria using the two disinfectants. But the differences could have occurred purely by chance.

Scientists use statistics to find the probability of any differences having occurred by chance. The lower this probability is, which is found out by statistical calculations, the more likely it is that it was (in this case) the disinfectant that caused the differences observed.

Statistical analysis can help to increase the confidence you have in your conclusions.

> **Tip**
> You have learnt about probability in your Maths lessons.

> **Definition**
> If there is a relationship between dependent and independent variables that can be defined, we say there is a **correlation** between the variables.

 ## Drawing conclusions

Observing trends in data or graphs will help you to draw conclusions. You may obtain a linear relationship between two sets of variables, or the relationship might be more complex.

### Example 6

**Conclusion from results A:** The higher the concentration of acid, the shorter the time taken for the magnesium ribbon to dissolve.

**Conclusion from results B:** The higher the concentration of acid, the faster the rate of reaction.

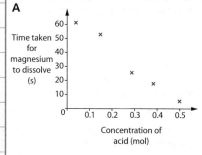

This graph shows **negative correlation**

When drawing conclusions, you should try to relate your findings to the science involved.

> In investigation A in Example 6, your discussion should focus on the greater possibility/increased frequency of collisions between reacting particles as the concentration of the acid is increased.

> In investigation B in Example 6, there's a clear scientific mechanism to link the rate of reaction to the concentration of acid.

But we sometimes see correlations between data in science which are coincidental, where the independent variable is not the cause of the trend in the data.

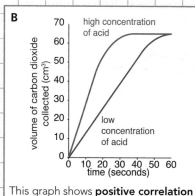

This graph shows **positive correlation**

## Carrying out practical work in GCSE Science

*Example 7*
Studies have shown that levels of vitamin D are very low in people with long-term inflammatory diseases. But there's no scientific evidence to suggest that these low levels are the cause of the diseases.

**Tip**
Scientists have to be careful not to mistake correlation for causation.

### ✺ Evaluating your investigation

Your conclusion will be based on your findings, but must take into consideration any uncertainty in these introduced by any possible sources of error. You should discuss where these have come from in your evaluation.

The two types of errors are:

✓ random error  ✓ systematic error.

Errors can occur when the instrument you're using to measure lacks sufficient sensitivity to indicate differences in readings. They can also occur when it's difficult to make a measurement. If two investigators measure the height of a plant, for example, they might choose different points on the compost, and the tip of the growing point to make their measurements.

The volume of liquid in a burette must be read to the bottom of the meniscus

Measurements can be either consistently too high or too low. One reason could be down to the way you are making a reading, for example taking a burette reading at the wrong point on the meniscus. Another could be the result of an instrument being incorrectly calibrated, or not being calibrated.

**Definition**
**Error** is a difference between a measurement you make, and its true value.

**Definition**
With **random error**, measurements vary in an unpredictable way.

**Definition**
With **systematic error**, readings vary in a controlled way.

**Tip**
What you shouldn't discuss in your evaluation are problems introduced by using faulty equipment, or by you using the equipment inappropriately. These errors can, or could have been, eliminated, by:
> the checking of equipment, and
> practising techniques beforehand, taking care and being patient when carrying out the practical.

### ✺ Accuracy and precision

When evaluating your investigation, you might mention accuracy or precision. But if you use these terms, it's important that you understand what they mean, and that you use them correctly. The terms accuracy and precision can be illustrated using shots at a dartboard.

**Definition**
When making measurements:
> the **accuracy** of the measurement is how close it is to the true value
> **precision** is how closely a series of measurements agree with each other.

precise but not accurate

precise and accurate

imprecise and inaccurate

Carrying out practical work in GCSE Science

## Improving your investigation

When evaluating your investigation, you should discuss how your investigation could be improved. This could be by improving:

- the reliability of your data. For example, you could make more repeats, or more frequent readings, or 'fine-tune' the range you investigate, or refine your technique in some other way
- the accuracy and precision of your data, by using more precise measuring equipment.

In science, the measurements you make as part of your investigation should be as precise as you can, or need to, make them. To achieve this, you should use:

- the most appropriate measuring instrument
- the measuring instrument with the most appropriate scale.

The smaller the scale divisions you work with, the more precise your measurements. For example:

- in an investigation on how your heart rate is affected by exercise, you might decide to investigate this after a 100 m run. You might measure out the 100 m distance using a trundle wheel, which is sufficiently precise for your investigation
- in an investigation on how light intensity is affected by distance, you would make your measurements of distance using a metre rule with millimetre divisions; clearly a trundle wheel would be too imprecise
- in an investigation on plant growth, in which you measure the thickness of a plant stem, you would use a micrometer or Vernier callipers. A metre rule would be too imprecise.

## Using secondary data

Another method of evaluation is to compare your data – primary data – with **secondary data**. One of the simplest ways of doing this is to compare your data with data from other members of your class who have carried out an identical practical investigation.

You should also, if possible, search through the scientific literature – in textbooks, the Internet, and databases, to find data from similar or identical practical investigations so that you can compare the data with yours.

Ideally, you should use secondary data from a number of sources, carry out a full analysis of the data you have collected, and compare the findings with your own. You should critically analyse any evidence that conflicts with yours, and suggest what further data might help to make your conclusions more secure.

You should review secondary data and evaluate it. Scientific studies are sometimes influenced by the **bias** of the experimenter.

- One kind of bias is having a strong opinion related to the investigation, and perhaps selecting only the results that fit with a hypothesis or prediction.
- Or the bias could be unintentional. In fields of science that are not yet fully understood, experimenters may try to fit their findings to current knowledge and thinking.

In other instances the 'findings' of experimenters have been influenced by organisations that supplied funding for the research.

You must reference secondary data you have used (see page 310).

### Definition

**Secondary data** are measurements/observations made by anyone other than you.

**Referencing methods**

The two main conventions for writing a reference are the:

✔ Harvard system
✔ Vancouver system.

In your text, the Harvard system refers to the authors of the reference, for example 'Smith and Jones (1978)'.

The Vancouver system refers to the number of the numbered reference in your text, for example '... the reason for this hypothesis is unknown[5]'.

Though the Harvard system is usually preferred by scientists, it is more straightforward for you to use the Vancouver system.

### Harvard system

In your references list a book reference should be written:

> Author(s) (year of publication). *Title of Book*, publisher, publisher location.

The references are listed in alphabetical order according to the authors.

### Vancouver system

In your references list a book reference should be written:

> 1 Author(s). *Title of Book*. Publisher, publisher location: year of publication.

The references are numbered in the order in which they are cited in the text.

**Tip**

Remember to write out the URL of a website in full. You should also quote the date when you looked at the website.

 **Do the data support your hypothesis?**

You need to discuss, in detail, whether all, or which of your primary data, and the secondary data you have collected, support your original hypothesis. They may, or may not.

You should communicate your points clearly, using the appropriate scientific terms, and checking carefully your use of spelling, punctuation and grammar. Your quality of written communication is important, as well as your science.

If your data do not completely match your hypothesis, it may be possible to modify the hypothesis or suggest an alternative one. You should suggest any further investigations that can be carried out to support your original hypothesis or the modified version.

It is important to remember, however, that if your investigation does support your hypothesis, it can improve the confidence you have in your conclusions and scientific explanations, but it can't prove your explanations are correct.

# Controlled assessment in 21st Century GCSE Science

## Introduction

*If you are following the GCSE Science course (and possibly going on to take Additional Science) there are two tasks that you will need to complete for the controlled assessment unit:*

✔ a Case Study of a topical issue in science, which gives you experience in evaluating evidence collected or reported by others

✔ a Practical Data Analysis task, designed as a stepping stone towards the wider range of skills required to carry out an independent, complete investigation. This focuses on interpreting your own first-hand data.

The controlled assessment task for GCSE Additional Science will be a Practical Investigation.

If you are using this book as part of your studies towards GCSE Biology, GCSE Chemistry and GCSE Physics, you will do a Practical Investigation in each of these; you won't do a Case Study or Data Analysis task.

### ✹ Case Study

In the Case Study you need to demonstrate your skills in evaluating science-related information of the type you might find in the media. You will be provided with some stimulus material as a starting point for the study. You will need to compare opposing views and use your understanding of science to decide on an appropriate course of action.

In addition to using the stimulus material, you are expected to use your research skills to locate and evaluate scientific evidence, claims and opinions in the media, in textbooks, and on the internet. You will need to consider this secondary data together with your own scientific knowledge and understanding, and take into account the quality of evidence and the apparent reliability of each source. See page 309 for hints on evaluating secondary data.

Finally, you will decide between the claims made and give recommendations for appropriate actions.

### ✹ Practical Data Analysis

In the Practical Data Analysis task you need to do an investigative practical activity to collect some primary data, then interpret the data and evaluate the activity.

You will be expected to:

✔ plan a test for a hypothesis

✔ record a risk assessment for the procedures you will use

✔ collect primary data

✔ draw evidence-based conclusions

✔ evaluate the quality of the data collected

✔ review the effectiveness of the practical procedures

✔ suggest improvements to your plan.

Pages 298–309 give guidelines for practical investigations.

A set of secondary data will be provided to help you to evaluate your own data. See page 309 for hints on comparing your own data with secondary data.

# How to be successful in your GCSE Science exam

## Introduction

*OCR uses assessments to test your understanding of scientific ideas, how well you can apply your understanding to new situations and how well you can analyse and interpret information you've been given. The assessments are opportunities to show how well you can do these things.*

To be successful in exams you need to:

- ✔ have a good knowledge and understanding of science
- ✔ be able to apply this knowledge and understanding to familiar and new situations, and
- ✔ be able to interpret and evaluate evidence that you've just been given.

You need to be able to do these things under exam conditions.

### ✸ The language of the external assessment

When working through an assessment paper, make sure that you:

- ✔ re-read the question enough times until you understand exactly what the examiner is looking for
- ✔ highlight key words in the question. In some instances, you will be given key words to include in your answer
- ✔ look at how many marks are allocated for each part of the question. In general, you need to write at least as many separate points in your answer as there are marks.

### ✸ What verbs are used in the question?

A good technique is to see which verbs are used in the wording of the question and to use these to gauge the type of response you need to give. The table lists some of the common verbs found in questions, the types of responses expected and then gives an example.

| Verb used in question | Response expected in answer | Example question |
|---|---|---|
| > write down<br>> state<br>> give<br>> identify | These are usually more straightforward types of question in which you're asked to give a definition, make a list of examples, or select the best answer from a series of options | 'Write down three types of microorganism that cause disease'<br><br>'State one difference and one similarity between radio waves and gamma rays' |
| calculate | Use maths to solve a numerical problem | 'Calculate the cost of supplying the flu vaccine to the whole population of the UK' |

# How to be successful in your GCSE Science exam

| estimate | Use maths to solve a numerical problem, but you do not have to work out the exact answer | 'Estimate the number of bacteria in the culture after five hours' |
|---|---|---|
| describe | Use words (or diagrams) to show the characteristics, properties or features of, or build an image, of something | 'Describe how antibiotic resistance can be reduced' |
| suggest | Come up with an idea to explain information you're given | 'Suggest why eating fast foods, rather than wholegrain foods, could increase the risk of obesity' |
| > demonstrate<br>> show how | Use words to make something evident using reasoning | 'Show how enzyme activity changes with temperature' |
| compare | Look for similarities and differences | 'Compare the structure of arteries and veins' |
| explain | Offer a reason for, or make understandable, information you're given | 'Explain why measles cannot be treated with antibiotics' |
| evaluate | Examine and make a judgement about an investigation or information you're given | 'Evaluate the evidence for vaccines causing harm to human health' |

##  What is the style of the question?

Try to get used to answering questions that have been written in lots of different styles before you sit the exam. Work through past papers, or specimen papers, to get a feel for these. The types of questions in your assessment fit the three assessment objectives shown in the table.

| Assessment objective | Your answer should show that you can… |
|---|---|
| **AO1** Recall the science | Recall, select and communicate your knowledge and understanding of science |
| **AO2** Apply your knowledge | Apply skills, knowledge and understanding of science in practical and other contexts |
| **AO3** Evaluate and analyse the evidence | Analyse and evaluate evidence, make reasoned judgements and draw conclusions based on evidence |

### Tip
Of course you must revise the subject material adequately. But it's as important that you are familiar with the different question styles used in the exam paper, as well as the question content.

**How to answer questions on: AO1 Recall the science**

These questions, or parts of questions, test your ability to recall your knowledge of a topic or a process. There are several types of this style of question:

✔ Fill in the spaces (you may be given words to choose from)
✔ Tick the correct statements or use lines to link a term with its definition or correct statement
✔ Add labels to a diagram or complete a table
✔ Describe a process
✔ Explain observations
✔ Write a full account or explanation of a topic or a process

To revise for these types of questions, make sure that you have learnt definitions and scientific terms. Produce a glossary of these, or key facts cards, to make them easier to remember. Make sure your key facts cards also cover important practical techniques.

### Tip

Don't forget that mind maps – either drawn by you or by using a computer program – are very helpful when revising key points.

*Example 8*

1 What is meant by the term *metabolic rate*?
   Tick (✓) **one** box.
   ☐ the amount of energy a person uses each hour
   ☐ the amount of exercise a person does each day
   ☐ the amount of food a person eats each day

2 Describe how to test the pH of a solution.

 **How to answer questions on: AO2 Apply skills, knowledge and understanding**

Some questions require you to apply basic knowledge and understanding in your answers.

You may be presented with a topic that's familiar to you, but you should also expect questions in your Science exam to be set in an unfamiliar context.

Questions may be presented as:

✔ experimental investigations
✔ data for you to interpret and analyse
✔ a short paragraph or article.

The information required for you to answer the question might be in the question itself, but for later stages of the question, you may be asked to draw on your knowledge and understanding of the subject material in the question.

You may be expected to describe patterns in data from graphs you are given or that you have drawn from given data.

Practice will help you to become familiar with some contexts that examiners use and common question styles. But you will not be able to predict all of the contexts used. This is deliberate; being able to apply your knowledge and understanding to different and unfamiliar situations is a skill the examiner tests.

### Tip

Work through the 'Preparing for Assessment: Applying your knowledge' tasks in this book as practice.

# How to be successful in your GCSE Science exam

Practise doing questions where you are tested on being able to apply your scientific knowledge and your ability to understand new and unfamiliar situations. In this way, when this type of question comes up in your exam, you will be able to tackle it successfully.

> *Example 9*
> 1 Measles is an infectious disease caused by a virus. Today, most children are vaccinated against measles when they are very young.
>
> The graph on the right shows the number of measles cases per year in the USA between 1950 and 2000. The graph also shows when vaccination against measles was first introduced in the USA.
>
> The use of the measles vaccine reduces the number of measles cases.
>
> Explain why this graph alone does not prove that the use of the measles vaccine reduces the number of measles cases.
>
> 2 Look at the graph on the right showing the resistance of the bacterium, *Streptococcus pneumoniae*, to three different types of antibiotic.
>  a Which antibiotic does there seem to be least resistance to, even when it has been used before?
>  b Can you explain why this might be the case?

You will also need to analyse scientific evidence or data given to you in the question. Analysing data may involve drawing graphs and interpreting them, and carrying out calculations. Practise drawing and interpreting graphs from data.

When drawing a graph, make sure you:

✓ choose and label the axes fully and correctly
✓ include units, if this hasn't been done for you already
✓ plot points on the graph carefully – the examiner will check individual points to make sure that they are accurate
✓ join the points correctly; usually this will be by a line of best fit.

When reading values off a graph you have drawn or one given in the question, make sure you:

✓ do it carefully, reading the values as accurately as you can
✓ double-check the values.

When describing patterns and trends in the data, make sure you:

✓ write about a pattern or trend in as much detail as you can
✓ mention anomalies where appropriate
✓ recognise there may be one general trend in the graph, where the variables show positive or negative correlation
✓ recognise the data may show a more complex relationship. The graph may demonstrate different trends in several sections. You should describe what's happening in each
✓ describe fully what the data show.

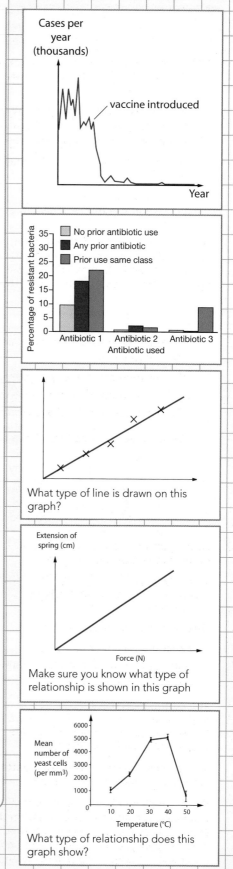

What type of line is drawn on this graph?

Make sure you know what type of relationship is shown in this graph

What type of relationship does this graph show?

315

How to be successful in your GCSE Science exam

## How to answer questions needing calculations

✔ The calculations you're asked to do may be straightforward, for example the calculation of the mean from a set of practical data.
✔ Or they may be more complex, for example calculating the yield of a chemical reaction.
✔ Other questions will require the use of formulae.

Remember, this is the same maths that you learnt in your Maths lessons.

### Example 10
1 Calculate the area on the agar plate, around the antibiotic disc, that is free from bacteria.
Use the formula:
area = $\pi r^2$
where $\pi$ = 3.14

**Tip**
Formulae are often given to you on the question paper, but sometimes you will be expected to recall and use these. Check the specification, or with your teacher, to make sure that you know the formulae that you have to learn and remember.

**Tip**
Remember, when carrying out any calculations, you should include your working at each stage. You may get credit for getting the process correct, even if your final answer is wrong.

**Tip**
When completing your calculation, make sure you include the correct units.

## How to answer questions on: AO3 Analysing and evaluating evidence

For these types of questions, in addition to analysing data, you must also be able to evaluate the information you're given. This is one of the hardest skills. Think about the validity of the scientific data: did the technique(s) used in any practical investigation allow the collection of accurate and precise data?

Your critical evaluation of scientific data in class will help you to develop the evaluation skills required for these types of questions.

### Example 11
1 In the experiment shown, on the testing of the inhibition of bacterial growth by a new antibiotic, explain why further investigation is required to confirm the effectiveness of the antibiotic.

**Tip**
Work through the 'Preparing for Assessment: Evaluating and analysing evidence' tasks in this book as practice.

You may be expected to compare data with other data, or come to a conclusion about its reliability, its usefulness or its implications. Again, it is possible that you won't be familiar with the context. You may be asked to make a judgement about the evidence or to give an opinion with reasons.

### Example 12
1 A catalytic converter reduces nitrogen oxide emissions from a car engine to nitrogen. Evidence shows that the reaction needs the catalyst to be hot. Explain the effect of this on air quality on housing estates where many people use their car to commute to work, and suggest the possible implications for health.

# How to be successful in your GCSE Science exam

 **The quality of your written communication**

Scientists need good communication skills to present and discuss their findings. You will be expected to demonstrate these skills in the exam. Questions will end with the sentence: The quality of written communication will be assessed in your answer to this question. It will be worth 6 marks.

✔ You must try to make sure that your writing is legible and your spelling, punctuation and grammar are accurate, so that it's clear what you mean in your answer. Examiners can't award marks for answers where the meaning isn't clear. When describing and explaining science, use correct scientific vocabulary.

✔ You must present your information in a form that suits its purpose: for example, a series of paragraphs, with lists or a table if appropriate, and a conclusion if required. Use subheadings where they will be helpful.

✔ You must use a suitable structure and style of writing: ensure that in continuous text you use complete sentences. Remember the writing skills you've developed in English lessons. For example, make sure that you know how to construct a good sentence using connectives.

Practise answering some 'quality of written communication (QWC)' questions. Look at how marks are awarded in mark schemes. You'll find these in the specimen question papers, and past papers.

> **Tip**
> You will be assessed on the way in which you communicate science ideas.

> **Tip**
> There are worked examples of these questions on pages 71, 135, 201 and 265.

### Example mark scheme

**For 5–6 marks:**
Ideas about the topic are correctly described and correctly used to explain it. All information in answer is relevant, clear, organised and presented in a structured and coherent format. Specialist terms are used appropriately. Few, if any, errors in grammar, punctuation and spelling.

**For 3–4 marks:**
Some aspects of the topic are correctly described, but only some are made use of in explaining it. For the most part information is relevant and presented in a structured and coherent format. Specialist terms are used for the most part appropriately. There are occasional errors in grammar, punctuation and spelling.

**For 1–2 marks:**
Some aspects of the topic are correctly described, but not used to explain it. Answer may be simplistic. There may be limited use of specialist terms. Errors of punctuation, grammar and spelling hinder communication of the science.

**0 marks:**
Insufficient or irrelevant science. Answer not worthy of credit.

How to be successful in your GCSE Science exam

##  Revising for your Science exam

You should revise in the way that suits you best. But it's important that you plan your revision carefully, and it's best to start well before the date of the exams. Take the time to prepare a revision timetable and try to stick to it. Use this during the lead up to the exams and between each exam.

When revising:

- ✓ find a quiet and comfortable space in the house where you won't be disturbed. It's best if it's well ventilated and has plenty of light
- ✓ take regular breaks. Some evidence suggests that revision is most effective when you revise in 30 to 40 minute slots. If you get bogged down at any point, take a break and go back to it later when you're feeling fresh. Try not to revise when you are feeling tired. If you do feel tired, take a break
- ✓ use your school notes, textbook and possibly a revision guide. But also make sure that you spend some time using past papers to familiarise yourself with the exam format
- ✓ produce summaries of each module
- ✓ draw mind maps covering the key information in a module
- ✓ set up revision cards containing condensed versions of your notes
- ✓ ask yourself questions, and try to predict questions, as you're revising modules
- ✓ test yourself as you're going along. Try to draw important labelled diagrams, and try some questions under timed conditions
- ✓ prioritise your revision of topics. You might want to allocate more time to revising the topics you find most difficult.

> **Tip**
> Try to make your revision timetable as specific as possible – don't just say 'science on Monday, and Thursday', but list the topics that you'll cover on those days.

> **Tip**
> Start your revision well before the date of the exams, produce a revision timetable, and use the revision strategies that suit your style of learning. Above all, revision should be an active process.

##  How do I use my time effectively in the exam?

Timing is important when you sit an exam. Don't spend so long on some questions that you leave insufficient time to answer others. For example, in a 60-mark question paper, lasting one hour, you will have, on average, one minute per question.

If you're unsure about certain questions, complete the ones you're able to do first, then go back to the ones you're less sure of.

If you have time, go back and check your answers at the end of the exam.

##  On exam day...

A little bit of nervousness before your exam can be a good thing, but try not to let it affect your performance in the exam. When you turn over the exam paper keep calm. Look at the paper and get it clear in your head exactly what is required from each question. Read each question carefully. Don't rush.

If you read a question and think that you have not covered the topic, keep calm – it could be that the information needed to answer the question is in the question itself or the examiner may be asking you to apply your knowledge to a new situation.

Finally, good luck!

# Data sheet

| Fundamental physical quantity | Unit(s) |
|---|---|
| length | metre (m); kilometre (km); centimetre (cm); millimetre (mm); nanometre (nm) |
| mass | kilogram (kg); gram (g); milligram (mg) |
| time | second (s); millisecond (ms); year (a); million years (Ma); billion years (Ga) |
| temperature | degree Celsius (°C); kelvin (K) |
| current | ampere (A); milliampere (mA) |

| Derived physical quantity | Unit(s) |
|---|---|
| area | $cm^2$; $m^2$ |
| volume | $cm^3$; $dm^3$; $m^3$; litre ($l$); millilitre (m$l$) |
| density | $kg/m^3$; $g/cm^3$ |
| speed, velocity | m/s; km/h |
| acceleration | $m/s^2$ |
| force | newton (N) |
| energy | joule (J); kilojoule (kJ); megajoule (MJ); kilowatt hour (kWh); megawatt hour (MWh) |
| power | watt (W); kilowatt (kW); megawatt (MW) |
| frequency | hertz (Hz); kilohertz (kHz) |
| information | bytes (B); kilobytes (kB); megabytes (MB) |
| voltage | volt (V) |
| distance (in astronomy) | light-year (ly) |

## Prefixes for units

| | | | |
|---|---|---|---|
| nano (n) | one thousand millionth | 0.000 000 001 | $\times 10^{-9}$ |
| micro (μ) | one millionth | 0.000 001 | $\times 10^{-6}$ |
| milli (m) | one thousandth | 0.001 | $\times 10^{-3}$ |
| kilo (k) | × one thousand | 1 000 | $\times 10^{3}$ |
| mega (M) | × one million | 1 000 000 | $\times 10^{6}$ |
| giga (G) | × one thousand million | 1 000 000 000 | $\times 10^{9}$ |
| tera (T) | × one million million | 1 000 000 000 000 | $\times 10^{12}$ |

## Symbols of some common elements

| | | | | | |
|---|---|---|---|---|---|
| H | hydrogen | O | oxygen | K | potassium |
| C | carbon | Na | sodium | Ca | calcium |
| N | nitrogen | S | sulfur | | |

## Formulae of some common molecules and compounds

| | | | | | | | |
|---|---|---|---|---|---|---|---|
| $H_2$ | hydrogen gas | $CO_2$ | carbon dioxide | $CH_4$ | methane | $Na_2CO_3$ | sodium carbonate |
| $O_2$ | oxygen gas | CO | carbon monoxide | $NH_3$ | ammonia | $H_2SO_4$ | sulfuric acid |
| $N_2$ | nitrogen gas | NO | nitrogen monoxide | NaCl | sodium chloride | $SO_2$ | sulfur dioxide |
| $H_2O$ | water | $NO_2$ | nitrogen dioxide | NaOH | sodium hydroxide | | |

## Useful equationa

| | |
|---|---|
| distance = wave speed × time | wave speed = frequency × wavelength |
| energy transferred = power × time | electrical power = voltage × current |
| efficiency = (energy usefully transferred ÷ total energy supplied) × 100% | |

# Bad Science for Schools

## When the evidence doesn't add up.

Sometimes people use what sound like scientific words and ideas to sell you things or persuade you to think in a certain way. Some of these claims are valid, and some are not. The activities on these pages are based on the work of Dr Ben Goldacre and will help you to question some of the scientific claims you meet. Read more about the work of Ben in his *Bad Science* book or at badscience.net.

## How much to look younger?

There are many ways to make yourself look younger if you're an adult. These include the style and colour of your hair, the texture of your skin, your body shape and the clothes you wear. Manufacturers and retailers know this and recognise where there's money to be made.

Which of these do you think is more effective?

Are there other ways for adults to make themselves look younger?

How are these age-defying products promoted?

## ✱ YOUNG SKIN FROM OLD?

Skin changes in appearance as people get older. These photographs show how older skin looks different to young skin.

> Examine the photographs. What are the differences?

> How might an anti-ageing skin cream work on the old skin? What would it need to do?

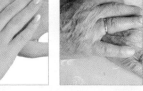

Young skin    Old skin

## ✱ THE SCIENCE BEHIND THE CLAIM

As you get older you may not like the appearance of wrinkles and crow's feet. You can spend quite a lot of money on anti-ageing skin creams. Creams are advertised with appealing images and lavish claims, but do they really work?

One immediate gain from a cream is rehydration. Dried-out skin doesn't look good so we can make it look better by moisturising it. This is easy and the active ingredients are really cheap. However something more is needed to make someone genuinely look younger.

These are three types of active ingredient commonly used:

> Alphahydroxy acids, such as vitamins A and C, are used to exfoliate the skin. Some of these work at high doses, but they are also irritants, so they can only be sold at low doses.

> Vegetable proteins, which are long chain molecules. As the cream dries on the skin the chain molecules tighten, applying tension and temporarily tightening it.

> Hydrogen peroxide, which is corrosive and will lightly burn the skin.

Why might someone who uses a cream with these ingredients think that it is working? Will the effects last?

**We are learning to:**
> find out how anti-ageing skin creams work
> examine claims that are made for them

# NEW AND IMPROVED! ADVERTISER'S CLAIMS

Many anti-ageing skin creams are sold on the basis that there is a scientific reason that they work. Some claims are justified but others are pretty dubious, even if they look persuasive at first glance. You should think critically about what you are told by the advertisers.

> Claims sound more convincing if they are based on tests and if scientists have been involved. Powerful scientific words include *conclusive tests*, *laboratory*, *cleanse*, *purify* and *health*.

> They may claim to make you feel better, look younger, have more energy and be healthier. Some of the claims may be difficult to prove; they should have been tested, the full results published and independently checked in a scientific way. There are very few cases of anti-ageing skin creams being proved to get rid of wrinkles. Why do you think this is?

> Watch out for claims such as 'eight out of ten users said that...' if it's not clear what kind of people and how many were asked. What do you think ten company employees might say about their product and would this be representative of all their consumers?

## ✲ THE PSYCHOLOGY OF COSMETICS

You can buy very cheap creams in the shops. You can even make your own skin cream using simple ingredients. If you did, it would be pretty good at moisturising, so your skin would feel soft and maybe a little smoother. It wouldn't, however, make you look younger for long.

Why do you think anti-ageing skin creams are sometimes quite expensive?

Do you think people who buy anti-ageing skin creams

   a) genuinely believe that they make them look younger?

   b) hope that they might but don't really believe it?

   c) do it because it makes them feel good?

# Bad Science for Schools

## When the evidence doesn't add up.

Sometimes people use what sound like scientific words and ideas to sell you things or persuade you to think in a certain way. Some of these claims are valid, and some are not. The activities on these pages are based on the work of Dr Ben Goldacre and will help you to question some of the scientific claims you meet. Read more about the work of Ben in his *Bad Science* book or at badscience.net.

## What we'll look like in the future

Here are some questions to get you thinking about the future of the human race.

> How do you think humans have evolved over the last few millions of years?

> What changes have taken place in the way we look and move?

> How might evolution affect us over the next few million years?

News stories that sound like they are about science can come from a number of different places. Sometimes they are fair reports of real scientific research, but sometimes they are just good stories.

The story opposite featured in a number of news reports, including The Times, where this version was printed, BBC, Daily Telegraph and The Sun. Your task is to work out if this story is good science or bad science.

You are going to investigate the predictions reported in this article. First you need to identify the predictions. This is the first one – that humans will evolve into two separate species. To decide whether you think this is good or bad science, think about what you have learnt about evolution. Do you think this likely to happen?

From **THE TIMES**
October 17, 2006

# The future ascent (and descent) of man

**Within 100,000 years the divide between rich and poor could lead to two human sub-species**

By Mark Henderson, Science Editor

The mating preferences of the rich, highly educated and well-nourished could ultimately drive their separation into a genetically distinct group that no longer interbreeds with less fortunate human beings, according to Oliver Curry.

Dr Curry, a research associate in the Centre for Philosophy of Natural and Social Science of the London School of Economics, speculated that privileged humans might over tens of thousands of years evolve into a "gracile" subspecies, tall, thin, symmetrical, intelligent and creative. The rest would be shorter and stockier, with asymmetric features and lower intelligence, he said.

**THE BRAVO EVOLUTION REPORT**

### We are learning to:

> analyse a piece of text about a scientific topic to identify the key features
> set these features against a background of accepted scientific evidence
> consider why some stories about science get written

---

*People today are taller and live longer than people a few hundred years ago. Why do you think this has happened? Do you think that trend will continue?*

*Dr Curry is a research associate, but is this story based on scientific research? Why do you think Bravo asked Dr Curry to write this piece? Do you know if Bravo usually take an interest in stories about science?*

---

Dr Curry's vision echoes that of H. G. Wells in *The Time Machine*. He envisaged a race of frail, privileged beings, the Eloi, living in a ruined city and coexisting uneasily with ape-like Morlocks who toil underground and are descended from the downtrodden workers of today.

Dr Curry also said that today's concept of race would be gone by the year 3000, relationships between people with different skin colours producing a "coffee-colour" across all populations.

*In Brazil, the black African, white European and native American populations have been having children together for hundreds of years but there is still a lot of diversity in physical appearance. Can you think of any other examples that you know of that either back up the idea about the whole population being "coffee-coloured" or make you think it might not happen that way?*

With improvements in nutrition and medicine, people would routinely grow to 6ft 6in and live to the age of 120, he said.

Genetic modification, cosmetic surgery and sexual selection — whereby mate preferences drive evolution — meant that people would tend to be better-looking than today.

Otherwise, humans will look much as they do now, with one exception: Dr Curry also suggested that increased reliance on processed food would make chewing less important, possibly resulting in less developed jaws and shorter chins. Ten thousand years from today this effect could be compounded as human faces grow more juvenile in appearance.

*This effect — neotony — is known from domestic animals: dogs resemble young versions of wild relatives such as wolves.*

Dr Curry raised the worrying possibility that reliance on technology could erode social skills and even health. As deaths from genetic diseases such as cancer are prevented, the genes themselves might become more common, no longer being "weeded out" of the gene pool. Increased use of medicine as a means of treating disease could lead to the deterioration of the body's immune system.

Dr Curry's predictions were commissioned by the television channel Bravo to celebrate its 21st anniversary on air.

"The Bravo Evolution Report suggests that the future of man will be a story of the good, the bad and the ugly," he said. "While science and technology have the potential to create an ideal habitat for humanity over the next millennium, there is the possibility of a genetic hangover due to an over-reliance on technology reducing our natural capacity to resist disease or get along with each other.

"After that, things could get ugly, with the possible emergence of genetic 'haves' and 'have-nots'."

*How have we ended up with a huge variety of breeds of domestic dog? Could the same happen with humans?*

*Do you think that the "genetic 'haves' and 'have-nots'" is a likely future for our race? Is a world with two species of humans likely? What evidence would you use to support your claim?*

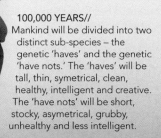

**1,000 YEARS//**
The peak of human enhancement – average height 6.5 feet, life expectancy of 120 years, coffee coloured, symetrical features, athleticism, large clear eyes and smooth hairless skin. Humans will have less developed jaws and shorter chins.

**100,000 YEARS//**
Mankind will be divided into two distinct sub-species – the genetic 'haves' and the genetic 'have nots.' The 'haves' will be tall, thin, symetrical, clean, healthy, intelligent and creative. The 'have nots' will be short, stocky, asymetrical, grubby, unhealthy and less intelligent.

# Bad Science for Schools

## When the evidence doesn't add up.

Sometimes people use what sound like scientific words and ideas to sell you things or persuade you to think in a certain way. Some of these claims are valid, and some are not. The activities on these pages are based on the work of Dr Ben Goldacre and will help you to question some of the scientific claims you meet. Read more about the work of Ben in his *Bad Science* book or at badscience.net.

## Keeping your brain fit

In science you learn about ideas that scientists have developed by collecting evidence from experiments; you are also learning to collect and evaluate evidence yourself. You can use this outside of the laboratory to weigh up information you come across every day. Let's look at this example about how to prepare for exams.

When it comes to exam season you will get lots of different tips from teachers, other students and companies that all claim to help you do better in exams but who is right? If you had an important exam coming up, what would be a good way of making sure that your brain was going to function well?

## ✱ GOOD ADVICE?

Here are three pieces of advice offered to students before exams. For each one:

> suggest why it might be true;

> suggest why you might be dubious about it;

> decide whether you think it's good advice and explain why.

Before sitting an aural exam (a listening test) spin round three times clockwise and three times anticlockwise. This stimulates the semicircular canals which are located in the inner ear, thus stimulating the cochlea.

A drink with caffeine in it is a good idea as it acts as a stimulant and will cause your brain to work quicker.

Before doing an exam in the morning make sure you have a good breakfast. Something like porridge is good as the energy is released slowly during the morning, so you don't get tired towards the end.

**We are learning to:**
> apply ideas about cells and body systems to evaluate advice on how to focus on mental activities
> examine scientific claims for accuracy
> function effectively in a group, developing ideas collaboratively

## ✱ THE SCIENCE BEHIND THE CLAIM

Let's look in more detail at some activities that some schools have used to try to improve students' concentration and learning. Your task is to work out which bits of science are good science and which are bad. To help you decide whether you think your activity is good or bad try discussing these questions:

> What *advice* is being given?
> What *claim* is being made?
> What *scientific ideas* are being used to justify that claim?
> What *scientific ideas* do you have that may tell you something about this topic?
> Is the advice *sound*?

*This is the advice.* *This is the claim.*

Interlock the fingers of both your hands, holding your elbows out at the sides. This completes a circuit and allows positive energy to flow. Positive energy creates positive thoughts, stimulating the brain, stilling anxieties and clearing the way for a free flow of logical thought.

| These are the scientific ideas used to back up this claim. Forming the arms in a loop creates no circuit that any kind of energy 'flows around' and 'positive energy' is a meaningless term. | So can you think of any reason why this might work? You know that regular exercise is good and could help to refocus on ideas and mental activities. | Is the advice sound? Well, it won't do you any harm and may even improve concentration, but not for the reasons claimed. |

**Now you have a go. Are these good or bad science?**

Water is a vital ingredient of blood and blood is essential to transport oxygen to the brain. For the brain to work well you have to ensure your blood is hydrated. This needs water, little and often. The best way of rehydrating the blood taking oxygen to the brain is to hold water in the mouth for up to half a minute, thus allowing direct absorption.

Your carotid arteries are vital to supplying your brain with richly oxygenated blood. Ensure their peak performance by pressing your brain buttons. These are just below the collar bone, one on either side. Make 'C' shapes with forefinger and thumb to place over the brain buttons and gently massage.

## ✱ WOULD YOU PAY MONEY FOR THIS?

Many products or services are sold on the basis that there is a scientific reason that they work. Some claims are justified but some are pretty dubious, even if they look persuasive at first glance. You should think critically about what you are told.

> Claims might sound more convincing if they use technical scientific terms, but sometimes they are used incorrectly, to make something sound scientific when it's not.

> Powerful scientific words include 'conclusive tests', 'energy', 'cleanse', 'purify' and 'health'.

> They may claim to make you feel better, look better, have more energy and be healthier.

> The claims may be true but they should have been tested, the full results published and independently checked.

# Glossary

**absorb** to take in energy from electromagnetic radiation; this is transferred to the particles of the material
**accuracy** how near a reading is to the true value
**acid** a chemical compound which when dissolved in water gives a pH reading of under 7 and turns litmus red
**acid rain** rain water which is made more acidic by pollutant gases
**adaptation** the way in which a species changes over time to become better able to survive in its environment
**adult stem cells** unspecialised body cells that can develop into other, specialised cells that the body needs
**alkali** a chemical compound which when dissolved in water gives a pH reading of over 7 and turns litmus blue
**alkanes** a family of hydrocarbons ($C_nH_{2n+2}$) found in crude oil
**alleles** different versions of a gene on a pair of chromosomes
**ampere** unit used to measure electrical current, often known as 'amp'
**amplitude** the maximum disturbance of a wave motion from its undisturbed position
**analogue** equipment that can display data with continuous values
**analogue signal** transmitted data that can have any value
**anode** positive electrode
**antibiotic** therapeutic drug acting to kill bacteria taken into the body
**antibody** protein normally present in the body, or produced in an immune response, which neutralises an antigen
**anti-diuretic hormone (ADH)** hormone which controls re-absorption of water in kidneys (and so water levels in the blood)
**antigen** harmful substance that stimulates the production of antibodies in the body
**antimicrobial** substance that acts to kill bacteria
**arteries** blood vessels that carry blood from the heart to other parts of the body
**asexual reproduction** reproduction (creation of offspring) involving only one parent; offspring are genetically identical to the parent
**asteroids** small objects in orbits in the solar system
**atmosphere** thin layer of gas surrounding a planet
**atom** the basic 'building block' of an element which cannot be chemically broken down
**bacteria** single-celled microorganisms, some of which may invade the body and cause disease
**base** solid alkali
**bias** intentional or unintentional selection of certain experimental results to fit an opinion or hypothesis
**Big Bang** the theoretical beginning of the Universe, when energy and matter expanded outwards from a point
**binary digit** a number that can only take the values 0 or 1
**binary fission** simple cell division
**biodegradable** a material that can be broken down by microorganisms
**biodiversity** the variety in terms of number and range of different life forms in an ecosystem
**biofuel** fuel such as wood, ethanol or biodiesel, obtained from living plants
**blood plasma** yellow liquid in blood, in which the blood cells are carried
**blood pressure** the pressure of blood against the walls of the blood vessels
**byte** a measure of digital data consisting of 8 binary digits
**capillaries** small blood vessels that join arteries to veins
**carbon** an element that combines with others, such as hydrogen and oxygen, to form many compounds in living organisms
**carbon cycle** the way in which carbon atoms pass between living organisms and their environment

**carbon dioxide** gas whose molecules consist of one carbon and two oxygen atoms, $CO_2$; product of respiration and combustion; used in photosynthesis; a greenhouse gas
**carbon monoxide** poisonous gas whose molecules consist of one carbon and one oxygen atom, CO
**carrier** someone who carries a gene but does not themselves have the characteristic
**carrier wave** electromagnetic wave on which a signal is superimposed for transmission
**catalyst** chemical that speeds up a chemical reaction but is not itself used up
**catalytic converter** a device fitted to vehicle exhausts to reduce the level of nitrogen oxides and unburnt hydrocarbons emitted
**cathode** negative electrode
**cell sampling** removal of a small number of fetal cells, e.g. from the placenta or amniotic fluid, for testing
**ceramics** non-metallic solids made by heating and cooling a material, such as clay to make pottery
**cholesterol** chemical needed by the body for the formation of cell membranes, but too much in the blood increases the risk of heart disease
**chromosome** thread-like structure in the cell nucleus that carries genetic information
**circulatory system** a transport system in the body that carries oxygen and food molecules
**classify** put things into groups according to their properties
**clinical trials** scientific testing of drugs, vaccines and medical processes
**clone** organism genetically identical to another
**combustion** process in which substances react with oxygen releasing heat
**comet** lump of rock and ice in a highly elongated orbit around the Sun
**competition** result of more than one organism needing the same resource, which may be in short supply
**component** part of an electric circuit
**composite** material consisting of a mixture of other materials
**compound** substance composed of two or more elements which are chemically joined together, for example $H_2O$
**compressive strength** a measure of resistance to squeezing or crushing forces
**condense** to turn from a gas into a liquid, as in steam (water vapour) which condenses to liquid water
**contaminated** having mixed with something harmful such as a pollutant or radioactive substance
**continental drift** slow movement of continents (land masses) relative to each other
**continuous variable** variable that can take any value
**continuous variation** variation in organisms of features that can take any value, for example height
**control group** in a drugs trial, the group that receives the placebo allowing researchers to assess whether the drug has an effect in the experimental group
**control variable** variable that is kept constant in an experiment in order to make a test fair
**convection** heat transfer in a liquid or gas, when particles in a warmer region gain energy and move into cooler regions, carrying this energy with them
**coronary arteries** blood vessels that carry blood away from the heart
**coronary heart disease** when arteries that supply the heart muscle gradually become blocked by fatty deposits, preventing the heart from working properly

# Glossary

**correlation** a link between two factors that shows they are related, but one does not necessarily cause the other; a positive correlation shows that as one variable increases, the other also increases; a negative correlation shows that as one variable increases, the other decreases
**cross-links** bonds that link one polymer chain to another
**crude oil** black substance extracted from the Earth, from which petrol and many other products are made
**crust** surface layer of Earth, made up of tectonic plates
**crystalline** a solid material with atoms, molecules or ions arranged in a regular repeating pattern
**current** flow of electrons in an electric circuit
**data** information, often in the form of numbers obtained from surveys or experiments
**decode** to extract information from a code
**decomposer** in a food chain, an organism such as a fungus that uses materials from dead or decaying matter
**decomposition** the action of bacteria and fungi to break down previously living material
**deforestation** the large-scale removal of trees from forested areas for building or farming
**denitrifying bacteria** bacteria vital to the nitrogen cycle, which change nitrates in the soil to nitrogen
**density** the mass of a substance per unit volume
**dependent variable** outcome variable that is being measured; it depends on the input variables
**detritivore** in a food chain, an organism such as an earthworm that breaks down dead or decaying matter into smaller particles
**differentiation** the change of an unspecialised body cell into a particular type of cell
**digital signal** transmitted information that can take only a small number of discrete values, usually just 0 and 1
**discrete variable** variable that can only take certain values and so changes in steps not continuously
**dissolve** to be soluble in water
**DNA** large molecule found in the nucleus of all body cells, the sequence of which determines genetic characteristics
**dominant (allele)** the allele that is always expressed, irrespective of the other allele in the pair
**dwarf planet** spherical object orbiting the Sun, smaller than a planet and larger than an asteroid
**effector** part of the body that responds to a stimulus
**efficiency** a measure of how effectively an appliance transfers the input energy into useful energy
**elastic** a material that returns to its original shape and size after a deforming force is removed
**electrode** solid electrical conductor through which the current passes into and out of the liquid during electrolysis
**electrolysis** process involving a chemical reaction when current is passed through a molten substance or a substance in solution
**electromagnet** a magnet which is magnetic only when a current is switched on
**electromagnetic radiation** energy transferred as electromagnetic waves
**electromagnetic spectrum** electromagnetic waves ordered according to wavelength and frequency – ranging from radio waves to gamma rays
**electron** small particle with a negative charge within an atom that orbits the nucleus
**element** substance made out of only one type of atom
**embryo** an organism in the earliest stages of development
**embryonic stem cells** cells in or from an embryo with the potential to become any other type of cell in the body
**energy input** the energy transferred into a device or appliance from elsewhere
**energy output** the energy transferred away from a device or appliance, which may be either useful or wasted
**environment** an organism's surroundings
**enzyme** biological catalyst that increases the speed of a chemical reaction
**epidemiological studies** studies of the patterns of health and illness in the population
**erosion** the wearing away of rock or other surface matter such as soil
**error** uncertainty in scientific data
**evaporate** turn from a liquid to a gas, such as when water evaporates to form water vapour
**evolution** change in a species over a long period of time
**excrete** to get rid of waste substances from the body
**experimental control** a method of controlling one or more variables to check that a result is not due to that variable
**extinct** a species that no longer survives
**extinction** the process or event that causes a species to die out
**family tree diagram** chart showing relationships between members of different generations of a family, which can be used to show inheritance of genetic characteristics
**fertilisation** when a sperm fuses (joins with) an egg
**fetus** a later-stage embryo of an animal; the body parts are recognisable
**fibre** a long thin thread or filament
**flue gas desulfurisation** industrial process whereby sulfur is removed from waste gases
**fold mountain** a mountain caused by folding of the Earth's crust when two tectonic plates push against one another
**food web** flow chart showing how a number of living things in an environment depend on one another for their food
**fossil** preserved evidence of a dead animal or plant
**fossil fuel** fuel such as coal, oil or natural gas, formed millions of years ago from dead plants and animals
**fossil record** the information obtained over the years from fossil collections
**fraction** group of substances with similar boiling points, produced by fractional distillation
**fractional distillation** process that separates the hydrocarbons in crude oil according to size of their molecules
**frequency** the number of waves passing a set point, or emitted by a source, per second
**functional protein** a protein such as an enzyme that speeds up a chemical reaction
**galaxy** group of billions of stars
**gamma rays** ionising high-energy electromagnetic radiation from radioactive substances, harmful to human health
**gas** state of matter in which atoms or molecules are spaced far apart and spread out to fill the available space
**gene** a section of DNA that codes for a particular characteristic
**gene pool** the complete set of alleles in a population; a larger gene pool results in greater genetic variation
**generator** equipment for producing electricity
**genetic diversity** the differences between individuals (because we all have slight variations in our genes)
**genetic screening** testing large numbers of individuals for a gene, such as a gene for a genetic disorder
**genetic testing** testing an individual for the gene for a genetic disorder
**genotype** an individual's genetic make up, such as whether they are homozygous or heterozygous for a particular gene
**geologist** scientist who studies rocks and the changes in the Earth
**global warming** gradual increase in the average temperature of Earth's surface
**greenhouse effect** the trapping of infra-red radiation by the Earth's atmosphere
**greenhouse gas** a gas such as carbon dioxide that reduces the amount of infrared radiation escaping from Earth into space, thereby contributing to global warming
**habitat** the physical surroundings of an organism

# Glossary

**hardness**  a measure of resistance to change in shape of a solid, for example by scratching or by impact
**hazard**  something that is likely to cause harm, e.g. a radioactive substance
**heart rate**  the number of heartbeats every minute
**hertz**  unit for measuring wave frequency; 1 hertz (Hz) = 1 wave per second
**heterozygous**  an individual who has two different alleles for an inherited characteristic
**high blood pressure**  blood pressure that is consistently abnormally high
**homeostasis**  the way the body keeps a constant internal environment
**homozygous**  an individual who has two identical alleles for an inherited characteristic
**hydrocarbon**  compound containing only carbon and hydrogen
**hydroelectric**  description of power station generating electricity from the energy of moving water
**hypothesis**  an idea that explains a set of facts or observations, and is the basis for possible experiments
**igneous rock**  rock formed by the solidification of molten magma or lava
**immune**  when a person has resistance to a particular disease
**immune system**  a body system which acts as a defence against pathogens, such as viruses and bacteria
**independent variable**  input variable that affects an outcome
**indicator**  in chemistry, a substance that shows the presence of an acid or an alkali by a change in colour; in biology, a measure of the quality of a natural environment, for example, the number of sensitive species present in an aquatic environment, or the level of pollutants in the air
**intensity**  a measure of the power of a beam of radiation
**intensive crop production**  large-scale farming using, for example, artificial fertilisers and pesticides to produce high yields
**interdependence**  relationship between several organisms that depend on one another
**inversely proportional**  when there is an increase in one variable and a proportionate decrease in another variable
**ion**  atom (or groups of atoms) with a positive or negative charge, caused by losing or gaining electrons
**ionisation**  the removal of electrons from atoms or molecules
**ionising radiation**  electromagnetic radiation that has sufficient energy to ionise the material it is absorbed by
**irradiated**  exposed to waves of radiation
**isolated**  separated, as in a strain of bacteria that can be separated from others, or as in an island that is remote from other land masses
**joule**  unit of energy
**kidney**  the organ in the body that controls water balance
**kilowatt**  unit of power equal to 1000 watts or joules per second
**kilowatt-hour (kWh)**  the energy transferred in 1 hour by an appliance with a power rating of 1 kW (sometimes called a 'unit' of electricity)
**lava**  molten rock (magma) from beneath the Earth's surface when it erupts from a volcano
**law of conservation of mass**  (in a chemical reaction) the total mass of the products is the same as the total mass of the reactants
**lichen**  small organism that consists of both a fungus and an alga
**Life Cycle Assessment**  an analysis of the environmental impact of a product, including the production of raw materials, its manufacture, packing, transport, use and disposal
**light pollution**  excessive artificial light that prevents us from seeing the stars at night and can disrupt ecosystems
**light-year**  the distance travelled by light in 1 year
**living indicator**  a species, the presence of which gives a measure of the quality of an environment; some species, such as the mayfly, are sensitive to pollutants and others are tolerant

**longitudinal**  a wave such as a sound wave in which the disturbances are parallel to the direction of energy transfer
**low blood pressure**  blood pressure that is consistently abnormally low
**magma**  molten (liquid) rock
**magnetic field**  a space in which a magnetic material feels a force
**malleable**  able to be beaten into a thin sheet; a common property of metals
**mantle**  semi-liquid layer of the Earth beneath the crust
**mass extinction**  the extinction of a large number of species at the same time
**mean**  an average of a set of data
**melting point**  temperature at which a solid changes to a liquid
**membrane cell**  electrolysis cell that uses a semi-permeable membrane to separate the reactions at the two electrodes, as in the electrolysis of brine
**memory cells**  white blood cells that form antibodies in response to a particular antigen and retain the ability to make that antibody should re-exposure to the antigen occur later in life
**metal**  a group of materials (elements or mixtures of elements) with broadly similar properties, such as being hard and shiny, able to conduct heat and electricity, and able to form thin sheets (malleable) and wires (ductile)
**methane**  a gas with molecules composed of carbon and hydrogen; a greenhouse gas
**microwave**  electromagnetic wave similar to radio waves but with higher energy
**microorganism**  very small organism (living thing) which can only be viewed through a microscope
**Milky Way**  the galaxy in which our Sun is one of billions of stars
**mixture**  one or more elements or compounds mixed together but not chemically joined, so they can be separated out fairly easily
**modulated**  an electromagnetic wave that has been altered by an information signal
**molecule**  two or more atoms held together by strong chemical bonds
**monoculture**  when a single crop is grown
**monomers**  small molecules that become chemically bonded to one another to form a polymer chain
**moon**  a large natural satellite that orbits a planet
**mutation**  a change in the DNA in a cell
**nanometre (nm)**  unit used to measure very small things (one-billionth of a metre, or $10^{-9}$ m)
**nanoparticles**  very small particles on an atomic scale
**nanotechnology**  technology making use of nanoparticles
**nanotube**  a carbon molecule in the form of a cylinder
**National Grid**  the network that distributes electricity from power stations across the country, using cables, transformers and pylons
**native species**  species that naturally inhabits a region
**natural materials**  materials made from plant and animal products
**natural selection**  process by which characteristics that can be passed on in genes become more common in a population over many generations (which are likely to give the organism an advantage that makes it more likely to survive)
**negative feedback**  information that causes a reversal in a control system, for example when we get too hot our body responds to bring our temperature back to normal through sweating and vasodilation
**neutralisation**  reaction between an acid and a base, to make a salt and water
**nitrates**  salts containing the nitrate ion (consisting of one nitrogen atom and three oxygen atoms); may be used as fertilisers, sometimes causing pollution of waterways
**nitrogen cycle**  the way in which nitrogen and nitrates pass between living organisms and the environment

# Glossary

**nitrogen-fixing bacteria** bacteria vital to the nitrogen cycle, which change nitrogen from the air to nitrates in the soil, needed by plants

**nitrogen oxides** gaseous molecules containing nitrogen and oxygen atoms according to the formula $NO_x$, where X = 1, 2 etc.; these pollutants are formed due to the high temperatures created by the combustion of fossil fuels

**noise** random alteration to a communication signal, possibly due to interference

**non-living indicator** a non-living measure of the quality of an ecosystem, such as water temperature

**non-native species** species that does not naturally inhabit a region; it has been introduced

**nuclear fission** a chain reaction employed in nuclear power reactors in which atoms are split, releasing huge amounts of energy

**nuclear fuel** radioactive fuel, such as uranium or plutonium, used in nuclear power stations

**nuclear fusion** nuclear reaction in which two small atomic nuclei combine to make a larger nucleus, with a large amount of energy released

**oceanic ridges** undersea mountain ranges formed by seafloor spreading and caused by the escape and solidification of magma where tectonic plates meet

**optical fibre** glass fibre that is used to transfer communication signals as light or infrared radiation

**orbit** near-circular path of an astronomical body around a larger body

**oscilloscope** laboratory equipment for displaying waveforms

**outlier** a measurement that does not follow the trend of other measurements

**oxidation** chemical process that increases the amount of oxygen in a compound; the opposite of reduction

**oxidised** a substance that has undergone oxidation

**ozone** gas found high in the atmosphere which absorbs ultraviolet rays from the Sun

**parallax** angle between two imaginary lines from two different observation points on Earth to an object such as a star or plant, used to measure the distance to that object

**particulates** pollution in the form of particles in the air, such as soot

**pathogen** harmful organism which invades the body and causes disease

**pH** a measure of the acidity or alkalinity of a substance

**phenotype** the physical expression of a gene; different genotypes can give the same phenotype

**photon** a 'packet' of electromagnetic energy, the amount of energy depending on the frequency of the electromagnetic wave

**photosynthesis** process carried out by green plants in which sunlight, carbon dioxide and water are used to produce glucose and oxygen

**phytoplankton** microscopic plant life, often forming the basis of aquatic food chains

**pixel** a tiny area (for example a dot or square) on a screen which conveys the data relating to a small part of a picture

**placebo** 'dummy' treatment given to some patients in a drug trial, that does not contain the drug being tested

**planet** large sphere of gas or rock orbiting a star

**plastic** a compound produced by polymerisation, capable of being moulded into various shapes or drawn into filaments and used as textile fibres

**plasticiser** small molecules which fit between polymer chains and allow them to slide over each other

**plate boundary** where two adjacent tectonic plates of the Earth's crust meet or are moving apart

**plate tectonics** the theory that explains how changes to the Earth's surface occur at tectonic plate boundaries

**pollutant** harmful substance in the environment

**polymer** large molecule made up of a chain of smaller molecules (monomers)

**polymerisation** chemical process that combines monomers to form a polymer

**power** amount of energy that something transfers each second, measured in watts (or joules per second)

**precision** how close together or spread out measurements are

**predate** to prey on (kill) another organism for food

**pre-implantation genetic diagnosis (PGD)** genetic testing of embryos created by in vitro fertilisation for a genetic disorder, so that healthy embryos can be transferred into the mother's uterus

**primary data** information obtained from the scientific experiment being carried out

**primary energy source** a source of energy before conversion to useful energy; examples include fossil fuels, wind, biomass and solar energy

**principal frequency** the main frequency of electromagnetic radiation emitted by an object; hotter objects have higher principal frequencies

**processing centre** a centre of control that acts in response to information, for example the hypothalamus in the brain which responds to changes in body temperature

**products** chemicals produced at the end of a chemical reaction

**protein** a type of chemical with important functions in living organisms

**pulse** in the body, a beat of the heart that may be felt in an artery close to the skin; in data transmission, a short on-phase

**pulse rate** a measure of the number of times per minute the heart is beating

**Punnett square** a diagram that can be used to work out the probability of outcomes resulting from a genetic cross

**PVC** a type of polymer (short for polyvinylchloride)

**P-waves** longitudinal shock waves following an earthquake that can travel through the molten core of the Earth; they change direction at the boundary between different layers of the Earth

**radiation** energy transfer by electromagnetic waves or fast-moving particles

**radioactive** a material that randomly emits ionising radiation from its atomic nuclei

**radiographer** medical worker who takes and processes body images

**range** in a series of data, the spread from the highest number to the lowest number

**rate** a measure of speed; the number of times something happens in a set amount of time

**rating** an assessment or classification according to a scale, as in electrical appliances that are rated in terms of power or energy efficiency

**reactants** chemicals that react together in a chemical reaction

**reactor** the part of a nuclear power station where energy is released from nuclear fuel

**real brightness** a measure of the light emitted by a star compared to the Sun, taking into account how far away it is

**receptor** nerve cell which detects a stimulus such as a hot surface

**recessive (allele)** an allele that is only expressed if the other allele in the pair is also recessive; it is hidden if the other allele in the pair is dominant

**redshift** the shift of lines in a spectrum towards the red (longer wavelength) end, due to the motion of the source away from us

**reduced** a substance that has undergone chemical reduction

**reduction** process that reduces the amount of oxygen in a compound – the opposite of oxidation

**reflect** in the case of light, re-direction of the light wave, usually back to the point of origin from a shiny surface

**relative brightness** the apparent brightness of a star as seen from Earth; a dim star close to Earth may appear brighter than a bright one that is further away

# Glossary

**reproducibility** the ability of the results of an experiment to be reproduced by another experimenter

**resistance** (in an electric circuit) a measure of how hard it is for an electric current to flow through a material; (in biology) ability of an organism to resist death/disease/harm; for example resistance may develop in some microorganisms against antimicrobials

**resolution** the clarity of an image, which for a digital image depends on the number of pixels used

**respiration** process occurring in living things where oxygen is used to release the energy in foods

**resting heart rate** a person's heart rate when inactive

**risk** the likelihood of a hazard causing harm

**rock cycle** the relationships between the different types of rock and the processes that occur to change these over long periods of time

**salt** generically, the dietary additive sodium chloride; in chemistry, an ionic compound formed when an acid neutralises a base

**Sankey diagram** diagram showing how the energy supplied to something is transferred into 'useful' or 'wasted' energy

**saturated fat** a component of the diet that, when eaten in excess, can contribute to coronary heart disease and other health problems

**seafloor spreading** an extension of the seafloor caused by tectonic plate movement and the extrusion of magma between two plates which solidifies to form rock

**secondary data** information obtained from other sources, not from the experiment being carried out

**secondary energy source** more convenient form of energy, such as electricity and refined fuels, produced from primary energy sources

**sediment** particles of rock etc. in water that settle to the bottom

**sedimentary rock** rock formed when sediments are laid down and compacted together

**sedimentation** the settling of particles in water to the bottom

**seismic waves** vibrations that pass through the Earth following an earthquake

**selective breeding** choosing organisms with desired characteristics to breed with one another

**sex-determining gene** a gene carried on the Y sex chromosome that causes a fetus to develop into a male

**sexual reproduction** reproduction of an organism that involves two parents

**shadow zone** an area on the Earth's surface where no earthquake waves can be detected because S-waves cannot pass through the Earth's core and P-waves are deflected at the inner/outer core boundary

**side effects** unwanted effects produced by medicines

**signal** information that is transmitted by, for example, an electrical current or an electromagnetic wave

**significant figures** the number of digits in a number; in data, depends on the precision of measurements

**smog** air pollution that is caused, for example by vehicle emissions and industrial fumes

**solar system** the planetary system around the Sun, of which the Earth is part

**soluble** able to dissolve (usually in water)

**species** basic category of biological classification, composed of individuals that resemble one another, can breed among themselves, but cannot breed with members of another species

**stem cells** unspecialised body cells that can develop into other, specialised cells

**stiffness** a measure of the resistance of a solid to bending forces

**structural protein** a protein, such as collagen, whose function is to build tissues

**sulfur dioxide** pollutant gas released from burning sulfur-containing fuels, which causes acid rain

**superbug** harmful microorganism that has become resistant to antimicrobials

**supercontinent** very large land mass

**supernova** explosion of a large star at the end of its life

**sustainability** measure of whether a resource or process we use now will still be able to be used by future generations

**sustainable** resource or process that will still be available to future generations

**S-waves** transverse shock waves following an earthquake that cannot travel through the molten core of the Earth

**synthetic material** material manufactured from chemicals

**tectonic plate** section of Earth's crust that slowly moves relative to other plates

**tensile strength** a measure of the resistance of a solid to a pulling or stretching force

**theory** a creative idea that may explain an observation and that can be tested by experimentation

**thermoplastic** plastic with a shape that can be changed by heating

**thermosetting** plastic with a shape that becomes permanent after heating and cooling

**toxins** poisons

**translucent** a material that lets some but not all light pass through

**transmitted** radiation that passes through a material

**transparent** a material that allows light to pass through

**transverse** a wave in which the disturbances are at right angles to the direction of energy transfer

**true value** a theoretically accurate value that could be found if measurements could be made without errors

**turbine** device which makes a generator spin to generate electricity

**Universe** the whole of space and all the objects and energy within it

**urea** waste product excreted by the kidneys

**vaccination** medical procedure, usually an injection, that provides immunity to a particular disease

**vaccine** weakened microorganisms that are given to a person to produce immunity to a particular disease

**vacuum** a space where there are no particles of any kind

**variation** differences between individuals belonging to the same species

**vasodilate** increase in diameter of small blood vessels near the surface of the body to increase the flow of blood

**veins** blood vessels that carry blood from parts of the body back to the heart

**volcano** landform from which molten rock erupts onto the surface

**voltage** a measure of the energy carried by an electric current

**watt (W)** unit of power, or rate of transfer of energy, equal to a joule per second

**wave** a periodic disturbance that transfers energy

**wave equation** the speed of a wave is equal to its frequency multiplied by its wavelength

**wave technology** in renewable energy, equipment that allows us to harness the power of ocean waves

**wavelength** distance between two successive wave peaks (or troughs, or any other point of equal disturbance)

**white blood cell** blood cell that defends the body against disease

**wind turbine** device that uses the energy in moving air to turn an electricity generator

**X-rays** ionising electromagnetic radiation

**zygote** the cell produced when egg and sperm fuse at fertilisation

# Index

accumulation of toxic chemicals 192
accuracy 309
acetylene 115
acid rain 326
 amount and spread 123, 124
 cause and sources 119, 135, 192
 environmental damage 112, 123
 health risks 113, 285
acids 180, 185, 326
 reaction with alkalis and bases 183
acrylic 141, 153
Adams, John Couch 208–9
adaptations of species to environment 74–5, 326
addition reactions 147
ADH (anti-diuretic hormone) 67, 326
adrenaline 58
adult stem cells 32, 326
advantageous genes 86–7, 89
advertising claims 320–1
air, composition 106–9
air quality
 human impacts 110, 128
 monitoring 85, 111, 122–3
 pollution and health 112–13
alchemists 212
alcohol
 consumption and heart disease 56
 effect on body water balance 66–7
 as fuel 117, 126, 129
Alkali Acts 185
alkalis 180–1, 326
 manufacture 184–5
 nineteenth-century demand 182
 reaction with acids 183
alkanes 147, 149, 326
alleles 14, 17, 326
 dominant and recessive 18–19
alum 181
aluminium 142–3
ammonia 109
 solution 181, 183
amniocentesis 28–9
amplitude, waves 226–7, 326
analogue signals 255, 258–9, 326
Andromeda galaxy 207, 215
anodes 191, 326
Antarctic, species adaptations 76, 87
antenatal testing 28–9
antibiotics 47, 326
 resistance, bacteria 47, 50–1, 90
antibodies 42–3, 326
antigens 43, 44–5, 326
antimicrobial substances 47, 50–1, 326

antibacterial properties of silver 162, 164
apparatus 299
aquatic ecosystems 74, 84–5
argon 106–7
arid environments 74
arteries 54, 55, 326
artificial cloning 31
asbestos 165
ASCII code 261
asexual reproduction 30–1, 41, 326
ash, source of alkalis 180, 181, 182
asteroids 204, 326
asthma 112, 113
athletes 58
atmosphere 326
 effects of human activities 110, 251
 formation and early changes 108–9
 and radiation 248
 thickness 107
atoms 145, 240, 326
 rearrangement in chemical reactions 116–17
 size 158–9

background radiation 243
bacteria 108, 162, 326
 causing diseases 40, 47, 188
 in nitrogen cycle 83
 reproduction and growth 30, 41
 useful (sewage treatment) 164
bad science, evaluation 320–5
bakelite 145
bar charts 37, 274, 306
bases, chemical reaction with acids 183
batteries 129, 270, 271
bias in scientific studies 245, 310
Big Bang theory 216–17, 326
binary fission 41, 326
biodegradable materials 99, 326
biodiversity 94–7, 326
biofuels 126–7, 129, 326
biological hazards 303
bladder 65, 66
block diagrams 282
blood
 plasma 64–7, 326
 spot tests, newborn babies 28
 white blood cells 42–3, 48
blood pressure 59, 60–1, 326
blood vessels 54–5, 63
body temperature, control 62–3
boiling points 148–9
bone marrow 32, 42
bones 64, 242
brain
 damage from genetic disorders 24, 28
 function, advice and claims 324–5

risk from nanoparticles 165
water balance control 67
brightness, stars 207, 210–11
brine
 in solution mining 177
 electrolysis 190–1
Britain
 history of rocks 172–3
 north-west, industry resources 174–5
broadcasting 255–7
bronze 144
brown coal 120
Brown tree snake 78–9
buckyballs 160
budding (hydra) 30
bulbs of plants (reproduction) 30
bullet-proof vests 140, 157
burning see combustion
bytes 260, 326

calcium carbonate 109, 183
calcium oxide (lime) 125, 181
calcium sulfate 125
calculations
 energy transfer efficiency 81, 276
 in exam questions 316
 mean (of measurements) 305
 power 271, 273
 wave equations 226, 229
camouflage 89
cancer 188, 241, 247
Cane toads 77
capillaries 55, 326
carbohydrates 250
carbon 120, 250, 326
carbonates 183
carbon cycle 82, 250, 326
carbon dioxide 183, 326
 amount in air 110–11, 249
 in carbon cycle 82, 250–1
 carried in blood 55
 emissions, reducing 126–7, 292–3
 in sea water 109
carbon fibre 152, 162
carbon footprint 82, 326
carbon monoxide 120–1, 128–9, 326
 in cigarette smoke 56
carbon nanotubes 159, 160, 162–3
cardiovascular fitness 58
carrier bags 99, 150, 155
carriers (genetic) 25, 26, 326
carrier waves 255, 259, 326
cars 128–9
 manufacture 142–3
catalysts 161
catalytic converters 128–9, 326
cathodes 191, 326
cattle breeding 88
cells
 damage from ionising radiation 241
 differentiation 32–3

genes and proteins 10–11
 membrane proteins 25, 43
 sampling, from fetus 28, 326
 virus infection 41
ceramics 144, 326
CFCs (chlorofluorocarbons) 193, 246
chain lengths, molecules 147
 polymers 150–1, 156
characteristics see traits
chemical hazards 192, 303
chemical industry 190
 industrial chemicals, safety 192–5
 natural resources, Britain 174–5, 182
 waste 184–5
chemical reactions 116–17, 182–3, 302
Cheshire salt mines 176, 177
chicken pox 41, 43
chlorides 183, 185
chlorination, drinking water 186–9
chlorine 185, 187, 190
chloroform 187, 188
cholesterol 57, 326
chorionic villus sampling 28
chromosomes 10, 326
 chromosomal disorders 15, 29
 inheritance from parents 17, 18
 number of pairs 14, 17
 sex chromosomes 20–1
circulatory system 54–5, 326
classification 92, 95, 326
cleaning products 50–1
climate change 252–3
 and extinction rate 94
climate models 253
clinical trials 52–3, 71, 326
clones 30–1, 326
cloth and clothing 141, 156, 276
 dyeing 181
clouds 106, 248
coal 114, 119
 formation 175
coal-fired power stations 282
 pollution 124–5, 135, 285
coding
 digital signal 259, 261
 genes, for proteins 10
collagen 11
collecting ducts, kidney 65
colour blindness, red-green 21
colours 236
 of stars 211, 215
 wavelength 215, 237
combined heat and power gas turbines 289
combustion (burning) 82, 114–15, 146–7, 326
comets 204, 326
common ancestors 92
common cold 41
communication 254–5
competition 75, 103, 326

331

# Index

with introduced species 76–7, 78
and natural selection 88–9
components, electric circuits 272, 326
composites 162, 326
composition of air 106–9
compounds 119, 185, 319, 326
compression (compressive) strength 138, 141, 326
computers 63, 259–61
    climate modelling 253
concrete 144, 243
condensation 107, 148, 326
confidence, large data sets 47
conservation (species/habitats) 94
conservation of mass 117
contamination (radioactive) 284–5, 326
continental drift 173, 220–1, 326
continuous variables 300
continuous variation (traits) 13, 326
control groups, clinical trials 53, 326
control systems 62–3
control variables 302
convection 221, 253, 327
coronary arteries 55, 56, 327
correlation 327
    air quality and lung diseases 113
    antibiotics and resistance 51
    compared with cause and effect 251
    heart disease and risk factors 57
    positive and negative 307
cosmic background radiation 216, 217
cosmic rays 216, 217, 243
cotton 144, 145, 153
counselling, genetic 19, 29
cracking 148
craters 218, 219, 327
crosses, genetic 18–19
cross-linked polymers 157, 327
crude oil (petroleum) 145, 146–9, 196, 327
crust 172, 218–19, 327
crystallinity of polymers 155, 157, 327
culling 78–9
current (electric) 271, 288, 327
cystic fibrosis 25, 26

dangerous chemicals 192–3
Darwin, Charles 93
data 327
    collection 304–5
    comparing data sets 307
    interpretation 37, 130–1, 315
    presentation 274, 306
    variation and range 107, 139
decay organisms 81
decoding 259, 327
decomposition 82–3, 99, 327
defences, body 42–3

deforestation 110, 251, 327
dehydration 66–7
denitrifying bacteria 83
density 138, 139, 327
deoxygenated blood 54
dependent variables 300, 301
desert adaptations 74
designer polymers 156–7
detectors, light 236
detritivores 81, 327
diesel 129, 149
diet
    amount of salt 178–9
    and heart disease risks 56–7, 71
differentiation, cells 32–3, 327
digital signals 258–9
dimples, genetic trait 12, 17
discrete variables 300
diseases
    body defences against pathogens 42–3
    correlated with air pollution 112–13
    epidemics 45
    immunisation programmes 45, 48–9
    stem cell therapy research 33
    see also genetic disorders
disinfectant by-products (DBP) 187
disposal of products 153, 196–7
distance
    effect on radiation intensity 239
    in space, measurement 205, 210–11
distillation columns 148–9
DNA (deoxyribonucleic acid) 10, 241, 327
    bases 10
    sequence analysis 11, 91, 92
Dolly the sheep 31
dominant alleles 18–19, 22, 327
dosimeters 243
double-blind trials 53
Down's syndrome 15
drilling, oil and gas 196, 269, 290
drinking water treatment 186–9
drugs
    medicines, testing 46, 52–3
    recreational 61, 66–7
dust 110, 213
dwarf planets 204, 327
dyeing cloth 181

Earth 205
    age 216, 219
    atmosphere 108–9
    crust (rocks) 172, 218–19
    interior structure 224–5
earthquakes 222, 224–5
ecosystems 96–7
Ecstasy 61, 66–7
effectors 62–3, 327
efficiency 80–1, 276–7, 327, 289
elasticity 141, 327
electrical devices
    appliance ratings 124, 271

components 272
electric cars 129
electricity
    consumption (use) 124, 297
    costs and units 272–3
    generators 280–1
    see also current
electricity supply
    choice of energy source 290–1
    future security 293
    meters 273, 297
    power cables and pylons 288–9
    standards 281
electrodes 191, 327
electrolysis of brine 190–1
electromagnetic radiation 327
    for communication 254–5, 259
    emission and temperature 249
    intensity 238–9
    spectrum, wave frequencies 237
electromagnets 280–1, 327
electrons 240, 327
elements 212–13, 233, 327
    symbols 319
embryos 327
    embryonic stem cells 32–3, 327
    implantation 26, 31
    research regulation 27, 31
    screening 26–7
    sex organ development 21
emissions
    legal limits 128
    reduction targets 292–3
endangered animals 76
energy
    demand 268, 292–3
    diagrams 274–5
    efficiency 272, 276–7
    of electromagnetic radiation 237–9
    sources 268–9, 290–1
    storage 80
energy transfer 272
    equation 270
    in food chains 80–1
    heat loss to surroundings 288–9
environment 327
    changes, survival of species 76, 90
    persistence of chemicals 192–3
    quality monitoring 84–5
    risks of nanoparticles 164
environmental variation 12–13, 31
enzymes 11, 62, 81, 327
epidemics 45
epidemiological studies 61, 327
equations 319
    for chemical reactions 114, 115, 183
    energy and electricity 270–1
    wave behaviour 226, 229

erosion 218–19, 327
errors 116, 123, 139, 327
    random and systematic 308
E. coli bacterium 40, 42
ethene 147, 150
ethical issues
    clinical trials 53
    embryo and stem cell research 27, 33
    genetic testing results 29, 37
    patenting genes 11
European Union laws 113, 192, 193
evaporation 149, 176, 327
evolution 87, 90–1, 327
    evidence for 92–3
    future humans 322–3
exam questions
    common and key words 312–13
    quality of written communication 317
    styles (assessment objectives) 313
    timing 318
    type AO1 (recall the science) 314
    type AO2 (apply your knowledge) 314–16
    type AO3 (evaluate/analyse evidence) 316
excretion 64–6, 192, 327
exercise 58, 64
    benefits 57
experimental controls 303
exponential growth 41
exposure to ionising radiation 242–3
extinction 327
    rate 94
    threats and effects 76–7
extreme weather 252, 253
eyes
    colour inheritance 13, 22
    light detection 236, 239

fair testing 302
family tree diagrams 19, 25, 327
Faraday, Michael 280
farming, intensive 97
    antibiotics use 47
fatty deposits, blood vessels 55, 56
feeding interactions 75
ferrofluids 163
fertilisation 14, 17, 327
    in vitro (IVF) 26–7
fertility treatment 26–7, 33
fetuses 26, 28, 327
fibres 138, 141, 145, 157
fire 114
fish, habitat adaptations 74
fitness 58
flu
    epidemics 45
    virus 41, 50
flue gas desulfurisation 124–5, 327
food
    fast and convenience 56

# Index

packaging information 179
salt content 176, 178–9
food chains
  accumulated toxins 192
  energy transfer efficiency 80–1
food webs 75, 77, 79, 327
forces of attraction, molecules 149, 154
formula of a compound 120, 319
fossil fuels 109, 127, 327
  burning, products 114, 120–1
  environmental effects of use 269
  increased use with Industrial Revolution 251, 268–9
fossils 86, 220, 327
  evidence for evolution 90, 92
  gaps in record 87
fractional distillation 148–9, 327
fraternal twins 13
freckles, inheritance 18
frequency (Hz) 228–9, 237, 327
fuel economy, cars 142–3
fuels 114–15
  biofuels 126–127, 129, 326
  carbon content 127, 129
  sulfur content 119, 124
functional proteins 11, 327
fusion reactions, stars 212–13

Galapagos finches 93
galaxies 205, 206–7, 214–15, 327
galligu 184
gamma rays 237, 240–2, 327
gases 327
  in air, recent changes 110–11
  greenhouse gases 249, 252, 292
  properties 106
  toxic 120, 121, 187
gas-fired power stations 127, 283
generators 280–1, 327
genes 10–14, 61, 327
  gene pool in populations 87, 327
  sex-determining and sex-linked 21, 330
  switching on/off 33
  see also mutations
genetic disorders
  dominant 19, 24
  recessive 25, 26–7
  testing for 25, 28–9, 37, 328
genetic (inherited) variation 12–13, 46
  and evolution 86–7, 88
  in offspring 16–17
genotype 13, 19, 328
geologists 172, 175, 218, 328
glass 229, 236, 255
  glassmaking 181, 182
  for storing radioactive waste 284
global dimming 110
global warming 252–3, 269, 328
gold nanoparticles 160, 161

Gossage, William 185
government
  advice on health 179
  decisions about global warming 253
  regulations for power generation 281
graphene 163
graphs 111
  choice, for displaying data 274, 306
  interpreting patterns 130–1, 315
gravity 217
greenhouse effect 248, 249, 328
greenhouse gases 249, 252, 292, 328

habitats 328
  adaptations of species 74, 90
  protection 96
haemophilia 21
hair
  colour, inheritance 22–3
  structure 11, 158
hangovers 67
hardness 138, 140–1, 328
Harvard system 310–11
hazards 165, 179, 328
  regulations and warnings 192–3
  symbols 303
head lice 103
health
  blood pressure monitoring 59, 60–1
  public health and sanitation 186, 188
health hazards
  chlorinated water 187, 188
  nanoparticles 165
  salty diet 179
heart
  disease 46, 56–7, 71
  rate 58, 328
  structure 54
heart attack 55, 56
helium 212, 213
herd immunity 45, 49
hertz (Hz) 228–9, 328
heterozygous individuals 15, 23, 328
high blood pressure 59, 60–1, 328
homeostasis 62–3, 328
homozygous individuals 15, 23, 328
hormones 26, 58, 65, 67
Hubble space telescope 207, 211
Human Genome Project 11, 328
humans
  cloning 31
  population 94, 277, 292
  taking part in clinical trials 52–3
Huntington's disease 19, 24
hydrocarbons 114, 146–9, 328
hydrochloric acid 185

hydroelectric power stations 283, 286, 287, 328
  Three-Gorges Dam, China 291
hydrogen 190, 212
  as fuel 115, 127
hydrogen chloride 184–5
hydrogen sulfide 118, 184
hydroxides 183
hypothalamus 63
hypothermia 62
hypotheses 207, 233, 298–9, 328

identical twins 13, 16, 31
igneous rocks 173, 328
illness, causes 40–1, 188
immune system 42–3, 44, 328
immunisation see vaccination
immunity 43, 45
independent variables 300
indicators 328
  for acids and alkalis 180, 183
  environmental, living and non-living 84–5
industrialisation
  in developing countries 293
  effect on carbon cycle 82, 251
  technology and social effects 278–9
Industrial Revolution 182, 251, 278
infections, preventing spread 44–5, 47
information
  storage, digital 260–1
  transmission 254–9
infrared radiation 237, 247, 248
  for communications 255
inheritance
  genetic crosses 18–19
  of genetic disorders 24–6
  patterns in families 16–17, 22–3
insulation (buildings) 126, 277
intensity, radiation 238–9, 328
intensive crop production 97, 328
interdependence 75, 328
Internet websites 298–9
introduced species 76–7, 78–9
investigations 298–9
  evaluation 308–11
in vitro fertilisation (IVF) 26–7
ionisation 240–1, 328
ionising radiation effects 242–3, 284
ions 240, 328
iron 109, 144
irradiation 284–5, 328
isolation and species 90, 328

Japan, diet and heart disease 57
Jenner, Dr Edward 47
joules (J) 270

karyotypes 14, 20, 29
keratin 11, 144
Kevlar 140, 153, 157

keystone species 97
kick sampling 85
kidneys 65, 66–7, 328
kilowatt-hours (kWh) 270, 273, 297, 328

Lamarck, Jean (evolution theory) 93
landfill 98
lasers 236, 237
Lavoisier, Antoine 114, 117
law of conservation of mass 117, 328
leaching 195
lead 243
Leblanc process 184–5, 191
lichens 85, 328
life, beginnings on Earth 86–7
Life Cycle Assessments (LCA) 99, 196–7, 201, 328
lifestyle
  and cancer risks 285
  and heart disease risk 56–7
light
  energy 80
  generation in stars 212
  pollution 211, 268, 328
  properties and behaviour 234, 236
  speed 206, 237
light bulbs, energy-efficient 124, 272, 276–7
light-emitting diodes (LEDs) 299
lightning 83
light-years 205, 206–7, 328
lime (calcium oxide) 125, 181
limestone 109, 144, 174
line spectra 215
litmus paper 180
living organisms
  environmental indicators 84–5, 328
  as source of materials 144–5
longitudinal waves 225, 328
loudness 227
lungs, blood supply 54

magma 222, 223
magnetic fields and electric current 280
magnetic patterns, rocks 173, 221
malleability 143, 328
mantle 221, 222–4, 328
mass
  law of conservation 116–17
  of solar system objects 205
  of the Universe 217
mass extinction events 94
materials
  choice, for products 140–3
  modern and old 152–3
  properties 138–9
Mauna Loa, atmosphere measurements 110–11
mayflies 84, 85
mean and true value 107, 123, 139, 328
measles 41, 45
measurement variation 116, 308

# Index

measuring equipment 299, 309
medical studies 60–1
medicine development 46, 52–3
melting points 138, 154, 155, 328
membrane cells 191, 328
memory cells (immunity) 43, 44, 328
meningitis 40
mercury, environmental contamination 191, 192
metals 144, 244, 328
meteorites 216
meter readings (electricity) 273, 297
methane 116, 146
  as greenhouse gas 249
microorganisms (microbes) 40–1, 328
  resistance to antimicrobials 47, 50–1
microscope, scanning tunnelling (STM) 158–9
microwaves 241
  in communications 254
  mobile phones, health risks 243, 245
  ovens 244–5
Milky Way 206, 328
mixtures 144, 147, 328
modulated waves 255, 328
molecules 116–17, 149, 240, 328
monitoring
  air quality 85, 111, 122–3
  blood pressure 59
  drug effects, long-term 52, 53
  environments 84–5
  radiation exposure 243
monocultures 97, 328
monomers 150–1, 328
moons 204, 328
MOT test 128
mountain building 218–19, 222–3
MRSA (superbug) 51
mutations 328
  chromosome abnormalities 15
  frequency, and adaptations 86–7, 89
  and microorganism resistance 51
naming of organisms 95
nanometres (nm) 159, 328
nanoparticles 160–1, 328
  safety 164–5
  uses 162–3
nanotechnology 159, 328
naphtha fraction 148–9
nappies, disposable and washable 98
National Grid 281, 288–9, 328
native species 76–7, 78
natural gas 118, 127
natural materials 144–5, 152, 328
natural selection 88–9, 88–90, 329

negative feedback 63, 329
neutralisation 180, 329
nicotine 56
nitrates 82–3, 329
  water pollution 84, 85
nitrogen cycle 82–3
nitrogen-fixing bacteria 83
nitrogen in air 106–7
nitrogen oxides 120, 121, 128–9, 329
noise (signal interference) 259, 329
nuclear fuels 268, 283, 284, 329
nuclear fusion 212–13, 329
nuclear power stations 283
nucleus
  atomic 212
  in cells 10, 17, 31
nutrient cycling 82–3
nylon 141, 145, 151, 153

obesity 46
oceans 220
  oceanic ridges 221, 329
  waves 226
offshore wind farms 287, 290
oil refineries 148–9
open-label trials 53
optical fibres 255, 329
orbits 204, 208–9, 329
origins of life 86
oscilloscopes 227, 329
outliers (data) 123, 139, 304–5, 329
oxidation 114–15, 185, 329
oxygen 108, 117
  amount in air 107
  carried in blood 54, 55, 58
  see also combustion
oxygenated blood 54
ozone 329
  layer and hole 162, 246–7
  for water treatment 187, 188

packaging 99
palm oil crops 126, 129
parallax 211, 329
particulates 110, 125, 130–1, 329
pascals (Pa) 139
pathogens 329
  illness and microbe growth 40–1
  immune responses 42–3
penguins 76
peppered moths (melanic) 89
PET (polyethenetetraphthalate) 141, 153
petrol 148
petroleum see crude oil
petroleum gases 148
pH 183, 201, 329
phenotype 13, 19, 329
photons 237, 238, 240–1, 329
photosynthesis 329
  energy absorption and storage, plants 80
  fixing carbon dioxide 82, 126
  source of oxygen in atmosphere 108

physical hazards 303
phytoplankton 84, 329
pie charts 274, 275
pitch (sound) 214, 227
pituitary gland 67
pixels 260, 329
placebos 53, 329
planets 204–5, 239, 329
  discovery 208–9
plant cells, water absorption 65
plants
  asexual reproduction 30
  photosynthesis 80
plasticisers 156, 194–5, 329
plastics 138, 140–1, 197, 329
plate tectonics theory 173, 222–3, 329
platinum catalyst 128–9
PM10 particulates 130
poisons see toxins
polio vaccine 45, 48–9
pollution
  air 112–13, 122–3, 128
  from alkali manufacture 184–5
  fossil fuel sources 120–1
  water 84, 85
polyethene 141, 156
  properties of LDPE and HDPE 154–5
  uses 150, 151
polymers 144, 329
  designed 156–7
  monomers and polymerisation 150–1
  properties 154–5
polystyrene 141
populations
  human 94, 277, 292
  of microorganisms 50–1
  spread of diseases 45
potash (potassium carbonate) 181, 183
power 270–1, 329
power stations 124–5
  costs and comparisons 290–1
  output and lifetime 291, 293
  types and structures 282–3
PP (polypropene) 141, 151
practical work 298–9
precision 309
pregnancy
  decisions 15, 29
  genetic testing 28–9
pre-implantation genetic diagnosis (PGD) 27, 329
prescription drugs 46
Priestley, Joseph 114
primary data 304
primary energy sources 268–9, 281, 329
principal frequency 249, 329
processing centres 62–3, 329
products
  Life Cycle Assessments 196–7, 201
  materials improvement 152–3
  sustainability 98–9
properties of materials 138–9
proteins 10–11, 43, 82–3, 329

public transport 126, 128
pulse rate 58, 329
Punnett squares 19, 21, 329
PVC 151, 156, 194–5, 329

quality of written communication (QWC) questions 317

radiation 207, 329
  absorption and heating 244
  intensity 238–9
  ionising 240–3
radioactive materials 242, 243, 329
  waste from nuclear power stations 284–5
radiographers 242–3, 329
radio waves 237, 254–5
radon 243
random errors 308
rating, power 124, 271, 277, 329
raw materials 145, 196–7
reactor fuel rods 283
receptors (homeostasis) 62–3, 329
recessive alleles 18, 21, 22, 329
recommended daily allowances 179
recycling 98, 276, 277
redshift 215, 329
reduction 114, 329
references 310–11
reflection 236, 248, 330
renewable energy sources 286–7, 290–1
repeatability 304–5
replication (viruses) 41
reproducibility 305
reproduction
  asexual 30–1, 41
  sexual 16
resistance 330
  head lice 103
  microorganisms 47, 50–1
resolution, image 260, 330
resources, competition and growth 41, 75
respiration 330
  in carbon cycle 82
  production of water 64
resting heart rate 58, 330
revision for exams 314, 318
risk assessments
  before experiments 303, 304
  industrial chemicals 192–3, 196–7
risks 187, 330
  difference from hazards 165, 179, 303
  health and treatments 46
  overestimated 45, 285
  willingness to accept risks 243
rockets 115
rocks 218–19
  formation 172, 174–5
  rock cycle 223, 330
rock salt 175
root systems 74, 82–3
rubber 138, 141, 169

# Index

synthetic (neoprene and silicone) 145

safety
  nanoparticles 164–5
  standards and testing 192–3, 195
  vaccination 46–7
salt (sodium chloride)
  extraction and uses 176–7
  in food 178–9
  solution (brine), electrolysis 190–1
  tax 174, 184
salts 180, 183, 330
sand grains in rock salt 175, 176, 177
Sankey diagrams 275, 289, 330
satellites 254, 256–7
saturated fats 56, 330
science writing and evidence 322–3
scientific claims and advice 324–5
seafloor spreading 221, 330
sea levels 252
sea otters 96
sea water 125
  evaporation for salt 176, 177
secondary data 310
secondary energy sources 268, 330
sedimentary rocks 109, 174, 330
sedimentation 219, 330
seismic waves 224–5, 330
selective breeding 88–9
sex cells (ovum and sperm) 14, 16, 17
sex determination 20–1
sex-linked traits 21
sexual reproduction 16, 330
shadow pictures, X-rays 242
shadow zone (seismic waves) 225, 330
side effects (medical treatment) 46, 53, 330
Signal crayfish 76
significant figures 305
silver nanoparticles 162, 164
single gene disorders 19, 24–7
skin
  anti-ageing creams 320–1
  cancer 247
  heat loss 63
  infection defence 42
smallpox 47
smog 122, 330
smoking 56, 57, 61
Snuppy the dog 30, 31
soap 180, 201
socks, de-odorising 162, 164
soda (sodium carbonate) 181, 183, 184
sodium chloride see salt
sodium hydroxide 190, 201
soil 82–3, 181
solar cells 238
solar system 204–5, 216, 330
solubility 85, 183, 185, 330
solution mining 177

Solvay alkali process 185
soot 110, 123
sound waves 227
  digital sampling 260–1
space see Universe
species 74, 90, 95, 330
speed
  of light 206, 237
  of sound 229
  waves (equations) 226–9
sphygmometers 59
starch 80
stars 206–7
  fusion and elements 212–13, 233
statistics (results analysis) 307
Steady State theory 217
steam and turbines 282–3
  early steam engines 278–9
stem cells 32–3, 330
stiffness 138, 139, 330
strength/stiffness units (Pa) 139
stress 56, 58, 71
strokes 59, 60, 179
structural proteins 11, 330
sugars 80
sulfur 118
sulfur dioxide 118–19, 125, 330
sulfuric acid 119, 184
Sun
  heat and light energy production 212, 238–9
  importance for life 80
  interior structure 213
  in solar system 204–5
sunburn 246
sunscreens 162
superbugs 50–1, 330
supercontinents 173, 330
supernova explosions 213, 330
surface area to volume ratio 74, 161
sustainability 330
  ecosystems 96–7
  energy sources 286–7, 290–1
  human actions 98–9
sweating 64
swim bladders 74
synthetic materials 145, 152–3, 330
systematic errors 308

tectonic plates 172, 222–3, 330
Teflon 151
telescopes 207, 211
Telstar 256–7
temperature
  effect on electromagnetic radiation 249
  regulation, in body 62–3
tennis racquets 152, 162
tensile strength 138, 141, 330
tetanus 40
theories 207, 330
  of continental drift and plate tectonics 220–1, 223
  of evolution 87, 93
thermoplastic and thermosetting polymers 157, 330
titanium oxide 162

toxins (poisons)
  from bacteria 40
  chemicals 192–3
traits (characteristics) 12–13
  alleles 14, 15
  in families 16–19, 22–3
transmitters 78, 254
transparent and translucent materials 236, 330
transport 126, 128–9
transverse waves 225, 330
tree of life 91
tropical rainforests 94
tuberculosis 40, 47
turbines 281, 282, 330
twin studies 13
typhoid 186

ultraviolet (UV) radiation 237
  Kevlar breakdown 153
  and ozone 246–7
  water disinfection 187, 188
umbilical cord stem cells 33
units 139, 319
  size and distance 158–9, 210–11
Universe (space) 205, 330
  expansion 214–15
  formation and age 216–17
uranium 268, 285
urea 65
urine
  excretion 64–7, 178
  used for dyeing 181
uses of materials 140–1

vaccination (immunisation) 44–6, 48–9, 330
  adverse reactions 47
  vaccine testing 46, 52–3
valves (blood vessels) 55
Vancouver system 310–11
variables 300–1
variation
  in data and measurements 116, 131, 139
  displaying on graphs 306
  genetic 12–13, 16–17, 86–8
vasodilation 63, 330
vehicle pollution 128–9
veins 54, 55, 330
viruses 41, 50
volcanoes 108, 219, 222–3, 330
voltage (volts, V) 271, 281, 288–9, 330
volunteers, clinical trials 52–3
vulcanisation of rubber 169

waste
  excretion by body 55, 58, 65–6
  from power stations 284–5
  waste disposal 98–9
water
  light absorption 239
  molecules, microwave absorption 244
  treatment 186–9
water balance 64–5
  plant adaptations 74

water vapour 106–7, 249
watts (W), power calculations 270–1
wave equation 228–9, 330
wavelength 214, 226–9, 330
wave power technology 286, 287, 330
waves 330
  carrying information 255
  frequency 228–9, 237
  longitudinal and transverse 224–5
  seismic 224–225, 330
  vibration measurements 226–7
weather 122, 252
weathering 218–19, 223
Wegener, Alfred 220–1
weight (body) 12, 46
white blood cells 42–3, 48, 330
wind turbines 286, 287, 330
wool 144
word equations (chemical reactions)
  acids and alkalis or bases 183
  combustion 114, 115
World Health Organisation (WHO) 45, 188

X and Y chromosomes 20–1
X-rays 237, 242–3, 265, 330

zygotes 13, 14, 330

# Acknowledgements

The publishers wish to thank the following for permission to reproduce photographs. Every effort has been made to trace copyright holders and to obtain their permission for the use of copyright material. The publishers will gladly receive any information enabling them to rectify any error or omission at the first opportunity. (t = top, b = bottom, c = centre, l = left, r = right)

cover & p1 Gustoimages/Science Photo Library, p8t Dr Gopal Murti/Science Photo Library, p8u Dr Stanley Flegler, Visuals Unlimited/Science Photo Library, p8l Francis Leroy, Biocosmos/Science Photo Library, p8b Smit/Shutterstock, p9t minifilm/Shutterstock, p9u Adrian T Sumner/Science Photo Library, p9l Monkey Business Images/Shutterstock, p9b Jeremy Sutton Hibbert/Rex Features, p10 Renee Lynn/Corbis, p11 Steve Gschmeissner/Science Photo Library, p12t Valentin Mosichev/Shutterstock, p12c Groge Blonsky/Alamy, p12b Rex Features, p13l Big Cheese Photo LLC/Alamy, p13r Gertjan Hooijer/Shutterstock, p14t Brandon Seidel/Shutterstock, p14c CNRI/Science Photo Library, p15 Diloute/iStockphoto, p16t Sam Vail, p16b Seth Joel/Corbis, p18 Radius Images/Alamy, p19 mypokcik/Shutterstock, p21 Biophoto Associates/Science Photo Library, p22 Hannamariah/Shutterstock, p24t Nicholas Sutcliffe/Shutterstock, p24bl Dr M O Habert, Pitie-Salpetriere, ISM/Science Photo Library, p24br ISM/Science Photo Library, p25 Simon Fraser, RVI, Newcastle-upon-Tyne/Science Photo Library, p26t Skip O'Donnell/iStockphoto, p26b Pascal Goetgheluck/Science Photo Library, p27 Suzanne Tucker/Shutterstock, p28 Ted Horowitz/Corbis, p29 Tomasz Markowski/Shutterstock, p30t Getty Images, p30r A.B. Dowsett/Science Photo Library, p30l Copit/Shutterstock, p30br Science Pictures Limited/Science Photo Library, p32 Professor Miodrag Stojkovic/Science Photo Library, p38t Steinhagen Artur/Shutterstock, p38c Sebastian Kaulitzki/Shutterstock, p38b oneclearvision/iStockphoto, p39t Leah-Anne Thompson/Shutterstock, p39c Wallenrock/Shutterstock, p39b Alexander Korobov/Shutterstock, p40t Brad Wynnyk/Shutterstock, p40b Dr Kari Lounatmaa/Science Photo Library, p41t Dr Steve Patterson/Science Photo Library, p41b Roger Harris/Science Photo Library, p42 Eye of Science/Science Photo Library, p43 Dagmara Ponikiewska/iStockphoto, p44 karnizz/Shutterstock, p45t Dmitry Naumov/Shutterstock, p45b Science Photo Library, p46t Margo Harrison/Shutterstock, p46b Luis Louro/Shutterstock, p47 Dario Sabljak/Shutterstock, p48 John Gay/English Heritage.NMR/Mary Evans Picture Library, p50 Thomas Perkins/iStockphoto, p51 les polders/Alamy, p52t Linda Bartlett, National Cancer Institute/Science Photo Library, p52b Brian Chase/Shutterstock, p53 Monkey Business Images/Shutterstock, p56 Joe Gough/Shutterstock, p57t oliveromg/Shutterstock, p57b Guy Erwood/Shutterstock, p58 Sipa Press/Rex Features, p59 Lana K/Shutterstock, p60 Lisa F. Young/Shutterstock, p61 annedde/iStockphoto, p62t louise murray/Alamy, p62b Dushenina/Shutterstock, p64 Scott Camazine/Alamy, p66 Corbis Premium RF/Alamy, p71 Alvey & Towers Picture Library/Alamy, p72t Elena Yakusheva/Shutterstock, p72c great_photos/Shutterstock, p72b Eco Stock/Shutterstock, p73t Marcio Jose Bastos Silva/Shutterstock, p73u Nick Biemans/Shutterstock, p73l Jasper_Lensselink_Photography/Shutterstock, p73b Christopher Meder - Photography/Shutterstock, p74t CCI Archives/Science Photo Library, p74l Frank Bach/Shutterstock, p74r Peter Leahy/Shutterstock, p75 cbimages/Alamy, p76t Keith Szafranski/iStockphoto, p76b Wildlife GmbH/Alamy, p77 Linda Hides/iStockphoto, p78t VisionsofParadise.com/Alamy, p78b John Mitchell/Science Photo Library, p80 AfriPics.com/Alamy, p82 Christopher Hudson/iStockphoto, p83 Paul Lampard/iStockphoto, p84t Awei/Shutterstock, p84b Knorre/Shutterstock, p85t Boris Shapiro/iStockphoto, p85b Goodluz/Shutterstock, p86t Pablo Romero/Shutterstock, p86b Erik Lam/Shutterstock, p87 John Foster, p88t goory/Shutterstock, p88tr Viorel Sima/Shutterstock, p88br dcwcreations/Shutterstock, p89 Michael W Tweedie/Science Photo Library, p90t Uryadnikov Sergey/Shutterstock, p90b Katarina Christenson/Shutterstock, p91l Alta Oosthuizen/Shutterstock, p91c Sam Dcruz/Shutterstock, p91r Yuri Arcurs/Shutterstock, p92l plampy/Shutterstock, p92r Emilia Stasiak/iStockphoto, p93l Gerald & Buff Corsi, Visuals Unlimited/Science Photo Library, p93r Gerald & Buff Corsi, Visuals Unlimited/Science Photo Library, p94t Scott Camazine/Science Photo Library, p94r Andre Nantel/Shutterstock, p95l Gary Unwin/Shutterstock, p95r IPK Photography/Shutterstock, p96t Mikhail Tchkheidze/Shutterstock, p96b Tom Mangel Sen/Nature Picture Library, p97t Danylchenko Iaroslav/Shutterstock , p97b Stephen Aaron Rees/Shutterstock, p98t Ralph125/iStockphoto, p98b Robin Phinizy/Shutterstock, p99 Lim Yong Hian/Shutterstock, p104t R.S.Jegg/Shutterstock, p104u charobnica/Shutterstock, p104l Martyn F. Chillmaid/Science Photo Library, p104b Juburg/Shutterstock, p105t Christian Jegou Publiphoto Diffusion/Science Photo Library, p105c Tim McCaig/iStockphoto, p105b GIPhotostock/Science Photo Library, p106 Detlev von Ravenswaay/Science Photo Library, p108t Stephen & Donna O' Meara/Science Photo Library, p108b John Reader/Science Photo Library, p109 Power & Syred/Science Photo Library, p110t Hank Morgan/Science Photo Library, p110b M D Baker/Shutterstock, p112t fstockfoto/Shutterstock, p112b FLPA/Alamy, p113 Ashley Cooper, Visuals Unlimited/Science Photo Library, p114t Louise Murray/Alamy, p114b svand/Shutterstock, p115 NASA, p116 European Southern Observatory/Science Photo Library, p118t repox/Shutterstock, p118b Leslie Garland Picture Library/Alamy, p119 Karol Kozlowski/Shutterstock, p120 mathieukor/iStockphoto, p122 Bettmann/Corbis, p123 Simon Fraser/Science Photo Library, p124t Smileus/Shutterstock, p124b Realimage/Alamy, p125 Colin Cuthbert/Science Photo Library, p126t Neil Lang/Shutterstock, p126b Kheng Guan Toh/Shutterstock, p127 Maksimilian/Shutterstock, p128 GQ/Shutterstock, p129 David Pearson/Alamy, p130 Guy Somerset./Alamy, p136t Can Balcioglu/Shutterstock, p136c abrakadabra/Shutterstock, p136b Joe Gough/Shutterstock, p137t Oliver Hoffmann/Shutterstock, p137u photobank.kiev.ua/Shutterstock, p137l Andrew Lambert Photography/Science Photo Library, p137b Victor Habbick Visions/Science Photo Library, p138t Hank Morgan/Science Photo Library, p138b Laurent davoust/iStockphoto, p139 Christophe Baudot/Shutterstock, p140t Edward Kinsman/Science Photo Library, p140b Pascal Goetgheluck/Science Photo Library, p141 Gusto Images/Science Photo Library, p142 Arno Massee/Science Photo Library, p144t Maugli/Shutterstock, p144b Margo Harrison/Shutterstock, p145r ansley johnson/iStockphoto, p145l Charles D. Winters/Science Photo Library, p146 Paul Rapson/Science Photo Library, p148 Paul Rapson/Science Photo Library, p150 Videowokart/Shutterstock, p152t Volker Steger/Science Photo Library, p152b Ted Foxx/Alamy, p153 MarVil/Shutterstock, p155l Martyn F. Chillmaid/Science Photo Library, p155r sevenke/Shutterstock, p156t Molly Borman/Science Photo Library, p156b Roman Sigaev/Shutterstock, p157 Craig DeBourbon/iStockphoto, p158t Steve Gschmeissner/Science Photo Library, p158b Philippe Plailly/Science Photo Library, p159t IBM, p159b Eye of Science/Science Photo Library, p160t David Freund/iStockphoto, p160b Dr Kostas Kostarelos & David McCarthy/Science Photo Library, p161 The Trustees of the British Museum, p162 Caroline Green, p163 JR Stock/Alamy, p164 Roger Harris/Science Photo Library, p165 Wojciech Krusinski/iStockphoto, p170t Paula Cobleigh/Shutterstock, p170u Andrey Armyagov/Shutterstock, p170l Charles D Winters/Science Photo Library, p170b Martyn F. Chillmaid/Science Photo Library, p171t Christian Darkin/Science Photo Library, p171u arteretum/Shutterstock, p171l James Stevenson/Science Photo Library, p171b Robert Brook/Science Photo Library, p172t Euro Color Creative/Shutterstock, p172b Steve Allen/Science Photo Library, p174 Leslie Garland Picture Library/Alamy, p175t Laurie O'Keefe/Science Photo Library, p175c David Scharf/Science Photo Library, p175b Dr Morley Read/Science Photo Library, p176t T.W. van Urk/iStockphoto, p176b Mike Dabell/iStockphoto, p177 Courtesy of Paul Hurley, p178 benicce/Shutterstock, p179t Monkey Business Images/Shutterstock, p179b Paul Rapson/Science Photo Library, p180t DNY59/iStockphoto, p180b CCI Archives/Science Photo Library, p181t Science Photo Library, p181b The Stapleton Collection/Bridgeman Art Library, p182t Martyn F. Chillmaid/Science Photo Library, p182b CCI Archives/Science Photo Library, p183 Martyn F. Chillmaid/Science Photo Library, p184t Courtesy of Paul Meera at Catalyst, p184b Courtesy of Halton Borough Council, p186 Joe Gough/Shutterstock, p187 Jim Varney/Science Photo Library, p188 Phase4Photography/Shutterstock, p190 Robert Brook/Science Photo Library, p192t Drake Fleege/Alamy, p192b Martyn F. Chillmaid/Science Photo Library, p193 3445128471/Shutterstock, p194t Steyno&Stitch/Shutterstock, p194l Gordon Ball LRPS/Shutterstock, p194l Li Wa/Shutterstock, p195 Ivonne Wierink/Shutterstock, p196t Ruslan Kudrin/Shutterstock, p196b photobank.kiev.ua/Shutterstock, p197 Stéphane Bidouze/iStockphoto, p202t Denis Tabler/Shutterstock, p202c NASA, p202b Scientifica, Visuals Unlimited/Science Photo Library, p203t NASA/Science Photo Library, p203c Vulkanette/Shutterstock, p203b Andrejs Pidjass/Shutterstock, p204 Walter Myers/Science Photo Library, p205 Detlev Van Ravenswaay/Science Photo Library, p206t Elenamiv/Shutterstock, p206b Lynette Cook, p207 NASA/ESA/F. Paresce/R. O'Connell/Wide Field Camera 3 Science Oversight Committee, p208 NASA/ESA/STScI/E. Karkoschka, U.Arizona/Science Photo Library, p210 Jerry Lodriguss/Science Photo Library, p212t NASA/CXC/StScI/JPL-Caltech/Science Photo Library, p212b William Radcliffe/Science Faction/Corbis, p214 NASA, p216 NASA, p218t LOOK Die Bildagentur der Fotografen GmbH/Alamy, p218b Leene/Shutterstock, p219 ollirg/Shutterstock, p224 Philip Hollis/Rex Features, p226t NASA/C.R.O, p226b Andrejs Pidjass/Shutterstock, p228 Fernando Jose Vasconcelos Soares/Shutterstock, p234t GIPhotostock/Science Photo Library, p234b mycola/Shutterstock, p235t vasakkohaline/Shutterstock, p235c Danicek/Shutterstock, p235b cbpix/Shutterstock, p236t Klaus Guldbrandsen/Science Photo Library, p236l Gusto Images/Science Photo Library, p236c Bork/Shutterstock, p236r Bliznetsov/Shutterstock, p238t Artur Synenko/Shutterstock, p238b Darren Baker/Shutterstock, p240 Brian Weed/Shutterstock, p242 dusan964/Shutterstock, p243 James Steidl/Shutterstock, p244 Feng Yu/Shutterstock, p245 CurvaBezier/iStockphoto, p246t Richard Clark/iStockphoto, p246b Suzanne Tucker/Shutterstock, p247 NASA/Science Photo Library, p249t Phil Degginger/Alamy, p249b Sergei Tarasov/iStockphoto, p251l Blazej Lyjak/Shutterstock, p251r jacus/iStockphoto, p252t Peter Arnold, Inc/Alamy, p252b Bart Pro/Alamy, p253 Lawrence Berkeley National Laboratory/Gary Strand, Ncar/Science Photo Library, p254t NASA/Science Photo Library, p254c Sergey Rusakov/Shutterstock, p254b Michael Ransburg/Shutterstock, p256t Science Museum/Science & Society Picture Library, p256b Stephen Meese/Shutterstock, p258 Mehau Kulyk/Science Photo Library, p260t Tatiana Popova/Shutterstock, p260c Kulish Viktoriia/Shutterstock, p261t taelove7/Shutterstock, p261r robootb/Shutterstock, p261b ifong/Shutterstock, p266t shirophoto/Shutterstock, p266c Brian A Jackson/Shutterstock, p266b Matthew Cole/iStockphoto, p267t Les Scholz/Shutterstock, p267c TebNad/iStockphoto, p267b Smileus/Shutterstock, p268t NOAA/Science Photo Library, p268b Topham Picturepoint/TopFoto, p270 Béatrice Nègre/iStockphoto, p271 Andrew Lambert Photography/Science Photo Library, p272 Mark Boulton/Alamy, p276 maigi/Shutterstock, p277t Christina Richards/Shutterstock, p277b Brasil2/iStockphoto, p278t The Print Collector/Alamy, p278b Mary Evans Picture Library/Alamy, p280 Science Museum/Science & Society Picture Library, p281 Rex Features, p282 Ted Clutter/Science Photo Library, p284t Christopher Pillitz/Alamy, p284b U.S. Dept. of Energy/Science Photo Library, p285t Will & Deni McIntyre/Science Photo Library, p285b Vladimir Caplinskij/Shutterstock, p286t Tony Craddock/Science Photo Library, p286b Aleksandr Kurganov/Shutterstock, p287 Martin Bond/Science Photo Library, p288 MrTwister/Shutterstock, p290 Thomas Barrat/Shutterstock, p292 Colin Cuthbert/Science Photo Library, p293 loong/Shutterstock.